Developments in Primatology: Progress and Prospects

Series editor

Louise Barrett, Lethbridge, Canada

More information about this series at http://www.springer.com/series/5852

Ulrich H. Reichard · Hirohisa Hirai
Claudia Barelli
Editors

Evolution of Gibbons and Siamang

Phylogeny, Morphology, and Cognition

 Springer

Editors
Ulrich H. Reichard
Department of Anthropology and Center
 for Ecology
Southern Illinois University Carbondale
Carbondale, IL
USA

Claudia Barelli
Sezione di Biodiversità Tropicale
MUSE—Museo delle Scienze
Trento
Italy

Hirohisa Hirai
Primate Research Institute
Kyoto University
Inuyama, Aichi
Japan

ISSN 1574-3489   ISSN 1574-3497   (electronic)
Developments in Primatology: Progress and Prospects
ISBN 978-1-4939-5612-8   ISBN 978-1-4939-5614-2   (eBook)
DOI 10.1007/978-1-4939-5614-2

Library of Congress Control Number: 2016942027

Printed on acid-free paper

This Springer imprint is published by Springer Nature
The registered company is Springer Science+Business Media LLC New York

Contents

About the Editors

Claudia Barelli is a primatologist and conservation scientist with a PhD in biology on female gibbons' reproductive strategies. She is currently a research fellow at MUSE—Science Museum in Trento, Italy. Her major research interests are integrated morphological and behavioral studies with genetics, endocrinology and parasitology to address questions relating to reproductive strategies, life history, signaling, sexual selection and evolution in primates. A second focus of her research involves conservation physiology with emphasis on developing multidisciplinary methods that integrate population ecology with metagenomics and physiological approaches for the rapid assessment of threatened populations to address questions concerning human/wildlife interactions and biodiversity conservation.

Hirohisa Hirai is a Professor at the Department of Cellular and Molecular Biology, and Former Director of the Primate Research Institute of Kyoto University, Japan. His primary research interests are in molecular cytogenetics and chromosome evolution in primates. Especially, he is interested in constitutive heterochromatin, rDNA genomic dispersion, centromere and telomere of hylobatids, hominids, and platyrrhines.

Ulrich H. Reichard is Associate Professor of Biological Anthropology at Southern Illinois University Carbondale, U.S.A. He co-authored *Monogamy: Mating Strategies and Partnerships in Birds, Humans and other Mammals* (2003). His research interests are wide, spanning topics related to the ecology, behavior, and cognition of primates, particularly small apes, with the purpose of finding answers to questions about what makes us human. Since nearly thirty years his empirical work focuses on the primate community of Khao Yai National Park, Thailand, where he and his team of students and colleagues study the life history, vocal communication, and spatial intelligence of white-handed gibbons (*Hylobates lar*). Current investigations also involve reproductive strategies of male and female northern pig-tailed macaques (*Macaca leonina*).

Part I
Introduction

Chapter 1
The Evolution of Gibbons and Siamang

Ulrich H. Reichard, Claudia Barelli, Hirohisa Hirai
and Matthew G. Nowak

Adult male white-handed gibbon grooms adolescent female, Khao Yai National Park, Thailand.
Photo credit: Ulrich H. Reichard

U.H. Reichard (✉)
Department of Anthropology and Center for Ecology,
Southern Illinois University Carbondale, Carbondale, IL 62901, USA
e-mail: ureich@siu.edu

C. Barelli
Sezione di Biodiversità Tropicale, MUSE—Museo delle Scienze, Trento, Italy
e-mail: Claudia.Barelli@muse.it; barelli.cla@gmail.com

H. Hirai
Primate Research Institute, Kyoto University, Inuyama, Aichi, Japan
e-mail: hirai.hirohisa.7w@kyoto-u.ac.jp

M.G. Nowak
Department of Anthropology, Southern Illinois University Carbondale,
Carbondale, IL 62901, USA
e-mail: nowak.mg@gmail.com

M.G. Nowak
Sumatran Orangutan Conservation Programme, Medan, Sumatra, Indonesia

© Springer Science+Business Media New York 2016
U.H. Reichard et al. (eds.), *Evolution of Gibbons and Siamang*,
Developments in Primatology: Progress and Prospects,
DOI 10.1007/978-1-4939-5614-2_1

Hylobatids as Hominoids

The independent evolution of small Asian apes[1] began about 16.26 million years ago (mya), early in the Miocene, based on molecular estimates (Thinh et al. 2010a; Fig. 1.1). Despite variation among studies that have used different molecular markers and samples to shed light on the origin of small Asian apes, the current consensus seems to be that the family's origin falls into the early to early middle Miocene, with an estimated first appearance by the end of the Oligocene 26 mya (Matsui et al. 2009) and a most recent appearance not after 15 mya (**16.8** mya [15.0–18.5 mya]: Raaum et al. 2005; **19.9** mya [16.7–23 mya] and **21.3** mya [17.3–26 mya]: Matsui et al. 2009; **19.2** mya: Chan et al. 2010; **21.8** mya [19.7–24.1 mya]: Israfil et al. 2011; **20.3** mya [17.5–23.6 mya]: Matsudaira and Ishida 2010; **20.3** mya [16.59–24.22]: Perelman et al. 2011; **16.2** mya [14.7–18.2 mya]: Thinh et al. 2010a; **15-18** mya: Meyer et al. 2012; **16.8** mya [15.9–17.6 mya]: Carbone et al. 2014). In the broader picture of hominoid evolution, the early to middle Miocene marks a crucial period when the ancestors of modern apes greatly diversified into many species (Begun et al. 1997; Begun 2007; McNulty et al. 2015). Following the establishment of the *Gomphotherium Landbridge*, at ∼19–18 mya (Rögl 1999; Harzhauser et al. 2007; Reuter et al. 2009; Fig. 1.1), *stem* hominoids (i.e., 'stem' meaning related to, but not uniquely linked to a modern ape lineage; sensu Donoghue 2005) left the African continent for the first time and then rapidly spread through the vast Eurasian rain forests (Andrews and Kelley 2007; Begun et al. 2003, 2012; Begun 2013). Therefore, the emergence of small apes in Asia and the nearly simultaneous (i.e., 16–14 mya) split of large Asian apes from the hominoid line (Locke et al. 2011) were not isolated events, but part of a series of significant changes happening in the superfamily Hominoidea. Probably these changes became possible by a northward expansion of forests and the success of flowering plants, which themselves were profiting from an ongoing warm and moist period that had started in the late Oligocene and that culminated in the middle Miocene climatic maximum 17–15 mya (Flower and Kennett 1994; Zachos et al. 2001; Reichard and Croissier 2016; Fig. 1.1).

In contrast to molecular clock estimates of the small Asian ape origin, a hylobatid phenotype is only traceable since as recent as the late Miocene, evidenced by the ∼8 my old *Yuanmoupithecus*, the oldest recognized ancestral hylobatid fossil (Harrison et al. 2008; Harrison 2016). Because of a lack of fossils to confirm the families' presence in the early Miocene, events during the earliest hylobatid evolutionary history from around 16–8 mya are still uncertain. Nevertheless, at present,

---

[1]The terms 'hylobatids', 'small Asian apes', 'small apes', and 'gibbons and siamang' are used synonymously in this chapter and volume to describe the family Hylobatidae (Gray 1870). The term 'gibbons' is used to mean the species of the genus '*Hylobates*'.

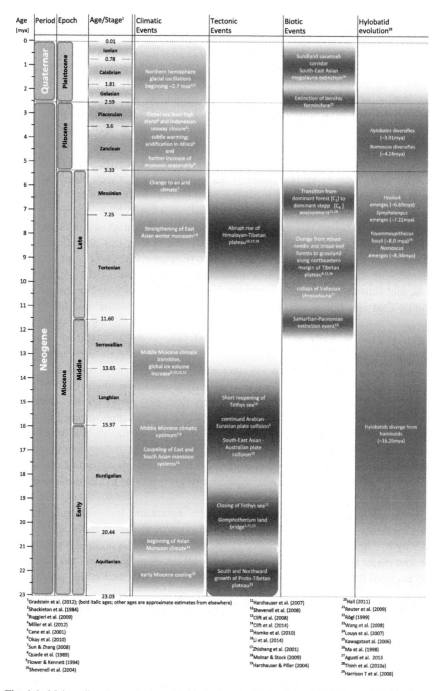

Fig. 1.1 Major climatic, tectonic, and biotic events in South and Southeast Asia from the Miocene to the Pleistocene

a hylobatid origin predating the middle Miocene seems unlikely given that the earliest dispersal of hominoids out of Africa probably occurred no earlier than the beginning of the middle Miocene (Begun 2005; Andrews and Kelley 2007). The first fossil evidence of any hominoid outside Africa represented by an upper molar fragment attributed to *Griphopithecus* or a kenyapithecine from the early Miocene assemblage of Engelswies, Germany was originally dated to 16.5–14.9 mya (Heizmann and Begun 2001) and more recently has been updated to 16.5–16.3 mya (Casanovas-Vilar et al. 2011). If one would still want to argue for an earlier appearance of hylobatids before ~16 mya, one would have to assume that an earlier, yet unrecognized wave of hominoid migration out of Africa had occurred or that the Engelswies fossil is incorrectly dated and is in fact much older. Based on circumstantial evidence and the fossil record of the earliest hominoids in Africa and Eurasia, it seems highly unlikely that the small apes could have emerged in Asia any time before hominoids first left Africa ~17 mya. Even an origin at 16.2 mya seems ambitious, because it assumes a nearly contemporaneous migration of stem hominoids, probably African kenyapithecids (Begun et al. 2003), out of Africa followed by their immediate, near simultaneous emergence in the Middle East, Asia, and Europe. While fossils support the early arrival of *Griphopithecus* in the Middle East and Europe, no paleontological support is yet available for an early arrival of a kenyapithecid or another hominoid in Asia. The closest geographic region to bear an early hominoid fossil that falls along an assumed migration corridor for African or western Eurasian hominoids into Asia are probably the Siwalik sediments of Northern Pakistan, where *Sivapithecus* remains have been found. However, *Sivapithecus* remains are only known from the middle Miocene 13–8.4 mya (Barry et al. 2002), which considerably postdates the arrival of *Griphopithecus* in Turkey (Begun et al. 2003) as well as the estimated origin of hylobatids. In summary, despite some molecular clock estimates that have placed the origin of hylobatids into the very early Miocene or even the terminal Oligocene, the currently available geophysical evidence, i.e., the *Gomphotherium Landbridge*, and Afro-Eurasian fossil evidence point toward an origin of hylobatids in Asia no earlier than 16 mya.

Great uncertainty about the families' early evolution does not end with the appearance of the *Yuanmoupithecus* fossil, however, because the time from 10.5 mya, when the gibbon radiation is estimated to have begun, to 8.3 mya, when the four major hylobatid genera *Nomascus, Hoolock, Symphalangus*, and *Hylobates* had emerged (Chatterjee 2006, 2009), is likewise undocumented in the fossil record and has so far only been reconstructed by molecular genetic methods. Since the early diversification of hylobatid genera was followed by a rather explosive speciation, particularly within concolor and lar gibbons during the Pleistocene (Carbone et al. 2014), we are left with a still unresolvable phylogeny of extant hylobatids, but also with the richest hominoid family in terms of species diversity. Unfortunately, until more fossils are found to guide our interpretations of early hylobatid evolution and the morphology of stem hylobatids, the emergence of small Asian apes must be considered the dark age of the families' evolutionary history, which can only be pieced together from nothing more than molecular clock divergence estimates.

Hylobatid evolution in Asia was unique, because different from the evolutionary roots of large Asian apes, the orangutans of Borneo (*Pongo pygmaeus*) and Sumatra (*Pongo abelii*) (Perelman et al. 2011), whose ancestry might be traceable to Eurasia, for example, through *Griphopithecus* and the proposed pongine *Ankarapithecus* (Harrison 2010; Fleagle 2013), hylobatids appear as an exclusive Asian branch of hominoid evolution with no known or suspected representation in Europe or the Middle East. This suggests that significant events were occurring in Asia during the early middle Miocene, when stem hominoids presumably had just arrived in Asia, which prompted the radiation of both a large-bodied as well as a small-bodied Asian ape.

The small-bodied hylobatids were the first apes to diverge from the hominoid line since stem hominoids, the proconsuloids, had diverged from the Old World Monkeys ∼25 mya (Gibbs et al. 2007; Han et al. 2007; Fleagle 2013; Stevens et al. 2013). As such, gibbons and siamang represent the most distant members of the closest living relative lineages of hominins (humans and their direct ancestors). It seems crucial that future research more forcefully aim to better understand small ape evolutionary history as a significant contribution to a comprehensive understanding of the forces hominoids faced and mastered at the Miocene-Pliocene transition, and how the presumably larger-than-hylobatid Miocene stem hominoids found adaptive solutions to these challenges. This volume aims to contribute to a more in-depth analysis and understanding of the uniqueness of hylobatids in the broader context of hominoid and great ape evolution.

Extant hylobatids are small-bodied apes weighing only between 5.3 and 11.9 kg (Smith and Jungers 1997). With such small body size, they resemble medium-size monkeys more than large-bodied great apes. At a glance, hylobatids may thus be interpreted to represent a morphological continuity from a small ancestral monkey *Bauplan* toward a larger bodied ancestral ape *Bauplan* in the hominoid lineage, in which hylobatids could represent an intermediate morphotype of an ape with a monkey-like body size. Thus, hylobatid morphology may represent a conserved, older trait of a moderately to small size ape of around 10 kg, or less (Harrison 2016). Alba et al. (2015) recently described a new, small (i.e., 4–5 kg) primate genus, *Pliobates*, from the Abocador de Can Mata stratigraphic series (Vallès-Penedès Basin, northeast Iberian Peninsula). The new fossil was named *Pliobates cataloniae* for its intermediate morphological features placing it between crown-hominoids (and even crown catarrhines) and hylobatids. A cladistic analysis, however, rejected that *Pliobates* was an ancestral hylobatid or small-bodied catarrhine, but supported that it represents a late-surviving stem hominoid lineage. Overall, in size and cranial morphology, *Pliobates cataloniae* was found to be more closely related to stem African hominoids than it appears to be related to *Proconsul* (Alba et al. 2015). Based on these new findings, Alba et al. (2015) suggested that the last common ancestor of extant hominoids might have been of small size and more gibbon-like instead of the larger size typical of extant great apes. However, most stem hominoids of the early Miocene slightly or clearly exceeded hylobatids in body size, ranging from 9 to 50 kg (Fleagle 2013; McNulty et al. 2015).

The small hylobatid body size has also been interpreted as a derived adaptation that became established before stem hylobatids split into the four genera known today, and it has been suggested that hylobatids emerged from a stock of moderately sized (i.e., >10 <30 kg) stem hominoids (Groves 1972; Tyler 1993) at a time when the global climate changed dramatically (Zachos et al. 2001; Westerhold et al. 2005; Lewis et al. 2008). The late Oligocene–early Miocene warming period had allowed hominoid habitable forests to expand into far northern latitudes (Janis 1993). The diverse Miocene hominoids successfully conquered not only tropical forests but various habitats such as moderately seasonal deciduous forests (e.g., *Proconsulidae* from Rusinga Island, Kenya), wet subtropical forests of low seasonality (e.g., *Dryopithecus* from central Europe), closed forests in swamp conditions (e.g., *Oreopitheucs* from Italy), and subtropical seasonal forests (e.g., *Griphopithecus* at Paşalar, Turkey) (Andrews et al. 1997). The early Miocene forest expansion was paralleled by the evolution of a diverse assemblage of large-bodied African apes reaching its acme toward the end of the early Miocene (Bernor 1983; Andrews 1992; Jablonski 2000, 2005; McNulty et al. 2015). While the early middle Miocene was still characterized by warm and moist conditions in the Eurasian forest belts, the late middle Miocene brought major global climatic changes, including ocean-level changes, leading to an abrupt high latitude cooling known as the mid-Miocene climatic transition (Miller et al. 1991; Wright et al. 1992; Flower and Kennett 1994; Zachos et al. 2001; Shevenell et al. 2004). In the aftermath of the middle Miocene climatic transition, the global climate gradually cooled while aridity increased significantly resulting in, for example, a global expansion of open habitats, grasslands, and consequently the emergence and spread of C_4 grasses (Kürschner et al. 2008; Eronen et al. 2009; Fig. 1.1). It is likely that changes following the post middle Miocene climatic optimum lead to an overall contraction of Asian forests (Cannon et al. 2009). What once may have been a vast, perhaps continuous hominoid habitable forest stretching all the way from West Europe to the Middle East, across India, into southern China until the South China Sea (Janis 1993; Grehan and Schwartz 2009) and south until Borneo (Morley and Flenley 1987), gradually shifted into a more fragmented landscape of closed forests interspersed with more open habitat types (Morley 2000), in which only isolated tropical forests (much like tropical rainforests today) preserved a stable microclimate of low fluctuating ecological conditions.

It is against a background of a fluctuating, mosaic landscape that the evolution of Asian apes must be understood and interpreted. Interestingly, the suggested rapid radiation of hylobatids at about 10.5 mya (Chatterjee 2006, 2009, 2016) nearly exactly falls into a period when elsewhere mammals and hominoids experienced a major extinction crisis (i.e., the mid-Vallesian crisis) due to altered oceanic and atmospheric circulation patterns that peaked at about 9.7 mya (Agustí et al. 2003, 2013; Eronen et al. 2009). As post early Miocene cool–dry conditions continued and coast line levels fluctuations became more dramatic (Haq et al. 1987; Woodruff and Turner 2009; Woodruff 2010), overall living conditions for ancestral Asian apes declined. To complicate survival even more, seasonality increased following the middle Miocene climatic optimum as Southeast Asian and East Asian monsoons began to grip Asia more tightly toward the end of the Miocene (Zhisheng et al.

2001; Nelson 2005). Thus, not only were the obligate arboreal Asian hominoids of the middle and late Miocene challenged by cycles of forest fragmentation, contraction, and expansion, but at a smaller scale challenges also came from an increasingly monsoon dictated, seasonal environment that curbed daily food supplies at least during part of the year. This would have been particularly problematic for Miocene apes who had by then already evolved a preference for sweet, succulent ripe fruits similar to that of living apes (Ungar and Kay 1995; Nelson 2007), which are relatively stable and predictable only in non- or slightly seasonal environments. A combination of these factors can explain the dearth of Asian ape fossils after about 8.5 mya, because some or many lineages had gone extinct, e.g., the *Sivapithecines* (Nelson 2007).

During the late Miocene with its increasing seasonality, aridity, and lower temperatures, large-bodied Asian apes at northern latitudes must have been particularly under threat as their habitat was shrinking most dramatically (Janis 1993). It is conceivable that in the late Miocene, the ancestral hylobatids and pongines became trapped or retreated to low seasonality forest refugia and mountain valleys such as, for example, localities like Shuitangba, the Lufeng basin, or the Yuanmou basin of Yunnan province, China, where numerous stem hominoid fossils have been found (Ji et al. 2013; Jablonski et al. 2014). Such a scenario of survival in forest refugia would also include the ancient Gaoligong Shan of the Hengduan mountains of Yunnan (Chaplin 2005), where *Lufengpithecus* survived at least until ~6.2 mya (Ji et al. 2013). But even these late-surviving stem hominoid perhaps responded to the worsening environmental conditions of the late Miocene, because *Lufengpithecus* may represent a 'downscaled' ancestral orangutan. Through the Miocene and into the early and middle Pleistocene, *Lufengpithecus* probably decreased in size (Jin et al. 2008) before going extinct in the late Pleistocene (Zhao and Zhang 2013). But the position and meaning of *Lufengpithecus* is controversial as it has also been suggested that it represents the most basal stem pongine (Harrison 2010), which implies that it perhaps survived due to its smaller size while larger pongines like *Sivapithecus* and *Khoratpithecus* perished during the late Miocene. Only the largest primate that ever lived, *Gigantopithecus*, seems to have withstood the Miocene-Pleistocene transition (Ciochon et al. 1996), perhaps because these 'giants' responded to seasonal shortage by switching to lower quality foods, which, for example, included bamboo, in a similar way robust African paranthropines (e.g., *Paranthropus boisei*) responded to increasing seasonality in East Africa (Cerling et al. 2011). With *Gigantopithecus* on one hand and hylobatids on the other, two primates at opposite ends of the hominoid size spectrum survived the Miocene-Pliocene transition. One interpretation could be that the climate change of the Miocene forced hominoids out of their comfortable medium to moderate body size equilibrium that early stem hominoids from Asia like *Sivapithecines* had occupied, because the ecological niche hominoids had been exploiting since their arrival in Asia was no longer sufficient, forcing a shift in the hominoid body size equilibrium to extremes of gigantism or dwarfism.

Despite the waxing and waning of tropical woodland and broadleaf forests after the middle Miocene as well as during the Plio-Pleistocene glaciation-interglaciation cycles, and in contrast to the general plight that gripped hominoids of the middle and late Miocene, hylobatid evolution is a success story, which is partly documented by a rich Plio-Pleistocene hylobatid fossil record (Jablonski and Chaplin 2009).

With as many as 19 surviving species (Anandam et al. 2013), gibbons and siamang are by far the most specious ape taxa. Contrary to common belief that hylobatids are uniform and socio-ecologically inflexible, because of their highly specialized anatomy, ecology, and behavior, the opposite conclusion seems more adequate considering the lineage's successful survival and diversification. Indeed, hylobatids are better characterized as flexible and adaptable, because they prospered through times of grave global and local climate and sea level oscillations while other hominoid lineages perished. Hylobatid evolution reflects adequate solutions to overcome the harsher, drier, and cooler climatic conditions even at northern latitudes where two hylobatid genera, *Hoolock* and *Nomascus*, are still found today (Fig. 1.2). Despite a common emphasis describing hylobatids as ripe fruit specialists, extant hylobatids show an ecological tolerance unmatched by other apes, which is probably the underlying reason for their much greater geographic distribution from equatorial to montane Asian forests, compared to the confined habitats of chimpanzees, bonobos, gorillas, and orangutans.

During the Plio-Pleistocene glaciation periods, hylobatids diversified as they radiated southward along the remaining or reestablished forest corridors and they colonized all habitable forest as far south as Java and the remote Mentawai islands where they survived even after some of these landmasses became islands. Despite a conserved, ape-typical slow life history (Reichard and Barelli 2008), hylobatids showed remarkable adaptability compared to their hominoid ancestors. We suggest that hylobatids adapted in three fundamental ways:

1. Perfection of forelimb suspensory locomotion (FSL) or brachiation.
2. Reduction of group size to small 2–4 adult groups; small group size in extant hylobatids includes: pair-living, small multi-male single female (Reichard et al. 2012) and small single-male/multi-female groups (Fan et al. 2006).
3. Reduction in body size.

Brachiation

Although the ability to move hand over hand is present in all apes (Gebo 1996; Hunt 2004; Manfreda et al. 2006; Thorpe and Crompton 2006; Fleagle 2013), only in hylobatids has *suspensory bimanual brachiation* i.e., using the forelimbs in an upright suspensory fashion for locomotion with the forelimb supporting over 50 % of the body weight, the elbow joint extended, and the brachium fully abducted (Hunt et al. 1996), become the dominant positional repertoire (Carpenter 1940; Fleagle 1976; Cannon and Leighton 1994). And while many hominoid anatomical traits of the upper torso and arms have been interpreted to reflect adaptations to brachiation (Gebo 1996; Chan 2007a, b, 2008), only the small Asian apes have perfected

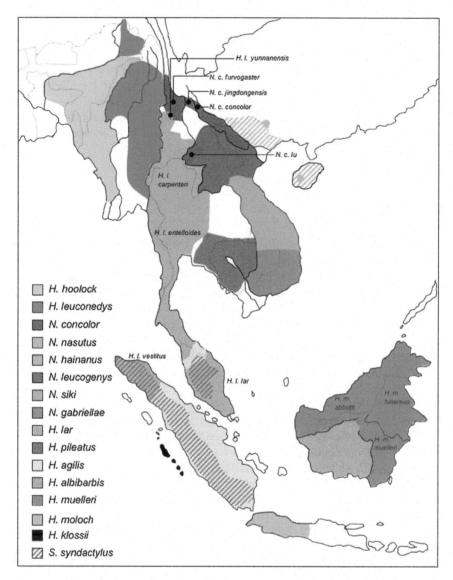

Fig. 1.2 Geographic distribution of gibbons. Dotted and solid lines indicate country borders and major rivers, respectively. Historical distribution of *Nomascus hainanus* and *N. nasutus* is hatched. Figure is reproduced with permission from Thinh et al. (2010a)

suspensory locomotion to include *ricochetal brachiation*, a form of locomotion where the body does not continuously contact the substrate during suspensory hand over hand movements (Hunt et al. 1996). The evolution of ricochetal brachiation was probably a progressive, milestone adaptation in early hylobatids' race to survive the Miocene, as it not only substantially reduced locomotion energy expenditure

(Usherwood and Bertram 2003; Bertram 2004; Preuschoft et al. 2016), but it also allowed hylobatids to outcompete heavier hominoids and emerging Asian monkeys in the harvest of fruit from the terminal branches of trees. The importance of brachiation in hylobatid evolution is supported by Carbone et al. (2014) recent findings of evidence that genes whose functions relate to anatomical specializations associated with brachiation (e.g., longer arms, powerful shoulder flexors, elbow flexors (Michilsens et al. 2009), and greatest craniodorsal shoulder mobility of all primates (Chan 2008)), underwent positive selection early in the hylobatid lineage, i.e., predating the split of the four hylobatid genera.

Group Size

A second adaptive revolution probably occurred when hylobatids shifted from a presumed, general ancestral Asian hominoid large group configurations, perhaps of promiscuous multi-male/multi-female groups as suggested, for example, by the degree of sexual dimorphism of *Sivapithecus* (Patnaik and Chauhan 2009), to living in small groups composed only of a few adults or just living in pairs. Reducing group size to a single-breeding female, which is the most common social organization in extant hylobatids (Bartlett 2011), was likely coupled with territoriality. Pair-living eliminates within-group female feeding competition, which associated with female priority of access to food resources as seen, for example, in white-handed gibbons (Barelli et al. 2008), can potentially increase a female's net caloric intake substantially compared to females living in larger multi-female groups. Similarly, territoriality reduces between-group feeding competition and in hylobatids presumably resulted in more efficient exploitation of resources by developing cognitive abilities to form mental representations of the environment, or a mental map, which has also been observed in white-handed gibbons (Asensio et al. 2011). Thus, both pair-living and territoriality were probably significant factors that contributed to a reduced net caloric need particularly for hylobatid females, which probably helped stem hylobatids to cope with an increasingly seasonal environment. Pair-living is rare among mammals, but represents about 29 % of species, thus is relatively common compared to mammals as a whole (Lukas and Clutton-Brock 2013). Lukas and Clutton-Brock (2013) analyzed an impressive sample of 2545 mammalian species, including 330 primates, and found that pair-living most often evolved under conditions where food competition drove females to live apart from each other. Increasing food competition is very likely the condition stem hylobatid females were facing as the Miocene climatic transition progressed. This scenario supports the notion that a change in social organization probably stemmed from ecological pressures forcing hylobatids early on to downscale group size and primarily live in single-breeding female groups, because when females live in separate territories, males tend to defend a single female (Lukas and Clutton-Brock 2013).

In summary, rather than appearing socio-ecologically inflexible, obligately monogamous brachiators doomed for extinction, hylobatids countered the ecological challenges of the late Miocene–early Pliocene with adaptive solutions allowing them to flourish when other ape lineages declined. They potentially increased their qualitative and quantitative food intake by further developing the hominoid predisposition for forelimb suspensory movement, i.e., brachiation, as their primary mode of locomotion, allowing them to exploit fruits in terminal branches more efficiently than their competitors. They optimized fast arboreal travel by reducing energy expenditure through ricochetal brachiation, and they reduced daily caloric need by adjusting adult group size to a single or very small breeding female group with only one–two males per group. The pivotal role of a third factor, the downscaling of body size as the ultimate cause of hylobatids astonishing success and current diversity, will be discussed in the next section.

Hylobatid Morphotype

Body Size

It has been suggested that hylobatids underwent a process of size reduction or dwarfing (Groves 1972; Tyler 1993; Pilbeam 1996). Dwarfing of lineages is a well-documented phenomenon in larger mammals that have become isolated on small islands (Foster 1964; van Valen 1973). A change in body size in response to an island environment may be adaptive for a colonizing mainland species, because the arriving species may be either too large (resulting in insular dwarfs) or too small (resulting in insular giants) relative to an island's landmass and its existing animal and plant communities (Burness et al. 2001). In biomass-limited situations, downscaled morphotypes will utilize available resources more completely than a larger mainland counterpart would and, at the same time, a population crash can be avoided by not overexploiting the smaller carrying capacity of the island compared to the mainland (Lomolino 1985). Also compared to the mainland, an island is less likely to support similarly large predators, because predators require larger areas and occur at an average lower density than their prey (McNab 1963). In the absence of large predators, selection pressure for large body size of prey species is relaxed and smaller body morphs may evolve through genetic drift or random mating of individuals from skewed size distributions (Lomolino 1985). Hence, island dwarfism may represent an adaptive response to a set of new socio-ecological conditions. In such a scenario, relaxed predation pressure is only one mechanism explaining body size adjustment on islands. This may be complemented by other factors such as competitive release, dispersal ability, and general physiological advantages of modal size (Perry and Dominy 2009), to which recently the combined effects of decreased extrinsic mortality and competition for resources on individual growth rates and body size have also been added (Palkovacs 2003; Raia and Meiri 2006). Although a balance between these factors will determine the strength of body size changes, resource limitation seems to be the strongest predictor of island dwarfism in large mammals (Lomolino 2005), because it results in an absolute reduction of caloric intake.

Theories developed for explaining island dwarfing fit well with the overall trend of downsizing of Miocene hominoids and in particular with the idea of dwarfing in hylobatids as a response to resource limitation. Processes similar to island dwarfing may also explain the potentially downscaled body size of *Lufengpithecus*. Although neither *Lufengpithecus* nor hylobatids evolved on oceanic islands, they may have originated in the middle to southern part of the Yunnan Plateau (Chatterjee 2006, 2009, 2016; Jablonski and Chaplin 2009), which may have resembled island conditions. Two of the earliest hylobatid fossils of the latest Miocene or earliest Pliocene, for example, have been found at two localities south of the eastward bend of the Yangzi River, north of the Pearl and Red Rivers and east of the Mekong River on the Yunnan Plateau in Yunnan Province, China (Jablonski and Chaplin 2009). These remote localities deep in the ancient valleys of the Hengduan Mountains may have represented island-like conditions under which the downsized hylobatid morphotype evolved. A similar scenario of forest island conditions leading to dwarfing has been suggested for several of the world's human pygmy groups (Perry and Dominy 2009). Perry and Dominy (2009) propose that the small pygmy body size represents an adaptive response to common ecological challenges posed by rainforests and rainforest edge habitats that include food limitation, high temperatures and humidity, structural properties of the forest itself, unusually high adult mortality or some combination thereof. In their analysis, they consider rainforest habitats (for humans) to be food-limited environments analogous to small islands. Similarly, Schmidt and Jensen (2003, 2005) have interpreted the decreasing body sizes of Denmark's birds and mammals across time as a result of increasing habitat fragmentation. Thus, it is possible that during the late Miocene the moderately elevated Yunnan Plateau (approximately 1000 m a.s.l.) had functionally become an island-like mosaic moist evergreen broadleaf forest ecosystem with deep, ecologically stable but resource limited and isolated localities, because these valleys were difficult to migrate into or disperse out of, not least because they were sealed against the north by the rising Himalayan Mountains. Such interpretation is consistent with the high endemism produced by both refugia and vicariance (Chaplin 2005). The remoteness of valleys of the Yunnan Plateau can explain why lineages like *Lufengpithecus* perhaps survived longer in localities like Lufeng and Shiutangba than hominoids elsewhere. As well as why the rapid radiation of hylobatids dating to the Plio-Pleistocene (Jablonski and Chaplin 2009) occurred only after stem hylobatids, who may have been similar to *Yuanmoupithecus* (Harrison 2016), who eventually left the sheltered Hengduan Mountain valleys and dispersed across a large geographic area, had already developed into the dwarfed morphotype we know today.

Interestingly, the first genus to split from the stem hylobatid lineage was probably crested gibbons (genus *Nomascus*) at ~8.3 mya (Roos and Geissmann 2001; Chatterjee 2006; Israfil et al. 2011; Thinh et al. 2010b). According to Jablonski and Chaplin (2009) proto-*Nomascus* hylobatids stayed east of the Mekong River, and initially spread northward into southern China, even though the negative impact of the late Miocene climate on northern latitude forests must have already been strong. Nevertheless, crested gibbons even during the later Pleistocene conquered northern habitats up to the Yangzi River (Jablonski and Chaplin 2009) and still today are able to cope with highly seasonal, harsh ecological conditions atypical for other hylobatid genera (Fig. 1.2). Extant crested gibbons are the second largest hylobatids with a body mass of between 7.3 and 7.8 kg (Smith and Jungers 1997). Inferring from molecular clock data, soon after the lineage of crested gibbons began its radiation, at about 7.2 mya (Thinh et al. 2010b), the second distinct hylobatid genus, the siamang (genus *Symphalangus*) has been recognized to have split off from the hylobatid lineage in a process of southward expansion (Chatterjee 2006), because their ancestry is only traceable to the west of the Mekong River (Jablonski and Chaplin 2009; Thinh et al. 2011). Although the phylogenetic position of the siamang is controversial (Meyer et al. 2012), it is possible that they came from the same stock of ancestral hylobatids who, like the proto-*Nomascus* gibbons, perhaps also emerged from the remote Hengduan Mountain valleys.

In trying to understand hylobatid body sizes, the siamang is enigmatic, because at 10.7–11.9 kg (Smith and Jungers 1997) they are 1½–2 times larger than other gibbons. If *Symphalangus* was not the basal hylobatid genus, siamangs underwent secondary size increase, because stem hylobatids had already undergone dwarfing probably to the size of crested gibbons. However, it seems also possible that stem hylobatids may have dwarfed only to the size of a siamang and thus, *Symphalangus* represents the ancestral hylobatid body size. Finally, the hylobatid lineage may have underwent dwarfing, but this trend continued differently in the different hylobatid genera with *Symphalangus* representing only the first downscaled step within the hylobatid dwarfing process. Which of these scenarios is most likely cannot be answered until more hylobatid fossils have been found. If, however, stem hominoids were of the size of *Pliobates* (Alba et al. 2015), then even stem hylobatids may have undergone size increase, with *Symphalangus* showing the largest positive change in body size.

In summary, to understand hylobatid body size(s) we are faced with a dilemma of two unknown size variables: first, the size of the last common ancestor of hylobatids and hominids is unknown and second the size of stem hylobatids is unknown. Perhaps the two primate groups were of similar size, although it is equally possible that there was always a size difference. With the recent discovery of *Pliobates* (Alba et al. 2015) as well as considering that some African Proconsuloids were of small size, it seems possible that the last hylobatid/hominid ancestor was of small size, perhaps smaller even than extant hylobatids. Alternatively, the last common ancestor may have been larger than extant hylobatids considering that most Miocene ape lineages, in particular the first Eurasian apes, are considerably larger than extant hylobatids. Depending on the assumed size of the last common hylobatid/hominid ancestor, stem hylobatids may already have underwent dwarfing or they were of equal size and only later hylobatid lineages became smaller. Thus, in each scenario the trajectory of body sizes of extant hylobatid genera is different.

We argue that stem hylobatids were already phyletic dwarfs as the stem hominoid that gave rise to the hylobatid lineage was most likely a primate of a size similar to the earliest Eurasian hominoids such as *Afropithecus* (30–40 kg), *Heliopithecus* (>20 kg), or *Griphopithecus* (20–40 kg) (Begun 2013). In the scenario we favor, stem hylobatids reduced in size in response to changing ecological conditions during the Miocene. Either crested gibbons or the siamang may represent a downsized stem hylobatid body morphology or either may represent already a second downsizing step if stem hylobatids had been of larger size but had been smaller than the last hylobatid/hominid ancestor. Although the sequence and extent of hylobatid dwarfing is highly speculative, dwarfing itself seems the most likely explanation for the hylobatid body morphology. Ancestral hylobatids may have responded to ecological limitations during the late Miocene-early Pliocene with a reduction in body size and later, during the radiation of hylobatids across the Sunda shelf, some hylobatid populations may have migrated into regions that eventually became islands when sea levels dropped during the Pleistocene (Reichard and Croissier 2016), where they experienced further island dwarfing. However, until fossil evidence of the last common hylobatids/Asian hominoid ancestor becomes available and we find more stem hylobatid fossils, the scenarios offered above to explain hylobatid body size evolution remain speculative.

Hylobatid Uniqueness

Extant hylobatids represent one of only five hominoid lineages (i.e., *Pan*, *Pongo*, *Gorilla*, *Hylobates*, and *Homo*) that survived the late Miocene-Pliocene environmental transition and Pleistocene glaciations. The hylobatid response to climate change by becoming smaller contrasts sharply with the hominoid trend of evolving larger body sizes that is already traceable in African proconsuloids and nyanzapithecids. Larger than hylobatid body sizes are also evident in European drypopithecids (Begun 2005) and Asian sivapithecids (Patnaik and Chauhan 2009). A large body size has been hypothesized to have protected hominoids and extant great apes against predation, but also against fluctuations in food availability (Pontzer et al. 2010). Small Asian apes, however, seem to have responded radically different but equally successful to ecological challenges of the late Miocene that lead to the evolution of a suit of unique traits, which have recently been recognized more prominently (Hayashi et al. 1995; Roos and Geissmann 2001; Chatterjee 2006; Chan et al. 2010, 2012, 2013; Israfil et al. 2011; Kim et al. 2011).

Understanding the unique morphological, behavioral, and genetic position of small Asian apes is important, because their adaptations reflect alternative solutions to lineage survival, as compared to that seen in great apes. As such, understanding the uniqueness of small apes is important for a comprehensive understanding of primate and particularly hominoid evolution. Key traits found in hylobatids not seen in large apes are:

1. perfection of forelimb suspensory locomotion (FSL) or **brachiation** and including ricochetal brachiation
2. reduction of **group size** to small 2–4 adult groups
3. small **body size**
4. rapid **diversification**
5. high **species richness**
6. high **genetic variation**
7. flamboyant **coat colors**, **hair patterns** and **vocalizations**
8. enhanced **habitat flexibility**
9. a current or recent **continuous distribution**

Traits (1)–(3) have been discussed earlier and will not be repeated. The goal of discussing traits (4)–(9) is to provide an overview of salient hylobatid features with the aim to create a comprehensive picture of the evolutionary event that the small Asian apes represent within the hominoid evolution.

Diversification

Rapid diversification occurred in hylobatids at two levels: First, at the genus level, because hylobatid genera diversified much faster than great ape genera. The four gibbon genera diverged from one another perhaps in as little as 2 million years (Thinh et al. 2010a; Israfil et al. 2011; Chan et al. 2013), whereas extant great ape genera *Homo*, *Pan*, *Gorilla*, and *Pongo* diverged over a time period of more than 5 million years (Hobolth et al. 2011; Locke et al. 2011; Wilkinson et al. 2011). Second, at the species level, because within the polyspecific hylobatid genera, the divergence of the nine *Hylobates* species occurred over a very short time period of about 1 million years or less (Whittaker et al. 2007; Thinh et al. 2010a; Chan et al. 2013) and divergence of the four Sundaic island species (*H. agilis*, *H. moloch*, *H. klossi*, *H. muelleri*) was presumably even faster (Chan et al. 2013).

Species Richness

Two mechanisms have been suggested to be responsible for the great species richness in hylobatids: gene flow and vicariance (Chatterjee 2006, 2009; Chan et al. 2013). Hylobatids are, with 19 recognized species, by far the most speciose of extant ape families (Andaman et al. 2013). The exact number of species and sub-species within each genus and phylogenetic relationships between them are fields of ongoing discussion (Whittaker et al. 2007; Israfil et al. 2011; Meyer et al. 2012; Chan et al. 2013; Carbone et al. 2014), except for the genus *Symphalangus* where broad agreement exists that this genus is monospeciose, i.e., the only species within the genus is the siamang (*Symphalangus syndactylus*) of Malaysia and Sumatra.

Table 1.1 Hylobatid chromosome numbers and body weights

Genus	DC	Body mass[a]	Species	Males (n)	Females (n)	Male mass (kg)	Female mass (kg)	Sexual dimorphism[b] (%)
Hoolock	38	6–7	H. hoolock	13	5	6.87	6.88	−0.15
Hylobates	44	5–6	H. pileatus	1	1	5.50	5.44	1.10
			H. lar	84	66	5.90	5.34	9.97
			H. agilis	19	10	5.88	5.82	1.03
			H. muelleri	20	19	5.71	5.35	6.51
			H. moloch	1	1	6.58	6.25	5.15
Symphalangus	50	10–12	S. syndactylus	7	10	11.90	10.70	10.63
Nomascus	52	7–8	N. concolor	28	13	7.79	7.62	2.21
			N. leucogenys	8	4	7.41	7.32	1.22

Data are taken from Smith and Jungers (1997)
DC diploid chromosomes; [a]data taken from Zihlmann et al. (2011); [b]sexual dimorphism = ln (average male mass/average female mass) * 100 (Smith 1999)

In contrast to small apes, great apes are only represented by approximately seven species (Groves 2001; Brandon-Jones et al. 2004). Thus, speciosity of small Asian apes contrasts astonishingly to that of other ape lineages, highlighting the families' great potential for comparative studies.

It is broadly agreed that the family Hylobatidae diverged into four monophyletic genera based on strikingly different chromosome numbers (diploid chromosomes: Hoolock n = 38, Hylobates n = 44, Symphalangus n = 50, and Nomascus n = 52; Prouty et al. 1983), and traceable also through distinct, genus-level body mass variation (Zihlmann et al. 2011). Although gibbons and siamang are generally portrayed as a morphologically homogenous group with no sexual dimorphism (Fleagle 2013), some authors have noticed variation in body mass between the sexes (Plavcan 2012). Based on Smith and Jungers (1997) admittedly small data set of wild hylobatids, body mass dimorphism is absent in Hoolock and probably negligible in Nomascus (males are on average only 1.7 % heavier than females), but is more pronounced in Hylobates where males are 5.6 % heavier than females, and is most expressed in Symphalangus where males weigh 10.1 % more than females (Table 1.1). Moreover, each genus exhibits a defined body mass class with adults of the genus Hylobates ranging at 5–6 kg, Hoolock ranging at 6–7 kg, Nomascus gibbons are again 1 kg heavier at 7–8 kg, and Symphalangus occupying a body mass category of around 10–11 kg, and systematic genus-level anatomical differences (Zihlman et al. 2011). Zihlman et al. (2011: 667) conclude that

> …the adult Hoolock has limb proportions of nearly equal mass, a pattern that differentiates it from species in the genus Hylobates, e.g., H. lar (lar gibbon), H. moloch (Javan gibbon), H. pileatus (pileated gibbon), Nomascus, and Symphalangus. Hylobates is distinct in having heavy hind limbs. Although Symphalangus has been treated as a scaled up version

of *Hylobates*, its forelimb exceeds its hind limb mass, an unusual primate pattern otherwise found only in orangutans.

With the recognition of genus-level anatomical differences future studies of hylobatid locomotor behavior may further reflect subtle anatomical variation. Although the meaning of intergenera anatomical variation has not yet been explored sufficiently, it seems plausible to hypothesize that morphological differences may have evolved in relation to ecological niche separation of hylobatid genera, which probably left traces also in variation in foraging strategies. Moreover, the graded body size classes of hylobatid genera may become informative for phylogenetic reconstruction in conjunction with graded molecular genetic markers.

Genetic Variation

Genetic differentiation among hylobatids exceeds the range of genetic variation found between other hominoids, for example, chimpanzees and humans (Roos and Geissmann 2001; Takacs et al. 2005; Whittaker et al. 2007). Molecular distances based on mtDNA sequences show that human and chimpanzee genomes differ by 9.6 %, but the genera *Hylobates* and *Hoolock* differ by 10.3 %, *Symphalangus* and *Hoolock* differ by 10.6 %, and *Nomascus* and other hylobatid genera differ even by 12.8 % (Roos and Geissmann 2001). Thus, although superficially similar in body size and endowed with long forelimbs compared to other hominoids, hylobatids differ significantly from each other (Zihlman et al. 2011). Extensive chromosomal rearrangement has occurred in each Hylobatid genus (van Tuinen and Ledbetter 1983; Groves 2001; Jauch et al. 1992; Müller et al. 2003; Capozzi et al. 2012), which was 10–20 times faster than in other mammals (Misceo et al. 2008). Genome evolution in hylobatids has been relatively rapid and complex, which makes molecular studies of genetic evolution particularly interesting.

Flamboyant Coat Colors, Hair Patterns, and Vocalizations

Small Asian apes have evolved an unmatched diversity in coat colors and hair patterns among apes. The eyebrows, hands, feet, cheeks, scrotum, chest, face, and head of several species are colored and patterned suspiciously different from the rest of the body (Table 1.2). The obviousness of small ape ornamentation is well reflected in their common names, which often point to body parts that have conspicuous colors or hair patterns or both, for example, crested gibbons, white-handed, and white-bearded gibbons (Roos 2016).

Some small ape species show sexual dichromatism, which is rare in mammals but common in birds (Bradley and Mundy 2008). In primates, it is only found in two lemur species and three New World primates, but it is common in all taxa of the hylobatid genera *Nomascus* and *Hoolock* and it is also found in pileated gibbons

Table 1.2 Gibbon and siamang pelage and hair patterns

Species/Subspecies[a,b,c,d]	SD[a]	PP	Pelage description[a]			Diagnostic morphological features[a,c,d,e]			Hybrids
			Male	Female	Notes	Male	Female	Notes	
Hoolock hoolock	Y	N	Black	Buffy brown or coppery tan		White brow streaks that turn up slightly at the ends; brow streaks are quite close together and connected by white hairs; little white on the chin or under the eyes; perputial tuft balck or only faintly grizzled	White face ring, continued around under the eyes as suborbital streaks; black fringe on the fingers, toes, and the edge of the hands	Both sexes have small laryngeal sac	N
Hoolock leuconedys	Y	N	Black	Buffy brown or coppery tan		White brow streaks well separated with no white hairs between; chin and suborbital zone often with white hairs; perputial tuft white; lighter silvery chest	White face ring, continued around under the eyes as suborbital streaks; somewhat lighter hands and feet		N
Hylobates pileatus	Y	N	Black	Silvery-gray or fawn-buff; crown hair directed fanwise from the front of the scalp.	Both sexes have white hands and feet; crown cap hair flat and long light to white hairs on sides of crown	Whitish border surrounding the face (sometimes just a brow band or a pair of long grayish fringes along each temple); white pubic tuft	White face ring, black cap, cheeks, and chest (spreading down the inner surfaces of limbs with age)		Y (H. lar)

(continued)

Table 1.2 (continued)

Species/Subspecies[a,b,c,d]	SD[a]	PP	Pelage description[a]			Diagnostic morphological features[a,c,d,e]			Hybrids
			Male	Female	Notes	Male	Female	Notes	
Hylobates lar	N	Y	Dark (brown or black) and pale (creamy-fawn to reddish-buff)	Dark (brown or black) and pale (creamy-fawn to reddish-buff)				Both sexes have face encircled with white face ring; crown hair directed fanwise from the front of the scalp	N
Hylobates lar lar	N	Y	Pale phase creamy, dark morph medium brown	Pale phase creamy, dark morph medium brown	Often with paler lumbar region and darker, browner legs				N
Hylobates lar carpenteri	N	Y	Pale morph creamy white	Pale morph creamy white	White face ring exceptionally broad with superolateral "angles" where the cheek and brow meet	Often white pubic tuft in dark morph			N
Hylobates lar entelloides	N	Y	Pale morph honey-colored, usually darker on legs	Pale morph honey-colored, usually darker on legs	White of hands and feet not extending on wrists and ankles				Y (*H. pileatus*)

(continued)

Table 1.2 (continued)

Species/Subspecies[a,b,c,d]	SD[a]	PP	Pelage description[a]			Diagnostic morphological features[a,c,d,e]			Hybrids
			Male	Female	Notes	Male	Female	Notes	
Hylobates lar vestitus	N	?	Fairly light brown from golden to fawn to gray-brown	Fairly light brown from golden to fawn to gray-brown					Y (*H. agilis*)
Hylobates lar yunnamensis	N	Y			Both sexes have very long hair	Hairs of pubic region dark brown or red-brown			N
Hylobates agilis	N	N	Highly variable: balck or maroon-brown or buff (gray or creamy, with a darker ventral surface)	Highly variable: balck or maroon-brown or buff (gray or creamy, with a darker ventral surface)	Hands and feet are the same as the body or somewhat darker; crown hair is directed fanwise from the front of the scalp	Light (white or russet-white) cheek whiskers; partial beard		Both sexes have white brow band; crown hair elongated over the ears	Y (*H. lar*)
Hylobates albibarbis	N	N	Highly colorful: light grayish-brown, becoming golden-toned on rump and with a blackened ventral zone, hand and feet	Highly colorful: light grayish-brown, becoming golden-toned on rump and with a blackened ventral zone, hand and feet	Crown hair directed fanwise from front of scalp	Light genital tuft (often)		Both sexes have crown hair of the cap dark although margined with buff	Y (*H. muelleri*)

(continued)

Table 1.2 (continued)

Species/Subspecies[a,b,c,d]	SD[a]	PP	Pelage description[a]			Diagnostic morphological features[a,c,d,e]			Hybrids
			Male	Female	Notes	Male	Female	Notes	
Hylobates muelleri	N	N	Mouse-gray or brownish	Mouse-gray or brownish	Crown hair directed fanwise from the front of the scalp	Pale, often incomplete, face ring; dark brown (as opposed to black) crown		Both sexes have dark cap and chest which varies in extent individually and with age; crown hair markedly elongated over ears	Y (*H. albibarbis*)
Hylobates abbotti	N	N	Medium gray; ventrum and crown are a little darker	Medium gray; ventrum and crown are a little darker	Crown hair directed fanwise from the front of the scalp	Crown hair markedly elongated over the ears	Crown hair markedly elongated over the ears		N
Hylobates funereus	N	N	Very dark brown or gray	Very dark brown or gray	Crown hair directed fanwise from the front of the scalp	Hands and feet same colored or lighter than limbs	Hands and feet same colored or lighter than limbs	Crown hair markedly elongated over ears	N
Hylobates moloch	N	N	Uniformly silvery-gray; traces of black on cap and chest	Uniformly silvery-gray; traces of black on cap and chest	Fur very long; crown hair directed fanwise from the front of the scalp	Poorly expressed white face ring; short white beard on chin	Poorly expressed white face ring; short white beard on chin	Eyebrows pronounced; crown hair elongated over ears	N
Hylobates klossi	N	N	Black	Black		Fur sparser than in other species of *Hylobates*; (nearly) bare area on throat	Fur sparser than in other species of *Hylobates*; (nearly) bare area on throat	Long pollex and hallux; skull resembles that of siamang	N

(continued)

Table 1.2 (continued)

Species/Subspecies[a,b,c,d]	SD[a]	PP	Pelage description[a]			Diagnostic morphological features[a,c,d,e]			Hybrids
			Male	Female	Notes	Male	Female	Notes	
Symphalangus syndactylus	N	N	Black	black	black	prominent black genital tuft		both sexes have large, inflatable laryngeal sac; second and third toes webbed up to second phalanx; crown hair directed fanwise from a whorl behind the brows	N
Nomascus hainanus	Y	N	Black	Brownish gray to brownish yellow		Crown hair erect; small laryngeal sac	White face ring; black crown streak; crown short and wide, oval with somewhat elongated, pointed posterior end; elongated clitoris		N
Nomascus nasutus	Y	N	Black	Yellow to beige brown	Transition from black female infant pelage occurs at subadulthood	Small laryngeal sac	Wide, white face ring; long, black crown streak extending past nape & may cross shoulders; chest sometimes with patch of gray, brown, or blackish hairs	Crown tuft; small laryngeal sac	N

(continued)

Table 1.2 (continued)

Species/Subspecies[a,b,c,d]	SD[a]	PP	Pelage description[a]			Diagnostic morphological features[a,c,d,e]			Hybrids
			Male	Female	Notes	Male	Female	Notes	
Nomascus concolor	Y	N	Black	Pale yellow to beige brown	Often female darker ventral area forms inverted triangle	Small laryngeal sac	Black-brown or black crown streak; black nape	Crown tuft; small laryngeal sac	N
Nomascus concolor concolor	Y	N	Black			Pronounced crown crest; small laryngeal sac		Small laryngeal sac	N
Nomascus concolor lu	Y	N	Black			Small laryngeal sac			N
Nomascus leucogenys	Y	N	Black	Pale yellow to orange-yellow	Both sexes have long hair and rather coarse	White cheek whiskers forming narrow streak starting under the chin and reaching the level of tops of ears	Thin white face ring that may or may not be complete; black or brown crown streak	Male crown hair erect and elongated in the middle, forming a high crown crest; small laryngeal sac	N
Nomascus siki	Y	N	Black	Pale yellow to orange-yellow		White cheek whiskers forming narrow streak starting under the chin and reaching halfway up to the ears with a pointed upper end and bracket the mouth, extending along the margin of the upper lips and onto sides of the chin	Thin white face ring; black or brown crown streak	Male crown hair erect, forming a high crown crest; small laryngeal sac	N

(continued)

Table 1.2 (continued)

| Species/Subspecies[a,b,c,d] | SD[a] | PP | Pelage description[a] | | | | Diagnostic morphological features[a,c,d,e] | | | | Hybrids |
			Male	Female	Notes		Male	Female	Notes	
Nomascus annamensis	Y	N	Black	Pale yellow to orange-yellow			Buff cheeks reaching less than halfway up the ears with rounded upper margin and brushed out away from the face; brown tinge on chest	Dark crown streak of variable size and a variable darker patch on the chest	Crown tuft	N
Nomascus gabriellae	Y	N	Black	Pale yellow to orange-yellow			Yellowish to orange cheeks reaching less than halfway up the ears with rounded upper margin and brushed out away from the face reaching the corner of the mouth	Dark crown streak of variable size, triangular and brushed sideways; variable darker patch on the chest; white face ring is rare	Male partial beard; rusty-colored chest; crown tuft	N

SD Sexual dichromatism; *PP* Pelage color polymorphism; *Y* Yes; *N* No; *?*: Unclear; [a]Anandam et al. (2013), [b]Thinh et al. (2010a, b), but see Roos et al. (2013) for subspecies (*N. concolor lu* and *N. concolor concolor*); [c]Groves (2001), [d]Mootnick (2006); [e]Mootnick and Groves (2005)

(*Hylobates pileatus*). Other small gibbons are asexually dichromatic, for example, white-handed gibbons, while only siamangs and Kloss' gibbons evolved a monomorphic black coat coloration that most closely resembles the uniform dark brown to black coat color of African apes. Although Bornean and Sumatran orangutans vary in the length, density, and redness of their fur, the species differences seem subtle compared to the striking species differences in hylobatids.

In general, primate coat colors are considered signals of rank, dominance, and attractiveness (Setchell and Wickings 2005), and also serve a function in individual status advertisement (Grueter et al. 2015). The conspicuous and, in some species, sexually dimorphic coat patterning and coloration of hylobatids conflicts with the idea that individual pelage badges evolved primarily in species living in large, relatively anonymous groups to facilitate distance identification (Kappeler and van Schaik 2004; Grueter et al. 2015). Besides social functions of coat colors it is also possible that color patterns evolved through natural selection varying with geography as darker colors in mammals seem to be associated with higher humidity (Kamilar and Bradley 2011) or different forms of camouflage aiding in crypsis against predators (Caro 2011). Overall, however, the patterning and function of hylobatid coat color variation is not well understood and needs to be studied in more detail in the future.

In contrast to coat colors, vocal communication has played a prominent role in hylobatid studies since the early days of field and captive observations (Carpenter 1940; Tembrock 1974). The loud vocalizations hylobatids produce are species-specific and in most species also sex-specific. Songs are usually organized into complex, often stereotyped notes and phrases that are combined into two primary types of songs: pair duets and soli (Marshall and Marshall 1976; Haimoff 1984; Geissmann 2002; Koda 2016; Table 1.3). Hylobatid loud calls surpass those of most primates (Delgado 2006), including the large apes, for example, in how frequently songs are performed, in amplitude, as well as in internal features, structural diversity, and organizational complexity (Geissmann 2002; Terleph et al. 2015).

Joint acoustic male–female displays in the form of duet singing (either as simultaneous or antiphonal singing) evolved independently numerous times in birds (Thorpe et al. 1972; Hall 2004, 2009) and some primate lineages (Haimoff 1986; Geissmann 2000a, b). Three nonmutually exclusive hypotheses have proven most promising in explaining the function(s) of duetting in hylobatids that have also gained some empirical support (Mitani 1987; Cowlishaw 1982): (1) a joint acoustic display in defense of resources, (2) a mutual signal of partner commitment in a newly established and/or existing partnership, and/or (3) a simultaenous individual acoustic display in defense of a sociosexual partner, i.e., a form of acoustic mate guarding.

The idea of duets as a joint acoustic display in defense of resources is suggested by the fact that, without exception, females of all hylobatid species produce great calls (Table 1.3), which are commonly part of a mated pairs' duet or are produced by females as a solo call. In most species, the female great call is the most prominent call phrase produced. The great call likely evolved for distance communication, as these calls are audible up to 1 km or further through the forest (Terleph et al. 2015). Because of features like the universal, conspicuous climax near the end of great calls, characterized by notes produced at the highest speed and/or pitch, and a peak in amplitude (Geissmann 1993), the great call is thought to primarily serve a spacing function in intergroup communication between females, aimed at excluding females from each others' territories (Raemaekers et al. 1984; Raemaekers and Raemaekers 1985; Mitani 1987; Cowlishaw 1982; Cheyne et al. 2007). Consistent with such interpretation, it has recently been shown that in white-handed gibbons the female great call displays acoustic properties suggesting that it evolved into a signal of female resources holding potential (Terleph et al. 2016). Because defending resources is more efficient when two individuals combine their efforts than when each defends on their own (Noë 2006), duetting is common in territorial species (Fedy and Stutchbury 2005).

Based on the general observation that duetting occurs in primarily pair-living species (Tilson 1976), duet singing has been suggested as an internal mechanism that communicates commitment to a partner and thereby contributes to strengthening the bond between adult pair members. For example, in the siamang it has been observed that pairs which have formed more recently produced more duet songs than pairs who had been paired for a longer time suggesting that duetting is related to pair formation and maintenance (Geissmann and Orgeldinger 2000). However, irrespective of these observations duets generally seem too loud to solely serve an intrapair function.

Under the acoustic mate guarding hypothesis, duetting is thought to be used in an effort to warn or keep away competitors of the same sex from the mate. This hypothesis differs from the joint territorial defense and mutual commitment hypotheses in that instead of assuming a cooperative nature of duetting, where individuals are guided by a common goal, acoustic mate guarding emphasizes a competitive nature of joint singing. Under this hypothesis, each individual performs a song to advertise (to potential rivals) that the partner already has a mate. However, duetting seems unnecessary for vocal mate guarding per se, because the simpler solo singing could serve the same function, unless one also assumes that not only is vocal mate guarding a necessary mechanism to prevent intruders from approaching the mate, but that the singing of the partner primarily serves to attract another/other partner(s). If one assumes that the partner calls to attract another mate, duetting may evolve as a form of vocal shadowing. When one individual begins to call the partner is obliged to also call in an attempt to communicate the mated status of the partner and perhaps even outperform the partner's call. The acoustic mate guarding hypothesis has not been tested in primates, but has received support from a study on a subspecies of the pain wren, *Thryothorus modestus zeledoni*, where it was found that both males and females used their contribution to the duet to mate guard the partner (Marshall-Ball et al. 2006). Overall, duetting likely serves multiple, sometimes overlapping and partially contradicting functions within and across duetting species, including hylobatids, which has made conclusive testing difficult.

Geissmann (2002) suggested an evolutionary scenario that explains vocal duetting in hylobatids. In this scenario, (1) duetting represents the ancestral state that was probably present in the lineage since before gibbons and siamang diversified into the four genera known today. This notion is supported by the general observation that duetting is present in all hylobatid species except two, and that in all species females sing great calls (Table 1.3). (2) Hylobatid duets probably evolved from a song which was common to both sexes and which later developed into discreet male-specific and female-specific parts. (3) By a process Geissmann (2002) termed 'duet-splitting' two gibbon species, the Moloch and the Kloss gibbons, lost duetting and secondarily returned to exlusive solo singing probably related to their isolated distribution on two islands.

Overall, although hylobatid vocal communication has been studied over many years, still many questions about interspecies, seasonal, and geographic variations remain. The lack of knowledge on hylobatid vocalizations is surprising given their outstanding diversity and sophistication, compared to the more limited vocal reprtoire of large apes, and aspects, such as a simple syntax, that relate hylobatid duetting to human language (Clarke et al. 2006).

Habitat Flexibility

Gibbons and siamang adapted to more diverse habitats within as well as across their geographical range compared to the great apes. Marshall (2009) provides an excellent, detailed example of habitat diversity for Bornean white-bearded gibbons (*Hylobates albibarbis*) at one locality, Cabang Panti Research Station (CPRS), Gunung Palung National Park. White-bearded gibbons are found across an impressive elevational range of 5–1100 m a.s.l and according to Marshall (2009: 257–258) inhabit the following habitats:

…(1) peat swamp forest on nutrient poor, bleached white soils overlaid with variable amounts of organic matter (5–10 m asl); (2) freshwater swamp forest on nutrient-rich, seasonally flooded, poorly drained gleyic soils (5–10 m asl); (3) alluvial forest on rich sandstone-derived soils recently deposited from upstream sandstone and granite parent material (5–50 m asl); (4) lowland sandstone forest on well drained sandstone derived soils with a high clay content and sparse patches of shale (20–200 m asl); (5) lowland granite forest on well drained, granite derived soils (200–400 m asl.); (6) upland granite forest on well drained, granite derived soils (350–800 m asl.); and (7) montane forest (750–1100 m asl.) […] mostly located on granite-derived soils; however, in many places these soils are overlaid with a substantial amount of bleached, sandy soil (derived from weathering of the granite substrate) similar to those found in the peat swamp.

Similarly, although less extreme, western Hoolocks (*Hoolock hoolock*) of Assam, India, range in altitude from 50–1400 m a.s.l. (Das et al. 2009) with a stunning reported maximum elevation for the genus *Hoolock* at 2600 m a.s.l. (Choudhury 2006). Also, the siamang (*Symphalangus syndactylus*) shows an incredible elevational flexibility ranging from near sea level to 2300 m a.s.l. (Yanuar 2009).

Table 1.3 Songs of gibbon species of the family hylobatidae

Genus	Species[a]	Common name	Types of songs				References
			Duet	Male solo	Female solo	Female GC	
Hoolock	H. hoolock	Western Hoolock gibbon	Y	N	N	Y	c
	H. leuconedys	Eastern Hoolock gibbon	Y	N	N	Y	d
Hylobates	H. pileatus	Pileated gibbon	Y	Y	Y	Y	c,e,f
	H. lar	Lar gibbon	Y	Y	N	Y	e,g
	H. agilis	Agile gibbon	Y	Y	Y	Y	c,e,f,h
	H. albibarbis	Bornean white-bearded gibbon	Y	Y	N	Y	i,j
	H. muelleri	Müller's gibbon	Y	Y	Y	Y	c,e,f,h
	H. abotti	Abbott's gray gibbon	DD	DD	DD	DD	
	H. funereus	East Bornean gibbon	Y	Y	N	Y	f,k
	H. moloch	Moloch gibbon	N	Y	Y	Y	c,l
	H. klossii	Kloss's gibbon	N	Y	Y	Y	c,m
Symphalangus	S. syndactylus	Siamang	Y	N	N	Y	c
Nomascus	N. hainanus	Hainan crested gibbon	Y	Y	N	Y	e,f,n
	N. nasutus	Eastern black crested gibbon	Y	N	DD	Y	c,n
	N. concolor	Western black crested gibbon	Y	Y	Y	Y	c,n,o
	N. leucogenys	Northern white-cheeked crested gibbon	Y	Y[b]	Y[b]	Y	c,n,p
	N. siki	Southern white-cheeked crested gibbon	Y	DD	DD	Y	n
	N. annamensis	Northern yellow-cheeked crested gibbon	Y	Y	N	Y	n
	N. gabriellae	Southern yellow-cheeked crested gibbon	Y	Y	DD	Y	c,n,o

[a]Taxonomy follows Anandam et al. (2013); *Y* Yes, present; *N* No, not present; *DD* data deficient; [b]only when aduld individuals are unpaired Ahsan (2001), [c]Geissmann (2002), [d]Skollar pers. comm., [e]Haimoff (1984), [f]Haimoff (1985), [g]Raemaekers et al. (1984), [h]Mitani (1984), [i]Cheyne et al. (2007), [j]Wanelik and Abdulazis (2013), [k]Gilhooly et al. (2015), [l]Dallmann and Geissmann (2009), [m]Keith et al. (2009), [n]Thinh et al. (2011), [o]Fan et al. (2009), [p]Ahsan 2001

More broadly, hylobatid habitats range from different types of lowland forests, for example, on Java, Sumatra, and the Mentawai Islands to temperate, high altitude montane forests of the Himalayas. Black crested gibbons (*Nomascus concolor*) are found in semihumid evergreen broad-leaved forests (1750–2500 m a.s.l.) and mid-montane humid evergreen broadleaf forests at elevations of 2200–2900 m a.s.l. in Yunnan province, China (Jiang et al. 2006). In comparison, orangutans (*Pongo pygmaeus*) are rarely found at altitudes >500 m a.s.l. on Borneo or >1500 m a.s.l. on Sumatra and commonly do not live at altitudes higher than 1200 m a.s.l. (Husson et al. 2009). Among apes, the elevation at which crested gibbons (*Nomascus*) occur is only trumped by mountain gorillas (*Gorilla beingei*), who can range up to heights of 4500 m a.s.l. in the Virunga Volcano National Park (Stewart et al. 2001). Because the altitudinal breadth of hylobatid habitats correlates with a corresponding floristic breadth, hylobatids are an ideal model group of primates to test predictions of socio-ecological theory, something that is not possible with the more ecology and specious limited African apes or orangutans.

Continuous Distribution

All hylobatid species currently are, or were until recently, continuously distributed across Southeast, South, and East Asian habitats either in close proximity to one another. Or in case of the siamang (*Symphalangus syndactylus*), in sympatric relationships by sharing their habitat with white-handed gibbons (*Hylobates lar*) in the North and agile gibbons (*Hylobates agilis*) in the South on the island of Sumatra, Indonesia, and the Malaysian Peninsula (Gittins and Raemaekers 1980; Thinh et al. 2010b; Chan et al. 2013; Fig. 1.2). Natural hybridization zones further support the notion of historic and current continuous geographic distributions of small Asian apes as contact zones between several species have long been recognized (Brockelman and Gittins 1984; Marshall and Sugardjito 1986; Mather 1992; Arnold and Meyer 2006). In contrast, the geographic distribution of living great apes is discontinuous. Again, because hylobatids share more continuous habitat, relative to the great apes, they are again an excellent model to address socio-ecological questions in comparative frameworks of hominin ecological flexibility that are more difficult to address in studies of extant great apes.

In conclusion, it is an extraordinary flexibility in many parameters that facilitated hylobatids to prosper and evolve into an astonishingly successful ape lineage at times when other apes struggled or even went extinct. Despite the fact that gibbons are sometimes interpreted as rather specialized, because of their unique brachiation locomotion and focus on a ripe fruit diet, which has lead to interpretations of reduced flexibility compared to large apes, a synthetic evaluation of traits that define hylobatids, such as forelimb dominated suspensory locomotion, small adult group size, and small body size, leads to the opposite conclusion and identifies hylobatids as astoundingly flexible.

References

Agustí J, Sanz de Siria A, Garcés M (2003) Explaining the end of the hominoid experiment in Europe. J Hum Evol 45:145–153

Agustí J, Cabrera L, Garcés M (2013) The Vallesian mammal turnover: a late miocene record of decoupled land-ocean evolution. Geobios 46:151–157

Ahsan F (2001) Socio-ecology of the hoolock gibbon (*Hylobates hoolock*) in two forests of Bangladesh. In: The apes: challenges for the 21st century. Conference Proceedings. Chicago Zoological Society, Chicago, pp 286–299

Alba DM, Almécija, DeMiguel D, Fortuny J, Pérez de los Ríos M, Pina M, Robles JM, Moyà-Solà S (2015) Miocene small-bodied ape from Eurasia sheds light on hominoid evolution. Science 350:6260

Anandam MV, Groves CP, Molur S, Rawson BM, Richardson MC, Roos C, Whittaker DJ (2013) Species accounts of Hylobatidae. In: Mittermeier RA, Rylands AB, Wilson DE (eds) Handbook of the mammals of the world, vol 3., PrimatesLynx Edicions, Barcelona, pp 778–791

Andrews P (1992) Community evolution in forest habitats. J Hum Evol 22:423–438

Andrews P, Kelley J (2007) Middle Miocene dispersals of apes. Folia Primatol 78:328–343

Andrews P, Begun DR, Zylstra M (1997) Interrelationships between functional morphology and paleoenvironments in Miocene hominoids. In: Begun DR, Ward CV, Rose MD (eds) Function, phylogeny, and fossils. Miocene hominoid, evolution and adaptation. Springer, US, pp 29–58

Arnold ML, Meyer A (2006) Natural hybridization in primates: one evolutionary mechanism. Zoology 104:261–276

Asensio N, Brockelman WY, Malaivijitnond S, Reichard UH (2011) Gibbon travel paths are goal oriented. Anim Cogn 14:395–411

Barelli C, Boesch C, Heistermann M, Reichard UH (2008) Female white-handed gibbons (*Hylobates lar*) lead group movements and have priority of access to food resources. Behaviour 145:965–981

Barry J, Morgan M, Flynn L, Pilbeam D, Behrensmeyer A, Raza S et al (2002) Faunal and environmental change in the late Miocene Siwaliks of Northern Pakistan. Paleobiology 28:1–72

Bartlett TQ (2011) The Hylobatidae: small apes of Asia. In: Campbell CJ, Fuentes A, MacKinnon KC, Panger M, Bearder SK (eds) Primates in perspective, 2nd edn. Oxford University Press, New York, pp 300–312

Begun DR (2005) *Sivapithecus* is east and *Dryopithecus* is west, and never the twain shall meet. Anthropol ScI 113:53–64

Begun DR (ed) (2007) Fossil record of Miocene hominoids. Handbook of Paleoanthropology, Springer, New York

Begun DR (2013) The Miocene hominoid radiations. In: Begun DR (ed) A companion to paleoanthropology. Blackwell Publishing Ltd, Oxford, pp 397–416

Begun DR, Ward CV, Rose MD (eds) (1997) Function, phylogeny, and fossils: miocene hominoids evolution and adaptations. Plenum Press, New York

Begun DR, Güleç E, Geraads D (2003) Dispersal patterns of Eurasian hominoids: implications from Turkey. Deinsea 10:23–39

Begun DR, Nargolwalla MC, Kordos L (2012) European Miocene hominids and the origin of the African ape and human clade. Evol Anthropol 21:10–23

Bernor RL (1983) Geochronology and zoogeographic relationships of Miocene Hominoidea. In: Ciochon R (ed) New interpretations of ape and human ancestry. Springer, US, pp 21–64

Bertram JEA (2004) New perspectives on brachiation mechanics. Am J Phys Anthropol 125:100–117

Bradley BJ, Mundy NI (2008) The primate palette: the evolution of primate coloration. Evol Anthropol Issues News Rev 17:97–111

Brandon-Jones D, Eudey AA, Geissmann T, Groves CP, Melnick DJ, Morales JC et al (2004) Asian primate classification. Int J Primatol 25:97–164

Brockelman WY, Gittins SP (1984) Natural hybridization in the *Hylobates lar* species group: implications for speciation in gibbons. In: Preuschoft H, Chivers DJ, Brockelman WY, Creel N (eds) The lesser apes: evolutionary and behavioural biology. Edinburgh University Press, Edinburgh, pp 498–532

Burness GP, Diamond J, Flannery T (2001) Dinosaurs, dragons, and dwarfs: the evolution of maximal body size. Proc Natl Acad Sci 98:14518–14523

Cane MA, Molnar P (2001) Closing of the Indonesian seaway as a precursor to east African aridification around 3–4 million years ago. Nature 411(6834):157–162

Cannon CH, Leighton M (1994) Comparative locomotor ecology of gibbons and macaques: selection of canopy elements for crossing gaps. Am J Phys Anthropol 93:505–524

Cannon CH, Morley RJ, Bush AB (2009) The current refugial rainforests of Sundaland are unrepresentative of their biogeographic past and highly vulnerable to disturbance. Proc Natl Acad Sci 106:11188–11193

Capozzi O, Carbone L, Stanyon RR, Marra A, Yang F, Whelan CW et al (2012) A comprehensive molecular cytogenetic analysis of chromosome rearrangements in gibbons. Genome Res 22:2520–2528

Carbone L, Harris RA, Gnerre S, Veeramah KR, Lorente-Galdos B, Huddleston J et al (2014) Gibbon genome and the fast karyotype evolution of small apes. Nature 513:195–201

Caro T (2011) The functions of black-and-white coloration in mammals: review and synthesis. In: Stevens M, Merilaita S (eds) Animal camouflage: function and mechanism. Cambridge University Press, Cambridge, pp 298–329

Carpenter CR (1940) A field study in Siam of the behavior and social relations of the gibbon (*Hylobates lar*). Comp Psych Monogr 16:1–201

Casanovas-Vilar I, Alba DM, Garcés M, Robles JM, Moyà-Solà S (2011) Updated chronology for the Miocene hominoid radiation in Western Eurasia. PNAS 108:5554–5559

Cerling TE, Mbua E, Kirera FM, Manthi FK, Grine FE, Leakey MG, Sponheimer M, Uno KT (2011) Diet of *Paranthropus boisei* in the early Pleistocene of East Africa. PNAS 108:9337–9341

Chan LK (2007a) Gelenohumeral mobility in primates. Folia Primatol 78:1–18

Chan LK (2007b) Scapular position in primates. Folia Primatol 78:19–35

Chan LK (2008) The range of passive arm circumduction in primates: do hominoids really have more mobile shoulders? Am J Phys Anthropol 136:265–277

Chan YC, Roos C, Inoue-Murayama M, Inoue E, Shih CC, Pei KJ, Vigilant L (2010) Mitochondrial genome sequences effectively reveal the phylogeny of hylobates gibbons. PLoS ONE 5:e14419

Chan YC, Roos C, Inoue-Murayama M, Inoue E, Shih CC, Vigilant L (2012) A comparative analysis of Y chromosome and mtDNA phylogenies of the hylobates gibbons. BMC Evol Biol 12:150

Chan YC, Roos C, Inoue-Murayama M, Inoue E, Shih CC, Pei KJ et al (2013) Inferring the evolutionary histories of divergences in *Hylobates* and *Nomascus* gibbons through multilocus sequence data. BMC Evol Biol 13:82

Chaplin G (2005) Physical geography of the Gaoligong Shan area of southwest China in relation to biodiversity. Proc Calif Acad Sci 56:527

Chatterjee HJ (2006) Phylogeny and biogeography of gibbons: a dispersal vicariance analysis. Int J Primatol 27:699–712

Chatterjee HJ (2009) Evolutionary relationships among the hylobatids: a biogeographic perspective. In: Lappan S, Whittaker DJ (eds) The hylobatids: new perspectives on small ape socioecology and population biology. Springer, New York, pp 13–36

Chatterjee HJ (2016) The role of historical and fossil records in predictingchanges in the spatial distribution of hylobatids. In: Reichard UH, Hirohisa H, Barelli C (eds) Evolution of gibbons and siamang. Springer, New York, pp 43–54

Cheyne SM, Chivers DJ, Sugardjito J (2007) Covariation in the great calls of rehabilitant and wild gibbons *Hylobates agilis albibarbis*. Raffles Bull Zool 55:201–207

Choudhury A (2006) The distribution and status of hoolock gibbon, *Hoolock hoolock* in Manipur, Meghalaya, Mizoram and Nagaland in Northeast India. Primate Conserv 20:79–87

Ciochon R, Long VT, Larick R, González L, Grün R, De Vos J et al (1996) Dated co-occurrence of *Homo erectus* and *Gigantopithecus* from Tham Khuyen cave, Vietnam. Proc Natl Acad Sci 93:3016–3020

Clarke E, Reichard UH, Zuberbühler K (2006) The syntax and meaning of wild gibbon songs. PLoS ONE 1:e73. doi:10.1371/journal.pone.0000073

Clift PD, Hodges KV, Heslop D, Hannigan R, Van Long H, Calves G (2008) Correlation of Himalayan exhumation rates and Asian monsoon intensity. Nat Geosci 1(12):875–880

Clift PD, Wan S, Blusztajn J (2014) Reconstructing chemical weathering, physical erosion and monsoon intensity since 25 Ma in the northern South China Sea: a review of competing proxies. Earth Sci Rev 130:86–102

Cowlishaw G (1982) Song function in gibbon. Behaviour 121:131–153

Dallmann R, Geissmann T (2009) Individual and geographical variability in the songs of wild silvery gibbons (*Hylobates moloch*) on Java, Indonesia. In: Lappan S, Whittaker DJ (eds) The gibbons. New prospectives on small apes socioecology and population biology, Springer, New York, pp 91–110

Das J, Biswas J, Bhattacherjee PC, Mohnot SM (2009) The distribution and abundance of hoolock gibbons in India. In: Lappan SM, Whittaker D (eds) The gibbons: new perspectives on small ape socioecology and population biology. Springer, Berlin, pp 409–433

Delgado RA (2006) Sexual selection in the loud calls of male primates: signal content and function. Int J Primatol 27:5–25

Donoghue PC (2005) Saving the stem groups—a contradiction in terms? Paleobiology 31:553–558

Eronen JT, Ataabadi MM, Micheels A, Karme A, Bernor RL, Fortelius M (2009) Distribution history and climatic controls of the late Miocene Pikermian chronofauna. Proc Natl Acad Sci 106:11867–11871

Fan PF, Jiang XL, Liu CM, Luo WS (2006) Polygynous mating system and behavioural reason of black crested gibbon (*Nomascus concolor jingdongensis*) at Dazhaizi, Mt Wuliang, Yunnan, China. Zool Res 27:216–220

Fan PF, Xiao W, Huo S, Jiang XL (2009) Singing behavior and singing functions of black-crested gibbons (Nomascus concolor jingdongensis) at Mt. Wuliang, Central Yunnan. China. Am J Primatol 71:539–547

Fedy BC, Stutchbury BJM (2005) Territory defence in tropical birds: are females as aggressive as males? Behav Ecol Sociobiol 58:414–422

Fleagle JG (1976) Locomotion and posture of the Malayan siamang and implications for hominid evolution. Folia Primatol 26:245–269

Fleagle JG (ed) (2013) Primate adaptation & evolution, 3rd edn. Academic Press Elsevier, Amsterdam

Flower BP, Kennett JP (1994) The middle Miocene climatic transition: east Antarctica ice sheet development, deep ocean circulation and global carbon cycling. Paleogeogr Paleoclimatol Paelaeocol 108:537–555

Foster JB (1964) Evolution of mammals on islands. Nature 202:234–235

Gebo DL (1996) Climbing, brachiation, and terrestrial quadrupedalism: historical precursors of hominid bipedalism. Am J Phys Anthropol 101:55–92

Geissmann T (1993) Evolution of communication in gibbons (hylobatidae). Dissertation, Zürich University

Geissmann T (2000a) Gibbon songs and human music in an evolutionary perspective. In: Wallin N, Merker B, Brown S (eds) The origins of music. MIT Press, Cambridge, Massachusetts, pp 103–123

Geissmann T (2000b) Duet songs of the siamang, *Hylobates syndactylus*: I. Structure and organization. Primate Rep 56:33–60

Geissmann T (2002) Duet-splitting and the evolution of gibbon songs. Biol Rev Camb Philos Soc 77:57–76

Geissmann T, Orgeldinger M (2000) The relationship between duet songs and pair bonds in siamangs, *Hylobates syndactylus*. Anim Behav 60:805–809

Gibbs RA, Rogers J, Katze MG, Bumgarner R, Weinstock GM, Mardis ER et al (2007) Evolutionary and biomedical insights from the rhesus macaque genome. Science 316:222–234

Gilhooly LJ, Rayadin Y, Cheyne SM (2015) A comparison of hylobatid survey methods using triangulation on Müller's gibbon (*Hylobates muelleri*) in Sungai Wain protection forest, East Kalimantan, Indonesia. Int J Primatol 36:567–582

Gittins SP, Raemaekers JJ (1980) Siamang, lar and agile gibbons. Malayan forest primates. Springer, US, pp 63–106

Gradstein FM, Ogg G, Schmitz M (2012) The geologic time scale 2012 2-volume set. Elsevier

Gray JE (ed) (1870) Catalogue of monkeys, lemurs, and fruit-eating bats in the collection of the British Museum. British Museum Trustees, London

Grehan JR, Schwartz JH (2009) Evolution of the second orangutan: phylogeny and biogeography of hominid origins. J Biogeogr 36:1823–1844

Groves CP (1972) Systematics and phylogeny of gibbons. In: Rumbaugh DM (ed) Gibbon and siamang. Karger, Basel, pp 1–89

Groves C (ed) (2001) Primate taxonomy. Smithsonian Institution Press, Washington

Grueter CC, Isler K, Dixson BJ (2015) Are badges of status adaptive in large complex primate groups? Evol Hum Behav 36:398–406

Haimoff EH (1984) Acoustic and organizational features of gibbon songs. In: Preuschoft H, Chivers D, Brockelman W, Creel N (eds) The lesser apes: evolutionary and behavioural biology. Edinburgh University Press, Edinburgh, pp 333–353

Haimoff EH (1985) The organisation of song in Müllers gibbon (*Hylobates muelleri*). Int J Primatol 6:173–192

Haimoff EH (1986) Convergence in the duetting of 2 primates. J Hum Evol 15:51–59

Hall ML (2004) A review of hypotheses for the functions of avian duetting. Behav Ecol Sociobiol 55:415–430

Hall ML (2009) A review of vocal duetting in birds. Adv Stud Behav 40:67–121

Hall R (2011) Australia–SE Asia collision: plate tectonics and crustal flow. In: Hall R, Cottam MA,Wilson MEJ (eds) The SE Asian gateway: history and tectonics of the Australia–Asia collision, vol 355. Geological Society, Special Publications, London, pp 75–109

Han K, Konkel MK, Xing J, Wang H, Lee J, Meyer TJ et al (2007) Mobile DNA in Old World monkeys: a glimpse through the rhesus macaque genome. Science 316:238–240

Haq BU, Hardenbol J, Vail PR (1987) Chronology of fluctuating sea levels since the Triassic. Science 235:1156–1167

Harrison T (2010) Apes among the tangled branches of human origins. Science 327:532–534

Harrison T (2016) The fossil record and evolutionary history of hylobatids. In: Reichard UH, Hirohisa H, Barelli C (eds) Evolution of gibbons and siamang. Springer, New York, pp 91–110

Harrison T, Ji X, Zheng L (2008) Renewed investigations at the late Miocene hominoid locality of Leilao, Yunnan. China. Am J Phys Anthropol 135(S46):113

Harzhauser M, Piller WE (2004) Benchmark data of a changing sea—Palaeogeography, Palaeobiogeography and events in the Central Paratethys during the Miocene. Palaeogeogr Palaeoclimatol Palaeoecol 253:8–31

Harzhauser M, Kroh A, Mandic O, Piller WE, Göhlich U, Reuter M, Berning B (2007) Biogeographic responses to geodynamics: a key study all around the Oligo-Miocene Tethyan Seaways. Zool Anz 246:241–256

Hayashi S, Hayasaka K, Takenaka O, Horai S (1995) Molecular phylogeny of gibbons inferred from mitochondrial-DNA sequences—preliminary report. J Mol Evol 41:359–365

Heizmann EP, Begun DR (2001) The oldest Eurasian hominoid. J Hum Evol 41:463–481

Hobolth A, Dutheil JY, Hawks J, Schierup MH, Mailund T (2011) Incomplete lineage sorting patterns among human, chimpanzee, and orangutan suggest recent orangutan speciation and widespread selection. Genome Res 21:349–356

Homke S, Vergés J, Van Der Beek P, Fernandez M, Saura E, Barbero L et al (2010) Insights in the exhumation history of the NW Zagros from bedrock and detrital apatite fission-track analysis: evidence for a long-lived orogeny. Basin Res 22(5):659–680

Hunt KD (2004) The special demands of great ape locomotion. In: Russon AE, Begun ER (eds) The evolution of thought: Evolutionary origins of great ape intelligence. Cambridge University Press, Cambridge, pp 172–189

Hunt KD, Cant JGH, Gebo DL, Rose MD, Walker SE, Youlatos D (1996) Standardized descriptions of primate locomotor and postural modes. Primates 37:363–387

Husson SJ, Wich SA, Marshall AJ, Dennis RD, Ancrenaz M, Brassey R et al (2009) Orangutan distribution, density, abundance and impacts of disturbance. In: Wich SA, Utami Atmoko SS, Setia TM, van Schaik CP (eds) Orangutans: geographic variation in behavioral ecology and conservation. Oxford University Press, Oxford, p 77–96

Israfil H, Zehr SM, Mootnick AR, Ruvolo M, Steiper ME (2011) Unresolved molecular phylogenies of gibbons and siamangs (Family: Hylobatidae) based on mitochondrial, Y-linked, and X-linked loci indicate a rapid Miocene radiation or sudden vicariance event. Mol Phylogenet Evol 58:447–455

Jablonski NG (2000) Micro-and macroevolution: scale and hierarchy in evolutionary biology and paleobiology. Paleobiology 26:15–52

Jablonski NG (2005) Primate homeland: forests and the evolution of primates during the Tertiary and Quaternary in Asia. Anthropol Sci 113:117–122

Jablonski NG, Chaplin G (2009) The fossil record of gibbons. In: Lappan SM, Whittaker D (eds) The gibbons: new perspectives on small ape socioecology and population biology. Springer, Berlin, pp 111–130

Jablonski NG, Su DF, Flynn LJ, Ji X, Deng C, Kelley J et al (2014) The site of Shuitangba (Yunnan, China) preserves a unique terminal Miocene fauna. J Vertebr Paleontol 34:1251–1257

Janis CM (1993) Tertiary mammal evolution in the context of changing climates, vegetation, and tectonic events. Annu Rev Ecol Syst 24:467–500

Jauch A, Wienberg J, Stanyon R, Arnold N, Tofanelli S, Ishida T, Cremer T (1992) Reconstruction of genomic rearrangements in great apes and gibbons by chromosome painting. Proc Natl Acad Sci USA 89:8611–8615

Ji XP, Jablonski NG, Su DF, Deng CL, Flynn LJ, You YS, Kelley J (2013) Juvenile hominoid cranium from the terminal Miocene of Yunnan, China. Chin Sci Bull 58:3771–3779

Jiang X, Luo Z, Zhao S, Li R, Liu C (2006) Status and distribution pattern of black crested gibbon (*Nomascus concolor jingdongensis*) in Wuliang Mountains, Yunnan, China: implication of conservation. Primates 477:264–271

Jin C, Qin D, Pan W, Tang Z, Liu J, Wang Y, Deng C, Zhang Y, Dong W, Tong H (2008) A newly discovered *Gigantopithecus* fauna from Sanhe Cave, Chongzuo, Guangxi, South China. Chin Sci Bull 54:1–10

Kamilar JM, Bradley BJ (2011) Interspecific variation in primate coat colour supports Gloger's rule. J Biogeogr 38:2270–2277

Kappeler PM, van Schaik CP (eds) (2004) Sexual selection in primates: new and comparative perspectives. Cambridge University Press, Cambridge

Kawagata S, Hayward BW, Gupta AK (2006) Benthic foraminiferal extinctions linked to late Pliocene-Pleistocene deep-sea circulation changes in the northern Indian Ocean (ODP Sites 722 and 758). Mar Micropaleontol 58:219–242

Keith SA, Waller MS, Geissmann T (2009) Vocal diversity of Kloss's gibbons (*Hylobates klossii*) in the Mentawai Islands, Indonesia. In: Lappan S, Whittaker DJ (eds) The gibbons. New prospectives on small apes socioecology and population biology, Springer, New York, pp 51–71

Kim SK, Carbone L, Becquet C, Mootnick AR, Li DJ, de Jong PJ, Wall JD (2011) Patterns of genetic variation within and between gibbon species. Mol Biol Evol 28:2211–2218

Koda H (2016) Gibbon songs: understanding the evolution and development of this unique form of vocal communication. In: Reichard UH, Hirohisa H, Barelli C (eds) Evolution of gibbons and siamang. Springer, New York, pp 347–357

Kürschner WM, Kvaček Z, Dilcher DL (2008) The impact of Miocene atmospheric carbon dioxide fluctuations on climate and the evolution of terrestrial ecosystems. Proc Natl Acad Sci 105:449–453

Lewis AR, Marchant DR, Ashworth AC, Hedenäs L, Hemming SR, Johnson JV et al (2008) Mid-Miocene cooling and the extinction of tundra in continental Antarctica. PNAS 105:10676–10680

Li J, Fang X, Song C, Pan B, Ma Y, Yan M (2014) Late Miocene-Quaternary rapid stepwise uplift of the NE Tibetan Plateau and its effects on climatic and environmental changes. Quat Res 81:400–423

Locke DP, Hillier LW, Warren WC, Worley KC, Nazareth LV, Muzny DM et al (2011) Comparative and demographic analysis of orang-utan genomes. Nature 469:529–533

Lomolino MV (1985) Body size of mammals on islands: the island rule reexamined. Am Nat 125:310–316

Lomolino MV (2005) Body size evolution in insular vertebrates: generality of the island rule. J Biogeogr 32:1683–1699

Louys J, Curnoe D, Tong H (2007) Characteristics of Pleistocene megafauna extinctions in Southeast Asia. Palaeogeogr Palaeoclimat Palaeoecol 243:152–173

Lukas D, Clutton-Brock TH (2013) The evolution of social monogamy in mammals. Science 341:526–530

Ma YZ, Li JJ, Fang XM (1998) Pollen assemblage in 30.6–5.0 Ma redbeds of Linxia region and climate evolution. Chin Sci Bull 43(3):301–304

Manfreda E, Mitteroecker P, Bookstein FL, Schaefer K (2006) Functional morphology of the first cervical vertebra in humans and nonhuman primates. Anat Rec B New Anat 289:184–194

Marshall AJ (2009) Are montane forests demographic sinks for Bornean white-bearded gibbons Hylobates albibarbis? Biotropica 41:257–267

Marshall JT, Marshall E (1976) Gibbons and their territorial songs. Science 193:235–237

Marshall JT, Sugardjito J (1986) Gibbon systematics. In: Swindler DR, Erwin J (eds) Comparative primate biology, vol 1., systematicsAlan R Liss, New York, pp 137–185

Marshall-Ball L, Mann N, Slater PJB (2006) Multiple functions to duet singing: hidden conflicts and apparent cooperation. Anim Behav 71:823–831

Mather R (1992) A field study of hybrid gibbons in Central Kalimantan Indonesia. Dissertation, University of Cambridge, Cambridge

Matsudaira K, Ishida T (2010) Phylogenetic relationships and divergence dates of the whole mitochondrial genome sequences among three gibbon genera. Mol Phylogenet Evol 55:454–459

Matsui A, Rakotondraparany F, Munechika I, Hasegawa M, Horai S (2009) Molecular phylogeny and evolution of prosimians based on complete sequences of mitochondrial DNAs. Gene 441:53–66

McNab BK (1963) Bioenergetics and the determination of home range size. Am Nat 97:133–140

McNulty KP, Begun DR, Kelley J, Manthi FK, Mbua EN (2015) A systematic revision of Proconsul with the description of a new genus of early Miocene hominoid. J Hum Evol 84:42–61

Meyer TJ, McLain AT, Oldenburg JM, Faulk C, Bourgeois MG, Conlin EM et al (2012) An Alu-based phylogeny of gibbons (Hylobatidae). Mol Biol Evol 29:3441–3450

Michilsens F, Vereecke EE, D'Août K, Aerts P (2009) Functional anatomy of the gibbon forelimb: adaptations to a brachiating lifestyle. J Anat 215:335–354

Miller KG, Wright JD, Fairbanks GG (1991) Unlocking the ice house: oligocene-miocene oxygen isotopes, eustasy, and marginal erosion. J Geophys Res 96:6829–6848

Miller KG, Wright JD, Browning JV, Kulpecz A, Kominz M, Naish TR et al (2012) High tide of the warm pliocene: implications of global sea level for Antarctic deglaciation. Geology 40:407–410

Misceo D, Capozzi O, Roberto R, Dell'oglio MP, Rocchi M, Stanyon R, Archidiacono N (2008) Tracking the complex flow of chromosome rearrangements from the Hominoidea ancestor to extant Hylobates and Nomascus gibbons by high-resolution synteny mapping. Genome Res 18:1530–1537

Mitani JC (1984) The behavioral regulation of monogamy in gibbons (*Hylobates muelleri*). Behav Ecol Sociobiol 15:225–229

Mitani JC (1987) Territoriality and monogamy among agile gibbons (*Hylobates agilis*). Behav Ecol Sociobiol 20:265–269

Molnar P, Stock JM (2009) Slowing of India's convergence with Eurasia since 20 Ma and its implications for Tibetan mantle dynamics. Tectonics 28(3)

Mootnick AR (2006) Gibbon (Hylobatidae) species identification recommended for rescue or breeding centers. Primate Conserv 21:103–138

Mootnick AR, Groves C (2005) A new generic name for the Hoolock gibbon (Hylobatidae). Int J Primatol 26:971–976

Morley RJ (ed) (2000) Origin and evolution of tropical rain forests. Wiley, New York

Morley RJ, Flenley JR (1987) Late Cainozoic vegetational and environmental changes in the Malay archipelago. In: Whitmore TC (ed) Biogeographical evolution of the Malay Archipelago. Clarendon Press, Oxford, pp 50–59

Müller S, Hollatz M, Wienberg J (2003) Chromosomal phylogeny and evolution of gibbons (Hylobatidae). Hum Genet 113:493–501

Nelson SV (2005) Paleoseasonality inferred from equid teeth and intra-tooth isotopic variability. Palaeogeogr Palaeoclimatol Palaeoecol 222:122–144

Nelson SV (2007) Isotopic reconstructions of habitat change surrounding the extinction of *Sivapithecus*, a Miocene hominoid, in the Siwalik Group of Pakistan. Paleogeogr Paleoclimatol Paleoecol 243:204–222

Noë R (2006) Cooperation experiments: coordination through communication versus acting apart together. Anim Behav 71:1–18

Okay AI, Zattin M, Cavazza W (2010) Apatite fission-track data for the Miocene Arabia-Eurasia collision. Geology 38:35–38

Palkovacs EP (2003) Explaining adaptive shifts in body size on islands: a life history approach. Oikos 103:37–44

Patnaik R, Chauhan P (2009) India at the cross-roads of human evolution. J Biosci 5:729–747

Perelman P, Johnson WE, Roos C, Seuánez HN, Horvath JE, Moreira MAM, Kessing B, Pontius J, Roelke M, Rumpler Y, Scheider MPC, Silva A, O'Brien SJ, Pecon-Slattery J (2011) A molecular phylogeny of living primates. PLoS Genet 7(3):e1001342

Perry GH, Dominy NJ (2009) Evolution of the human pygmy phenotype. Trends Ecol Evol 24:218–225

Pilbeam D (1996) Genetic and morphological records of the Hominoidea and hominid origins: a synthesis. Mol Phylogenet Evol 5:155–168

Plavcan JM (2012) Body size, size variation, and sexual size dimorphism in early *Homo*. Curr Anthropol 53:S409–S423

Pontzer H, Raichlen DA, Shumaker RW, Ocobock C, Wich SA (2010) Metabolic adaptation for low energy throughput in orangutans. PNAS 107:14048–14052

Preuschoft H, Schönwasser K-H, Witzel U (2016) Selective value of characteristic size parametersin hylobatids. A biomechanical approach to smallape size and morphology. In: Reichard UH, Hirohisa H, Barelli C (eds) Evolution of gibbons and siamang. Springer, New York, pp 227–263

Prouty LA, Buchanan PD, Pollitzer WS, Mootnick AR (1983) *Bunopithecus*: a genus-level taxon for the hoolock gibbon (*Hylobates hoolock*). Am J Primatol 5:83–87

Quade J, Cerling TE, Bowman JR (1989) Development of Asian monsoon revealed by marked ecological shift during the latest Miocene in northern Pakistan. Nature 342(6246):163–166

Raaum RL, Sterner KN, Noviello CM, Stewart CB, Disotell TR (2005) Catarrhine primate divergence dates estimated from complete mitochondrial genomes: concordance with fossil and nuclear DNA evidence. J Hum Evol 48:237–257

Raemaekers JJ, Raemaekers PM (1985) Field playback of loud calls to gibbons (*Hylobates lar*): territorial, sex-specific and species-specific responses. Anim Behav 33:481–493

Raemaekers JJ, Raemaekers PM, Haimoff EH (1984) Loud calls of the gibbon (*Hylobates lar*): repertoire, organization and context. Behaviour 91:146–189

Raia P, Meiri S (2006) The island rule in large mammals: paleontology meets ecology. Evolution 60:1731–1742

Reichard UH, Barelli C (2008) Life history and reproductive strategies of Khao Yai *Hylobates lar*: implications for social evolution in apes. Int J Primatol 29:823–844

Reichard UH, Ganpanakngan M, Barelli C (2012) White-handed gibbons of Khao Yai: social flexibility, complex reproductive strategies, and a slow life history. In: Watts DP, Kappeler PM (eds) Long-term field studies of primates. Springer, Berlin, pp 237–258

Reichard UH, Croissier MM (2016) Hylobatid evolution in paleogeographic and paleoclimatic context. In: Reichard UH, Hirohisa H, Barelli C (eds) Evolution of gibbons and siamang. Springer, New York, pp 111–135

Reuter M, Piller WE, Harzhauser M, Mandic O, Berning B, Rögl F, Kroh A, Aubry M-P, Wielandt-Schuster U, Hamedani A (2009) The Oligo-/Miocene Qom formation (Iran): evidence for an early Burdigalian restriction of the Tethyan Seaway and closure of its Iranian gateways. Int J Earth Sci 98:627–650

Rögl F (1999) Mediterranean and paratethys. Facts and hypotheses of an oligocene to Miocene paleogeography (short overview). Geol Carpath 50:339–349

Roos C, Geissmann T (2001) Molecular phylogeny of the major hylobatid divisions. Mol Phylogenet Evol 19:486–494

Roos C, Boonratana R, Supriatna J, Fellowes JR, Rylands AB, Mittermeier RA (2013) An updated taxonomy of primates in Vietnam, Laos, Cambodia and China. Vietn J Primatol 2:13–26

Roos C (2016) Phylogeny and classification of gibbons (hylobatidae). In: Reichard UH, Hirohisa H, Barelli C (eds) Evolution of gibbons and siamang. Springer, New York, pp 151–164

Ruggieri E, Herbert T, Lawrence KT, Lawrence CE (2009) Change point method for detecting regime shifts in paleoclimatic time series: application to δ18O time series of the Plio-Pleistocene. Paleoceanography 24:1

Schmidt NM, Jensen PM (2003) Changes in Mammalian body length over 175 years-adaptations to a fragmented landscape? Conserv Ecol 7:6

Schmidt NM, Jensen PM (2005) Concomitant patterns in avian and mammalian body length changes in Denmark. Ecol Soc 10:5

Setchell JM, Wickings EJ (2005) Dominance, status signals and coloration in male mandrills (*Mandrillus sphinx*). Ethology 111:25–50

Shackleton NJ, Backman J, Zimmerman H, Kent DV, Hall MA, Roberts DG et al (1984) Oxygen isotope calibration of the onset of ice-rafting and history of glaciation in the North Atlantic region. Nature 307:620–623

Shevenell AE, Kennett JP, Lea DW (1994) Middle Miocene southern ocean cooling and Antarctic cryosphere expansion. Science 305:1766–1770

Shevenell AE, Kennett JP, Lea DW (2004) Middle Miocene southern ocean cooling and Antarctic cryosphere expansion. Science 305:1766–1770

Shevenell AE, Kennett JP, Lea DW (2008) Middle Miocene ice sheet dynamics, deep-sea temperatures, and carbon cycling: a Southern Ocean perspective. Geochem Geophys 9(2)

Smith RJ, Jungers WL (1997) Body mass in comparative primatology. J Hum Evol 32:523–559

Stevens NJ, Seiffert ER, O'Connor PM, Roberts EM, Schmitz MD, Krause C, Gorscak E, Ngasala S, Hieronymus TL, Temu J (2013) Palaeontological evidence for an oligocene divergence between old world monkeys and apes. Nature 497:611–614

Stewart KJ, Sicotte P, Robbins MM (2001) Mountain gorillas of the virungas: a short history. In: Robbins MM, Sicotte P, Stewart KJ (eds) Mountain gorillas: three decades of research at Karisoke. Cambridge University Press, Cambridge, pp 1–26

Sun J, Zhang Z (2008) Palynological evidence for the mid-Miocene climatic optimum recorded in cenozoic sediments of the Tian Shan Range, northwestern China. Global Planet Change 64:53–68

Takacs Z, Morales JC, Geissmann T, Melnick DJ (2005) A complete species-level phylogeny of the hylobatidae based on mitochondrial *ND3-ND4* gene sequences. Mol Phylogenet Evol 36:456–467

Tembrock G (1974) Sound production of *Hylobates* and *Symphalangus*. In: Rumbaugh DM (ed) Gibbon and Siamang, vol 3. Karger, Basel, pp 176–205

Terleph TA, Malaivijitnond S, Reichard UH (2015) Lar gibbon (*Hylobates lar*) great call reveals individual caller identity. Am J Primatol 77:811–821

Terleph TA, Malaivijitnond S, Reichard UH (2016) Age related decline in female lar gibbon great call performance suggests that call features correlate with physical condition. BMC Evol Biol 16:4

Thinh VN, Mootnick AR, Geissmann T, Li M, Ziegler T, Agil M, Moisson P, Nadler T, Walter L, Roos C (2010a) Mitochondrial evidence for multiple radiations in the evolutionary history of small apes. BMC Evol Biol 10:74

Thinh VN, Rawson B, Hallam C, Kenyon M, Nadler T, Walter L, Roos C (2010b) Phylogeny and distribution of crested gibbons (genus *Nomascus*) based on mitochondrial cytochrome b gene sequence data. Am J Primatol 72:1047–1054

Thinh VN, Hallam C, Roos C, Hammerschmidt K (2011) Concordance between vocal and genetic diversity in crested gibbons. BMC Evol Biol 11:36

Thorpe SHS, Crompton K (2006) Orangutan positional behavior and the nature of arboreal locomotion in Hominoidea. Am J Phys Anthropol 131:184–401

Thorpe WH, Hall-Craggs J, Hooker B, Hooker T, Hutchins R (1972) Duetting and antiphonal song in birds: its extent and significance. Behaviour (Supplement No. 18:III, VII-XI):1–197

Tilson RL (1976) Monogamy and duetting in an Old World monkey. Nature 263:320–321

Tyler DE (1993) The evolutionary history of the gibbons. In: Jablonski NG (ed) Evolving landscapes and evolving biotas of East Asia since the mid-Tertiary. The University of Hong Kong, Hong Kong, Centre of Asian Studies, pp 228–240

Ungar PS, Kay RF (1995) The dietary adaptations of European Miocene catarrhines. Proc Natl Acad Sci 92:5479–5481

Usherwood JR, Bertram JE (2003) Understanding brachiation: insight from a collisional perspective. J Exp Biol 206:1631–1642

van Tuinen P, Ledbetter DH (1983) Cytogenetic comparison and phylogeny of three species of Hylobatidae. Am J Phys Anthropol 61:453–466

van Valen L (1973) A new evolutionary law. Evol Theor 1:1–33

Wanelik K, Abdulazis Cheyne SM (2013) Note-, phase- and song-specific acoustic variables contributing to the individuality of male duet song in the Bornean Southern gibbon (*Hylobates albibarbis*). Primates 54:159–170

Wang C, Zhao X, Liu Z, Lippert PC, Graham SA, Coe RS et al (2008) Constraints on the early uplift history of the Tibetan Plateau. PNAS 105(13):4987–4992

Westerhold T, Bickert T, Röhl U (2005) Middle to late Miocene oxygen isotope stratigraphy of ODP site 1085 (SE Atlantic): new constrains on Miocene climate variability and sea-level fluctuations. Palaeogeogr Palaeoclimatol Palaeoecol 217:205–222

Whittaker DJ, Morales JC, Melnick DJ (2007) Resolution of the *Hylobates* phylogeny: congruence of mitochondrial D-loop sequences with molecular, behavioral, and morphological data sets. Mol Phylogenet Evol 45:620–628

Wilkinson RD, Steiper ME, Soligo C, Martin RD, Yang Z, Tavare S (2011) Dating primate divergences through an integrated analysis of palaeontological and molecular data. Syst Biol 60:16–31

Woodruff DS (2010) Biogeography and conservation in Southeast Asia: how 2.7 million years of repeated environmental fluctuations affect today's patterns and the future of the remaining refugial-phase biodiversity. Biodivers Conserv 19:919–941

Woodruff DS, Turner LM (2009) The Indochinese-Sundaic zoogeographic transition: a description and analysis of terrestrial mammal species distributions. J Biogeogr 36:803–821

Wright JD, Miller KG, Fairbanks RG (1992) Early and middle Miocene stable isotopes: implications for deepwater circulation and climate. Paleoceanography 7:357–389

Yanuar A (2009) The population distribution and abundance of siamangs (*Symphalangus syndactylus*) and agile gibbons (*Hylobates agilis*) in west central Sumatra, Indonesia. In: Lappan SM, Whittaker D (eds) The gibbons: new perspectives on small ape socioecology and population biology. Springer, Berlin, pp 453–465

Zachos J, Pagani M, Sloan L, Thomas E, Billups K (2001) Trends, rhythms, and aberrations in global climate 65 ma to present. Science 292:686–693

Zhao LX, Zhang LZ (2013) New fossil evidence and diet analysis of *Gigantopithecus blacki* and its distribution and extinction in South China. Quat Int 286:69–74

Zhisheng A, Kutzbach JE, Prell WL, Porter SC (2001) Evolution of Asian monsoons and phased uplift of the Himalaya-Tibetan plateau since Late Miocene times. Nature 411:62–66

Zihlmann AL, Mootnick AR, Underwood CE (2011) Anatomical contributions to hylobatid taxonomy and adaptation. Int J Primatol 32:865–877

Chapter 2
The Role of Historical and Fossil Records in Predicting Changes in the Spatial Distribution of Hylobatids

Helen J. Chatterjee

Subadult female white-handed gibbon monitoring the surrounding, Khao Yai National Park, Thailand. Photo credit: Ulrich H. Reichard

H.J. Chatterjee (✉)
Department of Genetics, Evolution and Environment,
University College London, London WC1E 6BT, UK
e-mail: h.chatterjee@ucl.ac.uk

© Springer Science+Business Media New York 2016
U.H. Reichard et al. (eds.), *Evolution of Gibbons and Siamang*,
Developments in Primatology: Progress and Prospects,
DOI 10.1007/978-1-4939-5614-2_2

43

Introduction

Southeast Asia has experienced considerable palaeo-environmental and more recent environmental changes which have affected the spatial distribution of the region's fauna and flora. Interpreting and predicting changes in the spatial distribution of rare or endangered taxa is useful for both macro-evolutionary studies and for conservation action planning. Furthermore, quantifying these changes, spatially and temporally, provides valuable biogeographic information that can be used to understand the tolerance of species to environmental fluctuations (Stigall and Lieberman 2006). This is especially important for taxa such as hylobatids, where all but one of the species is listed as endangered or critically endangered by the IUCN (2015).

Environmental change in Southeast Asia continues to be a source of biogeographic interest. It is argued that the region is experiencing the highest relative rates of deforestation and forest degradation in the humid tropics (Koh and Sodhi 2010). Southeast Asia is widely acknowledged as one of the world's most significant biodiversity hotspots (e.g., Myers et al. 2000; Sodhi et al. 2004; Edwards et al. 2011). It is home to 5 of the 25 regionally defined biodiversity hotpots (comprising Sundaland, Wallacea, Philippines, Indo-Burma and South-Central China: Myers et al. 2000) and as such contains exceptional concentrations of endemic species which are undergoing extensive habitat loss. In light of the rapid rate of deforestation, and the high concentration of endemic species in the region, it is predicted that Southeast Asia could lose 13–42 % of local populations by the end of this century, at least 50 % of which could represent global species extinction (Brook et al. 2003). Given the relatively high diversity of hylobatids, with 19 recognized species (Anandam et al. 2013), the hominoids could be more severely affected compared to non-hominoid groups.

Habitat loss and fragmentation are likely to be exacerbated by other anthropogenic pressures including human-induced climate change and associated rapid industrialization, economic growth and human population migration in the future. South Asia and Southeast Asia have been identified as regions likely to experience more pronounced climate change and it has been argued that such ecoregions require more detailed impact assessments to inform effective conservation (Li et al. 2013). Southeast Asia has a long history of environmental change dating back many millions of years and these changes are responsible for the region's biogeographic uniqueness. It has been suggested that in order to achieve long-term success in protecting and conserving Southeast Asia's biodiversity it is vital to understand its historical biogeography (Koh and Sodhi 2010; Woodruff 2010).

Historical Biogeography and Species Ranges Changes in Southeast Asia

Dramatic plate tectonic and orogenic activity in Southeast Asia during the Cenozoic, gave rise to various land mass changes, which in tandem with glacial activity, affected the areas geomorphology, climate, flora and fauna (Chatterjee

2009). Palaeo-environmental changes dating back 15–20 million years impacted the extent of emergent land which provided both pathways and barriers to faunal and floral dispersal and vicariance events. In turn, climatic variations associated with glacial periodicity during the Pleistocene also affected sea levels and hence the extent of exposed land. Similarly, climatic changes associated with these fluctuations affected the type of ecosystems that could be sustained on exposed land (see Chatterjee 2009 and references therein).

Whilst climatic deterioration from the late Miocene onwards negatively affected some primate fauna, this does not appear to be the case regarding hylobatids (Jablonski 1993). Indeed, from both paleontological and palaeo-environmental records of China, in spite of increased seasonality and habitat fragmentation, hylobatids were among the most successful primates (Jablonski 1993; Jablonski et al. 2000). This scenario is in contrast to the hypothesis that in hominoids certain life history traits such as relatively long gestation times, long weaning periods, long inter-birth intervals, lower intrinsic rates of population increase and preferences for higher-quality fruits, imply higher vulnerability to environmental changes (Jablonski et al. 2000).

Hylobatids are predicted to have expanded their range from a putative gibbon ancestor in Eastern Indochina from approximately 10.5 million years ago, after which time hylobatids differentiated and radiated southwards (Chatterjee 2006). These radiation events occurred in tandem with considerable environmental changes, which raises interesting questions about hylobatid species tolerance. Regardless of the limitations of the sparse fossil record for hylobatids (Jablonski 1993; Jablonski et al. 2000; Jablonski and Chaplin 2009), gibbons and the siamangs were able to alter their spatial distribution in order to adapt to changing environmental conditions. This concept is supported when historical and fossil data are considered in the context of present and future predicted habitat ranges.

To predict the tolerance of Chinese hylobatids to changing environmental conditions, Chatterjee et al. (2012) used Ecological Niche Modelling (ENM) based on fossil, historical and present-day distribution data. The study developed a database of locality data for Chinese hylobatids from the published literature spanning three time intervals: fossil (Pliocene to earliest Holocene), historical (265 AD–1945) and modern (1945 AD to the present). The ENM software DIVA-GIS (Hijmans et al. 2005) was used to generate maps showing the distributional ranges for hylobatids in China in each of the three time intervals. Present and future habitat suitability was predicted using global climatic variables including precipitation, temperature and carbon dioxide emissions. According to recent models, mean temperature and precipitation in China are predicted to increase by c.0.71 °C and c.8.4 mm, respectively, over the next 30 years (Liu et al. 2010). These variables were manipulated in line with future climate change predictions to model the effects of these changes on future gibbon habitat suitability.

The results of this study show that Pliocene-Holocene gibbon fossils are distributed from southernmost China to the Yangtze River delta in eastern China (Fig. 2.1) Further, there are more gibbon fossils recorded from the south-western provinces of Yunnan, Guangxi and Hainan compared to more northern and eastern provinces (Guizhou, Guangdong, Hunan, Hubei, Chongqing,

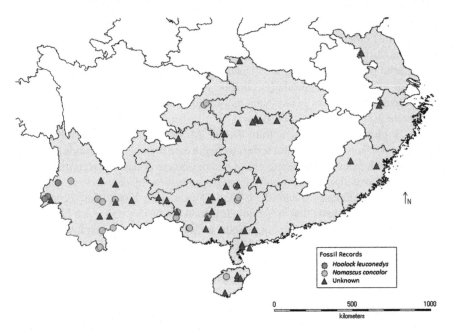

Fig. 2.1 Map of China showing the geographical distribution of fossil gibbons during the Pliocene-Holocene. Reproduced with permission from Chatterjee et al. (2012, Fig. 1). Copyright © 2012, Karger Publishers

Fujian, Zhejiang and Jiangsu), notwithstanding collection biases (Fig. 2.1). Similarly, records regarding the distribution of hylobatids during the Chinese historical period (265 AD–1945) indicate that hylobatids ranged across southern China to the north as far as the Yangtze region (Fig. 2.2). Given that such data are broadly congruent with the distribution of older Pliocene and Quaternary fossil gibbon records (Jablonski 1993; Jablonski et al. 2000; Jablonski and Chaplin 2009), hylobatids either remained widely distributed over southern China throughout Pleistocene-Holocene climatic cycles or were able to re-colonize this region relatively rapidly following periods of adverse climate (Chatterjee et al. 2012).

Present-day distribution of hylobatids in China, as elsewhere, has been dramatically affected by widespread habitat loss, along with other population stressors such as poaching. Although gibbon species diversity in China is relatively high with at least eight species and subspecies, their density is low with fewer than 300 individuals recorded by the IUCN and two species may already be extirpated from China (IUCN 2015). Current gibbon populations are restricted to the southernmost provinces of Yunnan, Guangxi and Hainan, suggesting that the greatest range shift in Chinese hylobatids took place between the late Holocene historical period and the modern era (from 1945 onwards). During these time periods northern populations had disappeared from China and remaining southern populations were considerably reduced and pushed to the far southwest of China (Chatterjee et al. 2012).

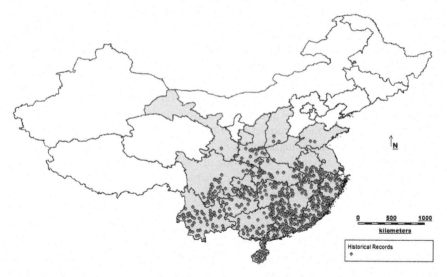

Fig. 2.2 Map of China showing the geographical distribution of gibbons during the historical period (AD 265–1945). Reproduced with permission from Chatterjee et al. (2012, Fig. 2). Copyright © 2012, Karger Publishers

The results of the ENM indicate that the areas suitable for hylobatids in Yunnan, Guangxi and Hainan are geographically restricted, with large parts of these provinces characterized by low-to-medium habitat suitability. Interestingly, the analysis shows that there are areas beyond modern gibbon ranges which are characterized by high-to-excellent habitat suitability, including parts of central China. This suggests that hylobatids in China are currently occupying sub-optimal locations. Moreover, it shows that not only recent historical events (e.g., habitat destruction) have affected population numbers, but that they are also responsible for causing adverse range shifts which may affect the success of remaining populations. When future climatic changes are considered, the study predicts that in the next 30 years, the suitability of habitats for hylobatids will be downgraded across south-western China, and Yunnan will be the only area to contain suitable habitats for hylobatids (Chatterjee et al. 2012, Fig. 7). Given the ongoing issues of habitat loss, poaching and other direct anthropogenic factors which are impacting gibbon population numbers and ranges, the added potential impact of anthropogenic climate change is likely to cause further population deterioration (Chatterjee et al. 2012).

Whilst the results of this study are only a proxy for the likely future spatial distribution of hylobatids in China, based on modelling predictions, they demonstrate the importance of considering past, fossil and historical distributions when studying range changes in endangered and critically endangered species. It also shows the value of incorporating historical biogeographic and environmental data in spatial distribution analyses.

The large-scale range contraction seen in Chinese hylobatids in fact seem to have occurred during a time period when environmental conditions were stable, compared to the fluctuations experienced in the Quaternary (Chatterjee et al. 2012). Although hylobatids' life history traits are expected to make them more vulnerable to environmental changes (Jablonski et al. 2000), historical and fossil records show the opposite (Jablonski 1993; Chatterjee et al. 2012). The ENM study reveals that the range changes observed in Chinese hylobatids can be attributed to past and ongoing anthropogenic activity rather than natural environmental change. The key finding of Chatterjee et al. (2012) is that the current-day distribution of hylobatids in China represents only their realized niche as a result of widespread recent habitat destruction, and not their fundamental niche of actual environmental tolerances. This is critically important for conservation planning since the findings of this study suggest that plans based on these modern distributions alone are likely to be artificially restrictive and pessimistic (Chatterjee et al. 2012).

To understand the history of biodiversity hotspots, refugia and biogeographic transitions, integrating historical biogeography into conservation planning is highly relevant (Willis et al. 2007; Woodruff 2010). Whilst hotspots capture present-day areas of exceptional species richness (Woodruff 2010), Pleistocene refugia are thought to have enabled these species to survive environmental challenges in the past. For example, during cooler Pleistocene glacial conditions rainforests retreated to the hills of peninsular Malaysia, western Sumatra, the Mentawi Islands and the centre of Borneo, and during warmer periods rainforests were replaced by savannah woodland or grassland on the emerged Sunda plains and elsewhere (Heaney 1991; Morley 2000, 2007).

By modelling changes in the distribution of major forest types during the last full 120,000-year glacial cycle, it was found that they expanded their ranges rather than contracted them during warm phases (Cannon et al. 2009). Thus, it is plausible that it is today's rainforests that are refugial and not those of the last glacial maximum (Cannon et al. 2009). In this regard, the changes experienced in Southeast Asian forests are the opposite of those in better-known temperate regions; rather than shrinking during cooler periods, lowland evergreen rainforests doubled in size (Woodruff 2010). Expansion of the forests may also have impacted the spatial distribution of the associated fauna, resulting in highly unusual present-day faunal and floral distribution patterns (Woodruff 2010). The biogeographic history of Southeast Asia could affect responses of species to land-use and climate change in the future, which could have considerable implications for conservation planning (Woodruff 2010).

Predicting Future Biogeographic Changes

Southeast Asia was almost entirely covered by rainforest 8000 years ago but it is predicted that the region could lose three quarters of its original forests by 2100 and up to 42 % of its biodiversity (Koh and Sodhi 2010; Sodhi et al. 2004). These biogeographic changes will occur in the backdrop of changes brought about by anthropogenic climate change, but the effects that climate change will have on

Southeast Asian biodiversity is difficult to predict. Whilst a species may appear to make an individualistic response to climate change, the species' dispersal abilities, niche breadth and ecological plasticity also play a role (Parmesan 2006). Biogeographic interdependence is also likely to be important. There are many examples of plants which are dependent upon specific seed dispersers. Hylobatids, for example, disperse many plant species, including fruit with hard covers and flesh attached to seeds that competitors, such as hornbills, are unable to eat (Kanwatanakid 2000; Kitamura et al. 2004). Further, it appears that some plant species, such as rambutans, may be entirely dependent upon hylobatids for dispersal (Brockelman 2011). Corlett (2009) notes that the survival of plants which respond to climate change through range shifts in their distribution will be tied to seed dispersing frugivorous birds and mammals. It is important therefore to understand the individualistic and ecosystem responses to climate change given the complex inter-relationships exhibited in such biotas.

Predicting the effects of climate change in a region such as Southeast Asia is challenging. The combination of a variety of different ecosystems, the region's varied and dramatic geomorphological history in association with rapid economic and human population growth, creates a complex and dynamic biogeographic landscape. Predictions based on a variety of studies, and discussed by Woodruff (2010), suggest a 2.4–2.7 °C rise in mean annual temperature (4 °C in subtropical China), a 7 % increase in wet season rainfall, a drier dry season, sea level rise of 1–2 m by 2150 and 2.5–5 m by 2300, higher frequency of El Nino Southern Oscillation (ENSO) events and higher frequency of fires. As Woodruff (2010) points out, such projections are not definitive end points but are based on the conditions expected when atmospheric CO_2 is double its pre-industrial concentration. If greenhouse gas emissions are not reduced and other climate change conditions not mitigated, temperatures and sea levels will continue to rise after this point, so most projections are likely to underestimate the effects on biota.

A considerable challenge with regard to future conservation planning is predicting how climate change will impact the distribution of species. The problem is further compounded in regions such as Southeast Asia where other direct anthropogenic factors, such as deforestation, continue at alarming rates. Woodruff (2010) argues that as land-use and climate change drive more people to become environmental refugees, displaced due to negative environmental impacts such as flooding, human biogeography and migration will also need to play a greater role in conservation planning. This is particularly significant for hylobatids living in areas which are predicted to undergo significant transformations as a result of climate change. For example, 14 million of the 28 million people living in the Mekong Delta in southern Vietnam are predicted to be displaced by a 2 m rise in sea level due to future climate change (Warner et al. 2009). Whilst some people will relocate to urbanized areas others will likely be forced into protected areas which provide habitats for numerous threatened species, including *Nomascus gabriellae* and *Nomascus annamensis*.

Understanding how and where species will shift their ranges in response to climate change can be implied by studying past biogeographic patterns, as outlined above. In keeping with the notion that it is important to incorporate historical bio-geography into present-day and future biogeographic studies, it has been found that species which occur on small islands tend to be smaller subsets of more diverse communities inhabiting larger islands (Okie and Brown 2009). Several authors have proposed that the mid-Pliocene (c.3 million years ago) is a useful model for pre-dicting faunal and floral range changes because global temperatures were on average 3 °C higher than today (Bonham et al. 2009; Haywood et al. 2009; Salzmann et al. 2009). Salzmann et al. (2009) for example, compared past data and models of Middle-Pliocene vegetation with simulations of vegetation distributions for the mid- and late twenty-first century to examine the extent to which the Middle Pliocene can be used as a 'test bed' for future climatic warming. Based on the premise that during the Middle Pliocene global temperatures were higher than today, as were higher atmospheric CO_2 concentrations, the study showed it was able to pinpoint specific future temperature and CO_2 levels which were concordant with those experienced in the Middle Pliocene. Their model simulations indicate a generally warmer and wetter climate, and afforded an opportunity to generate shifts in global vegetation patterns as a result of future climate change (Salzmann et al. 2009).

Many extant species living in Southeast Asia during the Pliocene have since survived multiple glacial/interglacial cycles. This ability to adapt may mean that species, such as hylobatids, will be less challenged by future environmental change such as temperature compared to seasonality and the length of the dry season (Woodruff 2010). It is possible that such species may have sufficient genetic variability and ecological plasticity to adapt to future climatic changes (Parmesan 2006). Although this confirms the possibility for hylobatids to adapt to environ-mental change (Chatterjee et al. 2012), further evidence is needed to fully under-stand hylobatids' ecological plasticity to future climate change.

Several studies have also highlighted the importance of considering within- and between-species genetic variability in conservation planning and management, for example Andayani et al. (2001) demonstrated at least two genetically differentiated lineages of *Hylobates moloch* on Java and suggested that the two lineages represent different management units. Given the well-documented challenges in reconstructing hylobatid phylogenetic relationships, revealed by whole genome sequencing (Carbone et al. 2014; Veeramah et al. 2015), understanding inter- and intraspecific genetic variability in gibbons in light of reduced population sizes is an important area of future research. In tandem, the relationships between species range shifts, population size and adaptive capacity are crucial in order to improve conservation planning for the future.

Hylobatid Historical and Fossil Records in Biogeographic Reconstruction

Southeast Asia has a long and complicated biogeographic past, but its history is relatively well studied and there is significant evidence that integrating historical biogeographic information improves our understanding of how species can be expected to respond to environmental changes (Woodruff 2010). The biggest challenge with regard to hylobatid historical biogeography is the sparseness of the fossil record, which is well documented (Chatterjee 2009; Jablonski and Chaplin 2009; Harrison 2016). This is particularly pertinent for the pre-Pleistocene time periods where the fossil record is very patchy; however, fossil data are especially useful when considered in the light of their associated locality and environmental information. Considerable efforts in documenting the gibbon fossil record have produced extremely valuable resources for investigating the spatial and temporal changes which occurred across hylobatids (Jablonski and Chaplin 2009). These meticulous records comprise species name, locality (including country, province, latitude and longitude) and time period for hylobatids, affording a time slice view of changes in the fossil record over time and space.

Historical records, in contrast to fossil records, have been relatively underexploited in spatial distribution analyses. In China, hylobatids formed a popular subject for paintings and sculptures and allude to the higher status attributed to hylobatids compared to other primates in Chinese culture. Geissmann (2008) was perhaps the first to investigate historical records for hylobatids and reported evidence from Van Gulik (1967) showing that as early as the Zhou dynasty (221–1027 BC) the Chinese singled out the gibbon as "the aristocrat among apes and monkeys" (van Gulik 1967, Preface). It is thought that this cultural interest in hylobatids stems from the belief that hylobatids were able to absorb the largest amount of "qi" (the key to longevity and immortality) and circulate it in the body. Other evidence that demonstrated the important role of *Nomascus* gibbons in Chinese culture includes literature and poems dating back to the Han dynasty (206 BC–220 AD) and Song dynasty (960–1279 AD). The cultural interest in hylobatids in China later spread to other Asian countries such as Japan and Korea. Historical Chinese poems, paintings and literature not only provide evidence of the cultural importance of hylobatids, but the localities of where these paintings and artistic compositions were found provide geographic information about their distribution. For example, based on historical records, Geissmann (2008) showed that in the tenth century, the distribution of the ancient Chinese hylobatids ranged over large parts of China, around 75 % of the country, down from the south all the way up to the Yellow River, which is concordant with Chatterjee et al. (2012). Geissmann[1] also points out that historical Chinese literature and art documents demonstrate that *Nomascus* gibbons lived in regions where winters were severe and notes that poets refer to seeing hylobatids in winter. Van Gulik refers to the below poem by Li Po (701–762 AD) about *Nomascus* gibbons in southern Anhui province:

The splendor of the mountains shivers under the accumulated snow.
Like shadows the gibbons are hanging from the cold branches... (cited in Van Gulik 1967, p. 61)

---

[1]http://www.hylobatids.de/main/index.html.

This evidence indicates that the hylobatids referred to in this poem inhabited considerably cooler environments than many extant hylobatid populations, most of which are restricted to tropical and subtropical forests. This exemplifies the intriguing possibility of using historical data to interpret past species ranges and indicates the possibility that hylobatids are able to adapt to different environments outside of their usual or preferred niche. In a similar vein to Jablonski's fossil record data, historical records for mammals in China collated by Wen (2009) provide details of species distributions. Based on the same records in some cases to those described by Geissmann (2008), Wen's data is based on historical books, manuscripts, illustrations and other texts. Aside from translation issues (Wen 2009 is in Chinese) one of the biggest challenges in using these data in spatial distribution analyses is interpreting the exact geographic locality to which the records pertain, and by extension therefore the exact species to which they refer. Most records refer to fairly broad geographic areas (c.10–100 km) within Chinese provinces which cover very large areas (many hundreds of kms) so pinpointing the exact location of a record can be problematic. However, by documenting these records and building up species distribution maps, studies such as Chatterjee et al. (2012) have shown that historical data can provide a useful tool for investigating the change in species distributions over time and can help to shed light on a species' adaptability when considered in the context of fossil and modern data.

Notwithstanding the limitations of gibbon fossil and historical data, including taphonomic and collecting biases, several studies have shown that incorporating these data into spatial distribution analysis affords a unique insight into how species respond to change over time and therefore how these data might be used to make predictions for the future (Woodruff 2010; Chatterjee et al. 2012). There are many inherent problems with historical and fossil data, not least the fact that the further back in time you go the poorer the spatial and temporal resolution of records is likely to be. The aforementioned studies have shown however, that not considering past species' responses to environmental change may provide a false or inaccurate view of its current and future success. Willis et al. (2007) have argued for the value of including historical records older than 50 years in conservation planning, despite the fact that this is not commonplace. They suggest that as well as using past data to interpret spatial range changes over time, these data also provide evidence of natural fluctuations in population size prior to the onset of anthropogenic activity. Furthermore, that data from the fossil record could be used to determine thresholds of natural variability and highlight those populations where the decline is real; in this way such studies could be used to identify conservation priorities (Willis et al. 2007).

In conclusion, given the dramatic population declines seen in many hylobatid species in the past 50–100 years and the continuing anthropogenic pressure these species face, including future climatic changes, the value of incorporating fossil and historical data in future conservation planning and studies of range shifts should not be underestimated.

References

Anandam MV, Groves CP, Molur S, Rawson BM, Richardson MC, Roos C, Whittaker DJ (2013) Species accounts of Hylobatidae. In: Mittermeier RA, Rylands AB, Wilson DE (eds) Handbook of the mammals of the world, vol 3. Primates. Lynx Edicions, Barcelona, pp 778–791

Andayani N, Morales JC, Forstner MRJ, Supriatna J, Melnick DJ (2001) Genetic variability in mtDNA of the silvery gibbon: implications for the conservation of a critically endangered species. Conserv Biol 15:770–775

Bonham SG, Haywood AM, Lunt DJ, Collins M, Salzmann U (2009) El Nino–southern oscillation, Pliocene climate and equifinality. Philos Trans R Soc A 367:127–156

Brockelman WY (2011) Rainfall patterns and unpredictable fruit production in seasonally dry evergreen forest and their effects on hylobatids. In: McShea W, Davies S, Phumpakphan N (eds) The unique ecology and conservation of tropical dry forests in Asia. Smithsonian Institution Scholarly Press, Washington, DC, p 195

Brook BW, Sodhi NS, Ng PKL (2003) Catastrophic extinctions follow deforestation in Singapore. Nature 424:420–423

Cannon CH, Morley RJ, Bush ABG (2009) The current refugial rainforests of Sundaland are unrepresentative of their biogeographic past and highly vulnerable to disturbance. Proc Natl Acad Sci USA 106:11188–11193

Carbone L, Harris RA, Gnerre S, Veeramah KR, Lorente-Galdos B, Huddleston J et al (2014) Gibbon genome and the fast karyotype evolution of small apes. Nature 513:195–201

Chatterjee HJ (2006) Phylogeny and biogeography of hylobatids: a dispersal-vicariance analysis. Int J Primatol 27:699–712

Chatterjee HJ (2009) Evolutionary relationships among the hylobatids: a biogeographic perspective. In: Lappan S, Whittaker DJ (eds) The hylobatids: new perspectives on small ape socioecology and population biology. Springer, New York, pp 13–36

Chatterjee HJ, Tse JS, Turvey S (2012) Using Ecological Niche Modelling to predict spatial and temporal distribution patterns in Chinese small apes (gibbons, Family Hylobatidae). Folia Primatol 83:85–99

Corlett RT (2009) Seed dispersal distances and plant migration potential in tropical East Asia. Biotropica 41:592–598

Edwards DP, Larsen TH, Docherty TDS, Ansell FA, Hsu WW, Derhe MA, Hamer KC, Wilcove DS (2011) Degraded lands worth protecting: the biological importance of Southeast Asia's repeatedly logged forests. Proc R Soc B 278:82–90

Geissmann T (2008) Gibbon paintings in China, Japan, and Korea: historical distribution, production rate and context. Gibbon J 4:1–38

Harrison T (2016) The fossil record and evolutionary history of hylobatids. In: Reichard UH, Hirohisa H, Barelli C (eds) Evolution of gibbons and siamang. Springer, New York, pp 91–110

Haywood AM, Dowsett HJ, Valdes PJ, Lunt DJ, Francis JE, Sellwood BW (2009) Introduction. Pliocene climate, processes and problems. Philos Trans R Soc A 367:3–17

Heaney LR (1991) A synopsis of climatic and vegetational change in Southeast Asia. Clim Change 19:53–61

Hijmans RJ, Cameron SE, Parra JL, Jones PG, Jarvis A (2005) Very high resolution interpolated climate surfaces for global land areas. Int J Climatol 25:1965–1978

IUCN (2015) The IUCN red list of threatened species. Version 2015.1. http://www.iucnredlist.org. Accessed 18 June 2015

Jablonski NG (1993) Quaternary environments and the evolution of primates in Eurasia, with notes on two new specimens of fossil Cercopithecidae from China. Folia Primatol 60:18–132

Jablonski NG, Chaplin G (2009) The fossil record of hylobatids. In: Lappan S, Whittaker DJ (eds) The hylobatids: new perspectives on small ape socioecology and population biology. Springer, New York, pp 111–130

Jablonski NG, Whitfort MJ, Roberts-Smith N, Xu Q (2000) The influence of life history and diet on the distribution of catarrhine primates during the Pleistocene in eastern Asia. J Hum Evol 39:131–157

Kanwatanakid C-O (2000) Characteristics of fruits consumed by the white-handed gibbon (*Hylobates lar*) in Khao Yai National Park, Thailand. Dissertation, Mahidol University, Bangkok, 136 pp

Koh LP, Sodhi NS (2010) Conserving southeast Asia's imperiled biodiversity: scientific, management, and policy challenges. Biodivers Conserv 19:913–1204

Kitamura S, Yumoto T, Poonswad P, Chuailua P, Plongmai K (2004) Characteristics of hornbill-dispersed fruits in a tropical seasonal forest in Thailand. Bird Conserv Int 14:S81–S88

Li J, Lin X, Chen A, Peterson T, Ma K et al (2013) Global priority conservation areas in the face of 21st century climate change. PLoS ONE 8(1):e54839

Liu Y, Li X, Zhang Q, Guo Y, Gao G, Wang J (2010) Simulation of regional temperature and precipitation in the past 50 years and the next 30 years over China. Quaterna Int 212:57–63

Morley RJ (2007) Cretaceous and Tertiary climate change and the past distribution of megathermal rainforests. In: Bush MB, Flenley JR (eds) Tropical rainforest responses to climate change. Springer, Berlin, pp 1–31

Morley RJ (ed) (2000) Origin and evolution of tropical rain forests. Wiley, New York

Myers N, Mittermeier RA, Mittermeier CG, da Fonseca GAB, Kent J (2000) Biodiversity hotspots for conservation priorities. Nature 403:853–858

Okie JG, Brown JH (2009) Niches, body sizes, and the disassembly of mammal communities on the Sunda Shelf islands. Proc Natl Acad Sci USA 106:19679–19684

Parmesan C (2006) Ecological and evolutionary responses to recent climate change. Annu Rev Ecol Evol Syst 37:637–669

Salzmann U, Haywood AM, Lunt DJ (2009) The past is a guide to the future? Comparing Middle Pliocene vegetation with predicted biome distributions for the twenty-first century. Philos Trans R Soc A 367:189–204

Sodhi NS, Koh LP, Brook BW, Ng PKL (2004) Southeast Asian biodiversity: an impending disaster. Trends Ecol Evol 19:654–660

Stigall AL, Lieberman BS (2006) Quantitative palaeobiogeography: GIS, phylogenetic biogeographical analysis, and conservation insights. J Biogeogr 33:2051–2060

Van Gulik RH (ed) (1967) The gibbon in China. An essay in Chinese animal lore. EJ Brill, Leiden

Veeramah KR, Woerner AE, Johnstone L, Gut I, Gut M, Marques-Bonet T, Carbone L, Wall JD, Hammer MF (2015) Examining phylogenetic relationships among gibbon genera using whole genome sequence data using an approximate bayesian computation approach. Genetics 200:295–308

Warner K, Erhart C, de Sherbinin A, Adamo S (2009) In search of shelter: mapping the effects of climate change on human migration and displacement. CARE international, 36 pp. http://www.careclimatechange.org

Wen RS (ed) (2009) The distributions and changes of rare wild animals in China. Chongqing Science and Technology Press, Chongqing

Willis KJ, Araujo MB, Bennett KD, Figueroa-Rangel B, Froyd CA, Myers N (2007) How can a knowledge of the past help to conserve the future? Biodiversity conservation and the relevance of long-term ecological studies. Philos Trans R Soc B 362:175–186

Woodruff DS (2010) Biogeography and conservation in Southeast Asia: how 2.7 million years of repeated environmental fluctuations affect today's patterns and the future of the remaining refugial-phase biodiversity. Biodivers Conserv 19:919–941

Chapter 3
Locomotion and Posture in Ancestral Hominoids Prior to the Split of Hylobatids

Matthew G. Nowak and Ulrich H. Reichard

Adult male white-handed gibbon foraging in understory, Khao Yai National Park, Thailand. Photo credit: Ulrich H. Reichard

M.G. Nowak (✉)
Department of Anthropology, Southern Illinois University Carbondale,
Carbondale, IL 62901, USA
e-mail: nowak.mg@gmail.com

M.G. Nowak
Sumatran Orangutan Conservation Programme (PanEco Foundation-YEL),
Medan, Sumatra 20154, Indonesia

U.H. Reichard
Department of Anthropology and Center for Ecology, Southern Illinois University
Carbondale, Carbondale, IL 62901, USA
e-mail: ureich@siu.edu

© Springer Science+Business Media New York 2016
U.H. Reichard et al. (eds.), *Evolution of Gibbons and Siamang*,
Developments in Primatology: Progress and Prospects,
DOI 10.1007/978-1-4939-5614-2_3

55

Introduction

Humans are the only living primate species that are truly bipedal. What triggered hominins to evolve an anatomy and behavior of walking on two legs with an upright torso is, despite decades of research, still controversial (Rose 1991; Niemitz 2010; Pawłowski and Nowaczewska 2015). Although walking on two legs is not unique to humans, as bipedalism has evolved in a number of animal lineages including extant Aves (Farlow et al. 2000; Abourachid and Höfling 2012) and is present at small frequencies in the positional repertoire of many primate species (Druelle and Berillon 2014), it is intriguing why of all extant primates only humans evolved habitual bipedalism. After all, hominoids share many postcranial morphological specializations (Harrison 1987, 2016), and interestingly, non-human hominoids are also known to share with humans a tendency to habitually position their torso in an upright manner during locomotion and posture, relative to Old World monkeys (Hunt 1991a; Thorpe and Crompton 2006), and perhaps this tendency may have been a necessary precondition upon which human bipedalism evolved (Tuttle 1974; Gebo 1996; Crompton et al. 2008).

Among living hominoids, upright or torso-orthograde (TO)-locomotion[1] and TO-posture,[2] and a suite of associated postcranial anatomical specializations[3] are suggested to have been significant adaptations that facilitate the acquisition of food resources in the arboreal canopy, and are thus thought to have played a major role in hominoid evolution and the cercopithecoid-hominoid divergence (Avis 1962; Grand 1972, 1984; Cartmill and Milton 1977; Ripley 1979; Temerin and Cant 1983; Cant 1992; Gebo 2004; Begun 2007, 2015; Ward 2015; Crompton et al. 2008; Hunt 2016). For instance, when the body size of a primate foraging in an above-branch pronograde fashion (e.g., standing or walking on four limbs) is large relative to the diameter of a given substrate, such as a branch (i.e., when the substrate to body size ratio is small), the possible diameter of the animal's support base becomes increasingly more constrained, leading to instability (Fig. 3.1). If an animal's support base becomes very narrow, even small lateral displacements will cause the body's center of gravity (CoG) to roll beyond the base of support (i.e., a

---

[1]Torso-orthograde locomotion includes bipedalism, brachiation/forelimb swing, clamber/transfer, and vertical climb, as outlined by Hunt et al. 1996.

[2]Torso-orthograde posture includes bipedalism and forelimb suspension, as outlined by Hunt et al. 1996.

[3]Extant apes share a number of postcranial anatomical features, including a relatively short stable lumbar region, a mediolaterally broad and dorsoventrally shallow thorax, relatively long forelimbs, a more dorsal positioning of the scapula, a mobile shoulder joint, an elbow that is fully extendable, a forearm that is stable during a wide range of pronation and supination, a mobile wrist, long curved phalanges, the lack of a tail, in addition to myological specializations associated with arm extension/flexion, forearm pronation/supination, and hand flexion (Keith 1923; Schultz 1930; Harrison 1987; 2016; Hunt 1991b, 2016; Rose 1993, 1994, 1997; Gebo 1996; Ward 1997a, 2015; Larson 1998; Stern and Larson 2001; Gibbs et al. 2002; Young 2003; Kagaya et al. 2008, 2009, 2010; Diogo and Wood 2011; Chan 2014; Williams and Russo 2015).

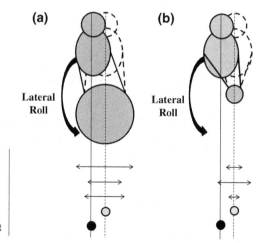

Substrate Diameter	
Chest Diameter	
Support Base Diameter	
GoG – Static Equilibrium	
CoG – After Displacement	

Fig. 3.1 A simple diagram of the substrate to body size ratio model, as related to balance. **a** Displays an animal with a large substrate to body size ratio. **b** Displays an animal with a small substrate to body size ratio. When the substrate to body size ratio is small, as in **b**, the animal's center of gravity (CoG) rolls beyond the base of support. This can induce a loss of balance. Adapted from Napier (1967)

rolling moment), resulting in a loss of balance (Fig. 3.1). Additionally, a small substrate to body size ratio can also increase the severity of branch deformation (i.e., the change of a branch's position under the effect of increasing weight), which can result in a disadvantageous forward displacement of the animal's CoG (i.e., a pitching moment), and in extreme instances branch failure (Fig. 3.2). Thus, maintaining an optimal substrate to body size ratio is a significant concern for animals foraging in the tree canopy (Napier 1967; Grand 1972, 1984; Cartmill 1974, 1985; Cant 1992; Preuschoft et al. 1995, 2016; Dunbar and Badam 2000; Preuschoft 2002).

TO-positional behaviors, such as brachiation/forelimb swing, clambering, forelimb suspension, and vertical climbing, can assist arboreal animals with a small substrate to body size ratio in circumventing problems associated with balance and branch deformation/failure, by placing the CoG below the substrate and thereby below the axis of rotation, maximizing the size of the base of support, and/or enhancing the spread of body mass among multiple substrates, assuming the animal can use multiple limbs to grasp adjacent substrates (Napier 1967; Grand 1972, 1984; Cartmill 1974, 1985; Cant 1992; Preuschoft et al. 1995, 2016; Dunbar and Badam 2000; Preuschoft 2002). As small-sized substrates are frequently encountered in the peripheral areas of a tree's crown, largely due to branch tapering and the presence of *first-order branches* (Bell and Bryan 2008; van Casteren et al. 2013), hominoid TO-positional behaviors have been considered beneficial for maximizing exploitation of the tree crown periphery, a resource-rich zone of tropical forest trees often unavoidably encountered during feeding and also while traveling between the

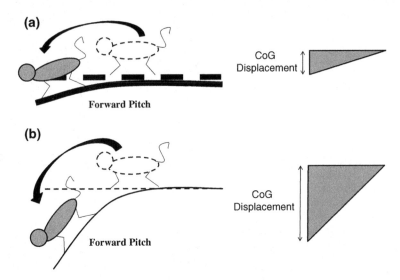

Fig. 3.2 A simple substrate to body size ratio model, as related to branch deformation. **a** Displays an animal with a large substrate to body size ratio. **b** Displays an animal with a small substrate to body size ratio. When the substrate to body size ratio is small, as in **b**, the pitching moment will create a disadvantageous displacement of the animal's center of gravity (CoG). This can induce a loss of balance and can restrict the path of movement

crowns of adjacent trees (Avis 1962; Grand 1972, 1984; Cartmill and Milton 1977; Ripley 1979; Temerin and Cant 1983; Preuschoft and Demes 1984; Sussman 1991; Cant 1992; Houle et al. 2007; Hunt 2016). Although currently untested, a working hypothesis has been that ape TO-behavioral specializations developed as a means to limit competition with sympatric cercopithecoids by enhancing apes' ability to access and acquire high-quality food resources (Ripley 1979; Temerin and Cant 1983; Lambert 2015; Hunt 2016).

Given the purported relationships between body size and arboreal canopy/substrate use, large body size has often been intimately linked to the evolutionary acquisition of TO-locomotion and posture among hominoids (Cartmill and Milton 1977; Gebo 2004; MacLatchy 2004; Begun 2007). However, the relatively small-bodied hylobatids (range: 5–12 kg; Smith and Jungers 1997) represent an enigmatic hominoid family, because their positional repertoire consists largely of TO-brachiation and forelimb suspension (Chivers 1972, 1974; Fleagle 1976, 1980; Gittins 1983; Srikosamatara 1984; Cannon and Leighton 1994; Sati and Alfred 2002; Fan et al. 2013; Fei et al. 2015; Nowak and Reichard 2016), yet their body sizes are far smaller than the closely related Asian and African great apes (range: 30–175 kg; Smith and Jungers 1997). Moreover, the majority of great apes have been reported to use TO-positional behaviors at reduced frequencies, relative to hylobatids, all of which questions suggested links between body size and TO-positional behaviors (Hunt 1991a, 2004, 2016; Thorpe and Crompton 2006).

Traditionally, distinct TO-positional behaviors were utilized to interpret the unique TO-postcranial anatomical specializations of living hominoids (Tuttle 1974; Gebo 1996; Crompton et al. 2008). For instance, early theorists (Keith 1891, 1899, 1903, 1923; Gregory 1916, 1928, 1934; Morton 1926; Washburn 1950; Washburn and Avis 1958; Avis 1962) had postulated that brachiation was the behavior that most succinctly differentiated hominoids from monkeys, and thus the observed postcranial similarities of living apes were thought to be associated with the biomechanical constraints of brachiation (i.e., the *brachiation hypothesis*). Due to their extensive use of forelimb-dominated below-branch orthogrady (e.g., brachiation and forelimb suspend), these early theorists saw the small-bodied hylobatids as a model organism for the earliest stage of hominoid anatomical and behavioral (i.e., locomotor/postural) evolution. Under the *brachiation hypothesis*, body size was considered to increase in the hominoid lineage, i.e., from a smaller-bodied hylobatid-like ancestor to body sizes more similar to that of the larger-bodied non-human hominids, made possible by a TO-positional repertoire that started to incorporate multiple supporting limbs during locomotion (e.g., TO-clambering) and/or posture (e.g., TO-quadrumanous suspension) (Tuttle 1974; Gebo 1996; Crompton et al. 2008).

Despite a longstanding popularity, empirical studies of wild ape positional behavior failed to produce quantitative support for the *brachiation hypothesis*, as the majority of nonhuman hominids relatively rarely utilize brachiation, especially when compared to the smaller-bodied hylobatids (Hunt 1991a, 2004; Thorpe and Crompton 2006), which ultimately necessitated a reevaluation of the *brachiation hypothesis*. Subsequent hominoid form-function concepts, such as the *slow-climbing hypothesis* (Stern 1971, 1975; Cartmill and Milton 1977), the *vertical climbing hypothesis* (Stern et al. 1977; Fleagle et al. 1981), or the combined *arm-hanging/vertical climbing hypothesis* (Hunt 1991a, b, 1992), which focused on synthesizing the available anatomical and behavioral data, effectively peripheralized brachiation as the focal positional behavior associated with hominoid postcranial similarities (but see, Gebo 1996; Chan 2007a, b, 2008).

Interestingly, however, these later hypotheses similarly singled out specific TO-positional behaviors in their attempts to understand hominoid anatomical specializations, and it was not until the recently developed *orthograde diversity hypothesis* that the notion of a combined use of all TO-positional behaviors was highlighted. Thorpe and Crompton (2006: 395) have been most explicit about the underlying principles of the *orthograde diversity hypothesis*, when they stated:

> Nevertheless it is possible we are mistaken in looking for a single unique mode to account for the morphological similarities of the apes. It is certainly plausible that what underlies ape anatomy is the exact opposite: the ability to respond to a complex array of ecological and environmental conditions with one of a broad range (if not a continuum) of orthograde-based postures and locomotion.

Consistent with the *orthograde diversity hypothesis*, a variable use of all TO-positional behaviors has already been documented among the majority of extant nonhuman hominid populations (Kano and Mulavwa 1983; Sugardjito and van

Hooff 1986; Cant 1987a, b; Hunt 1991a, 1992, 2004; Doran 1992, 1993a, b, 1996; Remis 1995, 1998; Thorpe and Crompton 2005, 2006, 2009; Manduell et al. 2011, 2012; Myatt and Thorpe 2011; Sarringhaus et al. 2014); however, TO-positional behavior flexibility has never been documented in the closely related extant hylobatids. The use of an expansive TO-positional repertoire among hylobatids similar to that of great apes has remained unclear until now, due to a lack of detailed data from the field (Nowak and Reichard 2016).

It was during the theoretical transition from the *brachiating hypothesis* to that of the *climbing/arm-hanging hypothesis* when Miocene hominoid fossils came under scrutiny and to the surprise of many scientists it was found that contrary to previous assumptions, the majority of purported Miocene fossil hominoids[4] (Table 3.1) apparently lack the postcranial specializations seen among living apes (Rose 1993, 1994, 1997; Pilbeam 1996; Ward 1997a, 2015; Pilbeam and Young 2004; Young and MacLatchy 2004; Begun 2007, 2015; Crompton et al. 2008). Even among Miocene fossil forms that more clearly exhibit postcranial affinities to extant

---

[4]We follow Harrison (2010a, 2013) in placing *Proconsul* within the subfamily Proconsulinae; *Afropithecus, Equatorius, Heliopithecus, Morotopithecus (Afropithecus?)*, and *Nacholapithecus* within the subfamily Afropithecinae; and *Mabokopithecus, Nyanzapithecus, Rangwapithecus, Turkanopithecus*, and *Xenopithecus* in the subfamily Nyanzapithecinae. *Rukwapithecus* is also tentatively placed in the subfamily Nyanzapithecinae, following Stevens et al. (2013). We recognize that McNulty et al. (2015) separate the fossil material commonly associated with the genus *Proconsul*, into two genera, *Ekembo* and *Proconsul*; however, given that the taxonomic changes are relatively recent and that the available postcranial material has been described as relatively homogeneous (Ward 1998), we tentatively place *Ekembo* as a separate genus within the subfamily Proconsulinae and utilize *Ekembo/Proconsul* to collectively designate the group, except in instances where a specific species is highlighted. Also similar to Harrison (2010a, 2013) we group the aforementioned subfamilies in the family Proconsulidae and superfamily Proconsuloidea (which may be paraphyletic); however, we place the Proconsuloidea as stem hominoids (Begun et al. 1997; Rae 1999; Fleagle 2013). Some recent studies prefer to place the Proconsuloidea into the superfamily Hominoidea (Zalmout et al. 2010; Stevens et al. 2013). The placement of *Otavipithecus* is unclear, and we therefore place this taxon *incertae sedis* among the Proconsuloidea. Following Harrison et al. (2008), *Yuanmoupithecus* currently remains the only fossil within the family Hylobatidae (Harrison 2016). Within the family Hominidae, we tentatively place the genera *Anoiapithecus, Dryopithecus, Graecopithecus, Hispanopithecus, Oreopithecus, Pierolapithecus, Rudapithecus*, and *Ouranopithecus*, within the subfamily Dryopithecinae (i.e., stem hominids) (Harrison 2010b; Fleagle 2013). Furthermore, the species *Ankarapithecus, Gigantopithecus, Khoratpithecus*, and *Sivapithecus* are tentatively grouped into the hominid subfamily Ponginae (i.e., stem pongines) (Harrison 2010b; Fleagle 2013). As *incertae sedis* within the family Hominidae, we place the species *Chororapithecus, Lufengpithecus, Nakalipithecus*, and *Samburupithecus* (Harrison 2010a; Ji et al. 2013). The taxonomic placement of *Griphopithecus* and *Kenyapithecus* is highly enigmatic (i.e., they may represent stem hominoids, early crown hominoids, or stem hominids); however, we tentatively group them *incertae sedis* in the family Hominidae (Harrison 2010a; Fleagle 2013). Finally, we highlight the recent work of Alba et al. (2015), who named a new fossil genus *Pliobates*, which they placed within the superfamily Hominoidea, in a newly named family Pliobatidae (i.e., stem hominoids). However, Benefit and McCrossin (2015), in response to the aforementioned publication have made a good case that *Pliobates* may actually be better placed within the early catarrhine superfamily Pliopithecoidea. Until subsequent studies confirm the placement of *Pliobates* within the Hominoidea, we follow Benefit and McCrossin's (2015) suggestion and view *Pliobates* as a pliopithecoid.

Table 3.1 Classification of hominoids used in this study

Order: Primates
Suborder: Anthropoidea
Infraorder: Catarrhini
Superfamily: Proconsuloidea or Hominoidea
Family: Proconsulidae
Subfamily: Proconsulinae
Genus: *Ekembo*[a]
Genus: *Proconsul*[a]
Subfamily: Afropithecinae
Genus: *Afropithecus*[a]
Genus: *Heliopithecus*[a]
Genus: *Morotopithecus* (*Afropithecus*?)[a]
Genus: *Nacholapithecus*[a]
Genus: *Equatorius*[a]
Subfamily: Nyanzapithecinae
Genus: *Nyanzapithecus*[a]
Genus: *Mabokopithecus*[a]
Genus: *Rangwapithecus*[a]
Genus: *Turkanapithecus*[a]
Genus: *Xenopithecus*[a]
Genus: *Rukwapithecus*[a]
Family: *incertae sedis*
Genus: *Otavipithecus*[a]
Superfamily: Hominoidea
Family: Hylobatidae
Genus: *Hylobates*
Genus: *Hoolock*
Genus: *Nomascus*
Genus: *Symphalangus*
Genus: *Yuanmoupithecus*[a]
Family: Hominidae
Subfamily: Dryopithecinae
Genus: *Dryopithecus*[a]
Genus: *Pierolapithecus*[a]
Genus: *Anoiapithecus*[a]
Genus: *Hispanopithecus*[a]
Genus: *Rudapithecus*[a]
Genus: *Ouranopithecus*[a]
Genus: *Graecopithecus*[a]
Genus: *Oreopithecus*[a]
Subfamily: Ponginae
Genus: *Pongo*
Genus: *Ankarapithecus*[a]
Genus: *Sivapithecus*[a]
Genus: *Gigantopithecus*[a]
Genus: *Khoratpithecus*[a]
Subfamily: Homininae
Genus: *Gorilla*
Genus: *Pan*
Genus: *Homo*
Genus: *Ardipithecus*[a]
Genus: *Orrorin*[a]
Genus: *Sahelanthropus*[a]
Genus: *Australopithecus*[a]
Genus: *Paranthropus*[a]
Genus: *Kenyanthropus*[a]
Subfamily: *incertae sedis*
Genus: *Kenyapithecus*[a]
Genus: *Griphopithecus*[a]
Genus: *Nakalipithecus*[a]
Genus: *Samburupithecus*[a]
Genus: *Chororapithecus*[a]
Genus: *Lufengpithecus*[a]

Adapted from Harrison (2010a, 2013), Zalmout et al. (2010), Fleagle (2013), Stevens et al. (2013), and McNulty et al. (2015)

[a]extinct members

hominoids (e.g., *Dryopithecus*, *Hispanopithecus*, *Morotopithecus*, *Oreopithecus*, *Pierolapithecus*, and *Rudapithecus*), only a mosaic of traits of the entire suite of extant ape anatomical specializations is evident (Begun 1992; Rose 1993, 1994, 1997; Moyà-Solà and Köhler 1996; Harrison and Rook 1997; Köhler and Moyà-Solà 1997; MacLatchy et al. 2000; Ward 1997a, 2015; Moyà-Solà et al. 1999, 2004, 2005a, b, 2009; Deane and Begun 2008; Alméija et al. 2007, 2009, 2012; Nakatsukasa 2008; Alba et al. 2010, 2011, 2012; Alba 2012; Hammond et al. 2013; Susanna et al. 2014). These observations, in addition to the fact that a number of more distantly related primates share some postcranial specializations with hominoids (e.g., atelins, dendropithecoids, lorises, palaeopropithecids, pliopithecoids, and some odd-nosed monkeys; Cartmill and Milton 1977; Rose 1993, 1994; Gebo 1996; Larson 1998; Begun 2002a; Godfrey and Jungers 2003; Young 2003; Hanna 2006; Su and Jablonski 2009; Harrison 2010a; Rossie and MacLatchy 2013), have forced some scholars to reconsider the homologous (sensu Hall 2007) nature of hominoid behavioral and postcranial morphological adaptations (Larson 1998; Young 2003; Begun 2007).

It has also become apparent that the assumption of a small-bodied hominoid ancestor, similar to living hylobatids, is not entirely consistent with the majority of Miocene fossil forms. For instance, proposed stem (sensu Donoghue 2005) hominoids (i.e., proconsuloids) are estimated to have weighed >9 kg (range: 9–50 kg; Fleagle 2013), while proposed stem hominids (e.g., *Dryopithecus*, *Hispanopithecus*, *Oreopithecus*, *Pierolapithecus,* and *Rudapithecus*) maintained body sizes of >17 kg (range: 17–48 kg; Fleagle 2013).

As a result of recent detailed studies of Miocene hominoid fossils, the small-bodied hylobatid-like hominoid morphotype that early theorists had assumed to have been basal to the hominoid lineage, is commonly replaced with a larger-bodied and behaviorally more generalized ape last common ancestor (LCA) (Pilbeam 1996; Young 2003; Gebo 2004; Pilbeam and Young 2004; Young and MacLatchy 2004; Thorpe and Crompton 2006; Crompton et al. 2008; Lovejoy et al. 2009; but see, Alba et al. 2015). Ironically, the unique suspensory-related anatomical and behavioral patterns that had originally placed hylobatids at the center of hominoid evolution, now marginalize extant hylobatids in contemporary ape evolutionary scenarios, and currently many scholars consider gibbons and siamang to be anatomically specialized outliers with morphological/anatomical traits of little value to the understanding of hominoid adaptations (Larson 1998; Young 2003; Thorpe and Crompton 2006; Crompton et al. 2008). The marginalization of hylobatids in ape evolutionary scenarios is however puzzling, because a significant caveat of current hominoid evolutionary theory is that hylobatids have evolutionarily undergone a reduction in body size from an assumed medium to large body-sized hominoid ancestor, and fine-tuned the generalized hominoid pattern of TO-positional behavior to that of below-branch forelimb-dominated TO-specialization (Groves 1972; Tyler 1991; Pilbeam 1996; Young 2003; Gebo 2004; MacLatchy 2004; Pilbeam and Young 2004; Young and MacLatchy 2004; Jablonski and Chaplin 2009; but see, Alba and Moyà-Solà 2009; Alba 2012; Alba et al. 2015).

Using recently published positional behavior data from a population of white-handed gibbons (*Hylobates lar*) from Khao Yai National Park, Thailand (Nowak and Reichard 2016), we are now for the first time in a position to compare in detail the positional behavior of all extant apes, including both small-bodied and large-bodied apes in relation to extant non-hominoid anthropoid primates. Using the results of the comparative behavioral dataset and adding to it a review of the hominoid fossil record, we reevaluate the hypothesis that the last common ancestor of hominoids possessed a diverse torso-orthograde positional repertoire.

An Expanded Hominoid Positional Behavior Dataset

By adding new, published data for adult white-handed gibbons (*Hylobates lar*) (Nowak and Reichard 2016) to previously published arboreal locomotor and postural data from 10 additional primate species (i.e., atelids, cercopithecoids, and hominoids), we have been able to expand and update the previous comparative datasets of Hunt (1991a and Thorpe and Crompton (2006), and reevaluate the uniqueness of the hominoid TO-positional repertoire (Tables 3.2 and 3.3). It is important to highlight that although earlier studies of hylobatid positional behavior exist (Chivers 1972, 1974; Fleagle 1976, 1980; Gittins 1983; Srikosamatara 1984; Cannon and Leighton 1994), either these studies predate the modified and extended locomotor- and posture-modes of Hunt et al. (1996) and are therefore difficult to compare to results of recent studies, or recent investigators of hylobatid locomotor and postural behavior chose to not integrate the extended and refined behavioral positional repertoires in their studies (Sati and Alfred 2002; Fan et al. 2013; Fei et al. 2015), which likewise prohibited including these studies in our comparative approach. Consequently, until now no comparative positional behavior data of wild hylobatids has been available, which prevented integrating hylobatids into a comprehensive hominoid TO-positional behavior evolutionary framework.

Despite the general advance that the now available white-handed gibbon behavioral repertoire (Nowak and Reichard 2016) provides, which incorporates all primary locomotor- and postural modes of Hunt et al. (1996), we acknowledge that our comparison rests on a single hylobatid species, while the positional behavior of the remaining 18 hylobatid species (Chivers et al. 2013) awaits to be studied in detail. However, because all hylobatid species share a similar postcranial morphological *bauplan* (Groves 1972; Schultz 1973; Andrews and Groves 1976; Jungers 1984, 1985), we believe that the diverse pattern of positional behavior described here for white-handed gibbons will also be observed in other hylobatid species.

Similar to Hunt (1991a) and Thorpe and Crompton (2006), we included in our dataset the olive baboon (*Papio anubis*) as an Old World monkey comparative outgroup to better understand similarities among hominoids; however, we have also incorporated positional behavior data from three additional populations of arboreal cercopithecoids (i.e., two African [*Cercopithecus mitis* and *Lophocebus albigena*]

Table 3.2 Comparison of primate postural behaviors

	Ateles belzebu[g]	Macaca fascicularis[h]	Cercopithecus mitis[i]	Lophocebus albigena[i]	Papio anubis[j]	Hylobates lar[k]	Pongo abelii[l]	Gorilla gorilla[m]	Gorilla beringei[n]	Pan paniscus[o]	Pan troglodytes[p]
Sit and Squat	26.1	82.0	63.9	66.0	77.0	67.3	58.0	79.5	86.6	90.0	86.7
TP Stand[a]	16.3	14.0	31.5	29.4	21.0	0.0	4.0	4.5	8.5	2.0	1.2
TO-Bipedal Stand[b]	p	4.0	2.6	1.3	0.0	0.1	6.0	4.4	0.7	p	0.2
TO-Suspend	13.9	0.0	0.0	0.1	1.0	23.5	5.0	1.4	1.4	5.0	9.6
TP-Suspend	–	–	0.1	0.1	–	1.1	13.0	–	–	–	–
Lie	0.0	0.0	1.9	2.5	0.0	8.0	12.0	10.3	2.8	3.0	2.1
Other[c]	43.7	0.0	<1.0	0.6	1.0	<0.1	1.0	0.0	0.0	0.0	0.0
Orthograde posture (%)[d]	13.9	4.0	2.6	1.4	1.0	23.6	11.0	5.8	2.1	5.0	9.7
Orthograde posture mode (#)[e]	2	1	1	2	1	2	2	2	2	2	2
Body Size (kg)[f]	8.1	4.5	6.1	7.1	19.2	5.6	56.8	121.0	130.0	39.1	41.1

– Unknown, p Present, but not distinguished in the cited publication

[a]TP stands for torso-pronograde

[b]TO stands for torso-orthograde

[c]Including cling, hind limb suspend, tripod, all tail-suspensory postural behaviors except tail-forelimb suspension, and postural bridge

[d]Sum of the frequencies of use for all torso-orthograde postural behaviors

[e]Total number of primary torso-orthograde postures utilized

[f]Body size averages were calculated from Smith and Jungers (1997)

[g]Recalculated from Youlatos (2008, Table 7.2). Tripod and all tail-suspensory postural modes, except tail-forelimb suspension, which is placed in TO-suspend, are categorized as other. Includes all arboreal behavioral contexts

[h]Recalculated from Cant (1988, Table 6). Foraging and feeding contexts are pooled. May include some terrestrial observations

[i]Recalcualted from Gebo and Chapman (1995, Table 6). Arboreal feeding and travel contexts are pooled

[i]Recalculated from Hunt (1991a, Table 1). Arboreal feeding contexts only

[k]Nowak and Reichard (2016). Includes all arboreal behavioral contexts

[l]Percentages taken from Thorpe and Crompton (2006, Table 6). Includes all arboreal behavioral contexts

[m]Recalculated from Remis (1995, Table 9). Includes all arboreal behavioral contexts during the wet season

[n]Percentages taken from Tuttle and Watts (1985, Table 4). Arboreal feeding contexts only, and is included in TO-suspend

[o]Percentages taken from Kano and Mulavwa (1984). Arboreal feeding contexts only

[p]Percentages represent the species averages for Pan troglodytes schweinfurthii and Pan troglodytes verus, which were recalculated from Hunt (1991a, Table 1) and Doran (1993a, Table 6). respectively. Arboreal feeding contexts only

Table 3.3 Comparison of primate locomotor behaviors

	Ateles belzebuth[h]	Macaca fascicularis[i]	Cercopithecus mitis[j]	Lophocebus albigena[j]	Papio anubis[k]	Hylobates lar[l]	Pongo abelii[m]	Gorilla gorilla[n]	Gorilla beringei[o]	Pan paniscus[p]	Pan troglodytes[q]
Quadrupedal/tripedal walk	32.8	84.0	87.3	76.7	68.3	0.0	14.4	18.9	49.0	18.4	29.4
TO-vertical climb[a]	13.6	7.0	1.6	5.5	21.3	15.9	22.5	47.0	29.9	26.1	52.6
TO-bipedal walk	0.6	0.0	0.0	0.1	0.0	6.9	6.2	5.0	12.8	1.2	5.0
TO-suspend locomotion[b]	36.8	0.0	0.2	0.6	0.0	58.1	37.5	20.6	7.0	20.0	10.5
TO clamber and transfer	13.7	0.0	0.0	0.0	0.0	10.1	23.6	17.4	5.8	1.6	4.1
TO brachiation and forelimb swing	23.1	0.0	0.2	0.6	0.0	48.0	13.9	3.2	1.3	18.4	6.3
Drop and leap	2.3	8.0	10.5	17.1	10.4	15.8	1.4	p	0.0	11.4	0.6
TP-suspend locomotion[c]	–	–	<1.0	<1.0	–	0.7	3.7	–	–	–	–
Bridge	10.1	0.0	0.4	0.3	0.0	2.6	2.5	p	p	13.5	1.0
Other[d]	3.8	0.0	0.0	0.0	0.0	0.0	12.3	8.1	0.0	9.4	3.3
Orthograde locomotion (%)[e]	51.0	7.0	1.8	6.2	21.3	80.9	66.1	72.6	49.8	47.3	68.1
Orthograde locomotor mode (#)[f]	4	1	2	3	1	4	4	4	4	4	4
Body size (kg)[g]	8.1	4.5	6.1	7.1	19.2	5.6	56.8	121.0	130.0	39.1	41.1

– Unknown, p Present, but not distinguished in the cited publication

[a]TO stands for torso-orthograde

[b]The percentage is the sum of all TO suspensory locomotor behaviors, including both TO clamber/transfer and TO brachiation/forelimb swing

[c]TP stands for torso-pronograde

[d]Including bipedal hop, tail swing, forelimb-hind limb swing, tree-sway, and ride

[e]Sum of the frequencies of use for all torso-orthograde locomotor behaviors

[f]Total number of primary torso-orthograde locomotor modes utilized

[g]Body size averages were calculated from Smith and Jungers (1997)

[h]Recalculated from Cant et al. (2001, Tables 1 and 10). Includes all arboreal behavioral contexts

[i]Recalculated from Cant (1988, Table 1). Foraging, feeding, and traveling contexts are pooled. May include some terrestrial observations

^aRecalculated from Gebo and Chapman (1995, Tables 6 and 13). Arboreal feeding and travel contexts are pooled

^kRecalculated from Hunt (1991a, Table 3). All arboreal behavioral contexts

^lNowak and Reichard (2016). Includes all arboreal behavioral contexts

^mSpecies percentages for *Pongo abelii* were recalculated from Thorpe and Crompton (2006, Table 7) and Manduell et al. (2012, Table 8). Includes all arboreal behavioral contexts

ⁿRecalculated from Remis (1995, Table 11). Includes all arboreal locomotor contexts during the wet season. TO brachiation/forelimb swing includes frequencies for drop, whereas the other category is comprised of frequencies for tree-sway, leaping, fireslide, and bridging

^oRecalculated from Carlson (2005, Table 2). Females and males are pooled. Arboreal locomotor contexts only. The vertical climbing percentages include climbing on both vertical and inclined substrates. TO clamber/transfer includes scramble, tree-sway, and bridging. Does not sum to 100 %

^pRecalculated from Doran (1993a, Tables 2 and 4). Includes all arboreal behavioral contexts for. Males, females without infants, and females with infants are pooled

^qPercentages represent the species averages for *Pan troglodytes schweinfurthii* and *Pan troglodytes verus*, which were recalculated from Hunt (1991a, Table 3) and Doran (1993b, Table 3). Includes all arboreal locomotor contexts

and one Asian population [*Macaca fascicularis*]) and we have expanded the outgroup comparative sample of Thorpe and Crompton (2006), by also including both the locomotor and postural behavior of a population of convergently adapted white-bellied spider monkeys (*Ateles belzebuth*) (e.g., Larson 1998; Young 2003). These additions provided a more diverse comparative outgroup, relative to Hunt (1991a) and Thorpe and Crompton (2006), which in addition to the original moderate-sized semi-terrestrial olive baboon outgroup, also includes smaller-bodied more arboreal quadrupeds from both Africa and Asia and the complete positional behavior repertoire of the white-handed spider monkey. In addition to the white-handed gibbon sample population, our ape comparative sample included recalculated values for eastern gorillas (*Gorilla beringei*), western gorillas (*Gorilla gorilla*), chimpanzees (*Pan troglodytes*), bonobos (*Pan paniscus*), and Sumatran orangutans (*Pongo abelii*), (Tables 3.2 and 3.3).

To avoid biases related to behavioral categorization, we only included studies that closely followed the standardized positional modes described in Hunt et al. (1996) and we have indicated where studies of locomotion and/or posture may have deviated from these standardized locomotor and postural modes (Tables 3.2 and 3.3). Similar to Hunt (1991a) and Thorpe and Crompton (2006), we restrict our comparisons to arboreal observations, in order to allow for more accurate comparisons with the habitually arboreal Asian apes included in this dataset. Because the studies we included in the comparative dataset used various sampling methods (i.e., bout versus instantaneous sampling) and also presented locomotor/postural data from different behavioral contexts, we were by default restricted to broader comparisons in our conclusions. Nevertheless, the modifications we made to the earlier analyses of Hunt (1991a) and Thorpe and Crompton (2006), have allowed for a more comprehensive and detailed survey of hominoid positional behavior than has previously been possible.

Torso-Orthograde Positional Behavior Among *Extant* Hominoidea

In general, both cercopithecoids (64–82 %) and hominoids (58–90 %) frequently utilize **sit/squat**[5] (Table 3.2). The low frequency of **sit/squat** in white-bellied spider monkeys (26 %) is likely an artifact of their extensive use of tail-suspensory postures (Youlatos 2008). Compared to both hominoids (2–24 %) and white-bellied spider monkeys (14 %), our comparative cercopithecoid sample exhibits a low range of frequencies of total TO-postural behavior (1–4 %). Only eastern gorillas

---

[5]From this point on, we highlight in bold the specific modes detailed in Hunt et al. (1996) to maintain consistency and to differentiate these modes from expansive and/or vague locomotor and postural categories used in previous studies of positional behavior or in discussions of fossil taxa.

overlap in their use of TO-postural behavior with the cercopithecine range, but if the primarily terrestrial eastern gorilla (Tuttle and Watts 1985; Doran 1996) is excluded from comparisons, the hominoid (5–24 %) and cercopithecoid (1–4 %) values no longer overlap in their ranges of total TO-postural behavior and the two primate superfamilies can be separated based on TO-postural behavior alone. Furthermore, hominoids are capable of utilizing both **TO-bipedal stand** and **TO-suspend**, whereas the majority of cercopithecoids only utilize a single TO-postural mode, with the exception of gray-cheeked mangabeys (*Lophocebus albigena*) (Gebo and Chapman 1995). When TO-postural behaviors are analyzed separately, hominoids (and white-bellied spider monkeys) utilize far greater frequencies of **TO-suspend** (hominoids: 1–24 %; white-bellied spider monkeys: 14 %), relative to cercopithecoids (0–1 %), whereas cercopithecoids utilize **TO-bipedal** stand at frequencies (0–4 %) that overlap with hominoids (<0.1–6 %). An additional difference between cercopithecoids and hominoids is the extensive use of **TP-stand** by all cercopithecoid species (Table 3.2).

While Thorpe and Crompton (2006) and Myatt and Thorpe (2011) have drawn attention to the unique use of **TP-suspend** among Sumatran orangutans, we note that **TP-suspend** is also utilized by blue monkeys (*Cercopithecus mitis*), gray-cheeked mangabeys, and the white-handed gibbon population included here (Gebo and Chapman 1995; Nowak and Reichard 2016). The appearance of a more widespread use of **TP-suspend** among primates suggests that primates other than orangutans are capable of utilizing **TP-suspend**, which is therefore unlikely a behavioral trait unique to orangutans. However, we agree that orangutans may be unique in that they most frequently utilize this postural mode (Table 3.2; Thorpe and Crompton 2006; Myatt and Thorpe 2011). The reason that only Asian apes but not African apes were documented to utilize **TP-suspend** within the current dataset may be a methodological side effect of the postural behavior categorization from previous studies of African apes; however, Sarringhaus et al. (2014)[6] have documented **TP-suspend** in the postural repertoire of infant, juvenile, and adolescent chimpanzees (all ≤ 1 % of postural time) at Ngogo, but not in adult individuals from the same population. Thus, younger and presumably smaller chimpanzees from Ngogo are capable of utilizing **TP-suspend**, suggesting that size/developmental stage may also account for the lack of use of **TP-suspend** in previous studies of African ape populations, as all of the sampled populations in the comparative dataset are based on data from adult individuals. Given that adult orangutans are as large, if not sometimes larger, e.g., flanged adult male orangutans, than adult chimpanzees (Smith and Jungers 1997), further data are needed to substantiate if and why this postural mode is absent in African apes but uniquely utilized by all age–sex classes of Asian ape species, especially large-bodied orangutans. Perhaps habitual use of the arboreal canopy and extreme morphological

---

[6]The study by Sarringhaus et al. (2014) was not included in the comparative sample presented here, as their study combined data from both arboreal and terrestrial contexts, whereas in this study, the focus was strictly arboreal locomotion and posture.

specializations related to suspensory positional behavior (Ward 2015) may permit Asian apes to utilize a more diverse suspensory component in their positional behavior repertoires, relative to that of African apes.

Consistent with their unique suspensory-related anatomical specializations (Erikson 1963; Groves 1972; Schultz 1973; Andrews and Groves 1976; Jungers and Stern 1980; Jungers 1984; Preuschoft and Demes 1984, 1985; Larson 1988, 1998; Takahashi 1990, 1991; Gebo 1996; Young 2003, 2008; Michilsens et al. 2009; Diogo and Wood 2011; Diogo et al. 2012; Arias-Martorell et al. 2015), lar gibbons utilize the greatest frequencies of **TO-suspend** (23 %), followed by white-bellied spider monkeys (14 %) and chimpanzees (10 %). Nevertheless, lar gibbons and the other hominoid taxa in our dataset are distinguishable from cercopithecoids in a similar manner. Overall, the postural dataset indicates that in contrast to cercopithecoids, hominoids collectively utilize greater frequencies of TO-postural behavior, especially **TO-suspend**, consistently utilize all TO-postural modes, and infrequently utilize **TP-stand**.

Similar to our postural dataset, both white-bellied spider monkeys (51 %) and hominoids (47–81 %) possessed a greater range of total TO-locomotion, relative to that of cercopithecoids (2–21 %). However, differences were more marked and there was no overlap between ranges (Tables 3.2 and 3.3). Both white-bellied spider monkeys and the hominoid sample utilized all TO-locomotor modes, whereas cercopithecoid species utilized between one and three of the primary TO-locomotor modes, with gray-cheeked mangabeys again utilizing the greatest number (i.e., a total of three TO-locomotor modes). Hominoids are additionally differentiated from cercopithecoids by the hominoids' relative lack of use of **quadrupedal/tripedal walk/run** (i.e., above-branch pronogrady; <1–49 %) compared to that of cercopithecoids (68–87 %).

When TO-locomotor modes were analyzed separately, white-bellied spider monkeys and hominoids utilized all TO-locomotor modes at greater frequencies (i.e., there was no overlap in ranges), except **TO-vertical climb** (Table 3.3). Furthermore, no cercopithecoid from our comparative sample was documented to utilize **TO-clamber/transfer** (Table 3.3). Similar to figures for **TP-suspend**, only lar gibbons (<1 %) and Sumatran orangutans (4 %) have been documented to utilize **TP-suspensory locomotion** among the hominoids sampled here. However, two cercopithecoid species have also been documented to utilize **TP-suspensory locomotion**, albeit both at frequencies <1 %. Similar to the situation noted above for **TP-suspend**, Sarringhaus et al. (2014) did document **TP-suspensory locomotion** in both infant and juvenile chimpanzees at Ngogo (both ≤ 1 % of locomotor time), but not in adolescent or adult individuals from the same population. Nevertheless, further data are required to clarify if and why **TP-suspensory locomotion** is uniquely used by all age–sex classes amongst Asian hominoids, with an emphasis on analyzing the contextual use of TP-suspensory positional behaviors at different developmental stages and/or for different body size classes.

Among hominoids, lar gibbons are differentiated by their lack of use of **quadrupedal/tripedal walk/run**, reduced use of **TO-vertical climb** (albeit they utilize **TO-vertical climb** at greater frequencies than the majority of cercopithecoids included in this sample), and frequent use of **TO-brachiation/forelimb swing** (Table 3.3). Only bonobos (18 %) and the convergently adapted white-bellied spider monkey (23 %) approach the frequency of **TO-brachiation/forelimb swing** utilized by Khao Yai lar gibbons (48 %), which as noted above for **TO-suspend** is consistent with their unique suspensory-related anatomical specializations (Erikson 1963; Groves 1972; Schultz 1973; Andrews and Groves 1976; Jungers and Stern 1980; Jungers 1984; Preuschoft and Demes 1984, 1985; Larson 1988, 1998; Takahashi 1990, 1991; Gebo 1996; Young 2003, 2008; Michilsens et al. 2009; Diogo and Wood 2011; Diogo et al. 2012; Arias-Martorell et al. 2015). Nevertheless, lar gibbons and all other hominoids are collectively differentiated from cercopithecoids by their extensive use of all TO-locomotor behavioral modes and infrequent use of **quadrupedal/tripedal walk/run**.

Torso-Orthograde Positional Behavior Among *Extinct* Hominoidea

Despite the fact that differences do exist in how frequently specific TO-positional behaviors are used, i.e., each hominoid taxon utilizes a relatively unique combination of primary TO-positional modes, it is clear that in an arboreal setting all living hominoids differ from Old World monkeys by their collective and diverse TO-positional repertoires and lack of use of above-branch TP-positional behaviors. While this supports the hypothesis that extant hominoids may have shared a last common ancestor (LCA) who also expressed a diverse set of TO-behavioral specializations, the importance of evaluating this hypothesis in relation to the catarrhine fossil record must be underscored (Harrison 1987, 1991; Donoghue et al. 1989; Ward 2002; Andrews and Harrison 2005; Wood and Harrison 2011).

The molecular divergence dates for the cercopithecoid/hominoid lineage split range between ~38 and 20 mya (range median: 29 mya; Raaum et al. 2005; Steiper and Young 2006; Chatterjee et al. 2009; Matsui et al. 2009; Chan et al. 2010; Thinh et al. 2010; Israfil et al. 2011; Perelman et al. 2011; Wilkinson et al. 2011; Steiper and Seiffert 2012; Finstermeier et al. 2013; Springer et al. 2013; Carbone et al. 2014), and a recent publication has suggested that stem cercopithecoids (i.e., *Nsungwepithecus*) and stem hominoids (i.e., *Rukwapithecus*) had potentially already diverged by ~25 mya (Stevens et al. 2013). Compared to these early fossil taxa, the next oldest fossils for either of these two catarrhine stem groups are considerably younger at ~23 mya for stem hominoids (i.e., *Proconsul meswea*; Harrison and Andrews 2009; Harrison 2010a) and ~19 mya for stem cercopithecoids (i.e., *Victoriapithecus* from Napak; Miller et al. 2009). The molecular divergence dates for hylobatids/hominids ranges between ~24 and 13 mya (range median: 19 mya; Raaum et al. 2005; Steiper and Young 2006; Chatterjee et al. 2009; Matsui et al. 2009; Chan et al. 2010; Thinh et al. 2010; Israfil et al. 2011; Matsudaira et al. 2010; Perelman et al. 2011; Wilkinson et al. 2011;

Steiper and Seiffert 2012; Finstermeier et al. 2013; Springer et al. 2013; Carbone et al. 2014). The oldest potential crown hominoid is represented by a molar commonly associated with the genus *Griphopithecus*, from Engelswies, Germany, that has been dated to ∼17 mya (Casanovas-Vilar et al. 2011). Given the enigmatic taxonomic affiliations of *Griphopithecus* and *Kenyapithecus*, the best evidence for early stem hominids are the fossil taxa *Dryopithecus fontani* and *Pierolapithecus catalaunicus* (Harrison 2010b), which have their earliest appearances at ∼13 mya and ∼12 mya, respectively (Casanovas-Vilar et al. 2011). Finally, the oldest definitive stem hylobatid (i.e., *Yuanmoupithecus*) is dated to ∼7–9 mya (Harrison et al. 2008; Harrison 2016).

Because stem cercopithecoids can be differentiated (based primarily on partial bilophodont molars) at ∼25 mya, it is possible that TO-postcranial specializations could have similarly developed in the hominoid lineage by this time. However, the lack of fossil crown catarrhines with extant hominoid postcranial specializations from the Oligocene strongly contradicts this possibility (Rasmussen 2002; Fleagle 2013; Godinot 2015), although it is important to keep in mind that a lack of fossils is not necessarily evidence for a lack of the presence of a trait (Steiper and Young 2008). Nevertheless, the ∼25 mya time mark seems currently the best possible fossil upper bound, based on the earliest presence of crown catarrhines, available to evaluate the presence of an early crown hominoid with a TO-morphological pattern. Conversely, *Dryopithecus* and *Pierolapithecus* provide provisional evidence for the presence of stem hominids by at least ∼13 mya, and *Pierolapithecus* also conveniently displays relatively well-developed TO-postcranial morphological adaptations (Moyà-Solà et al. 2004, 2005a; Deane and Begun 2008; Almécija et al. 2009, 2012; Alba et al. 2010; Hammond et al. 2013; Pina et al. 2014). As only a few postcranial remains are currently attributed to *Dryopithecus fontani* (e.g., Moyà-Solà et al. 2009; Alba et al. 2011), ∼12 mya appears as a logical, yet tentative fossil lower bound to evaluate the occurrence of an early crown hominoid with a TO-body plan.

If the hypothesis that extant hominoids evolved homologous TO-specializations, fossil crown hominoids with TO-postcranial specializations are predicted to occur in the fossil record between ∼25 and 12 mya, a very broad time period, but one that is nevertheless relatively consistent with the currently suggested absolute minimum and maximum molecular divergence dates of hylobatids and hominids (i.e., ∼24–13 mya).

A considerable number of publications have addressed in detail the postcranial osteology of known Miocene fossil catarrhines (Rose 1993, 1994, 1997; Ward 1997a, 1998, 2015; Crompton et al. 2008), and therefore we provide a brief review of more salient details. Previous comparative studies have concluded an undeniable pronograde quadrupedal body plan in most early Miocene stem hominoids, i.e., the proconsuloids, despite some variation in postcranial locomotor osteology. Nevertheless, compared to the pronograde quadrupedal body plan of earlier stem catarrhines, a generalized yet more advanced morphological pattern appears in early Miocene stem hominoids, with an incipient emphasis on elbow stability and forearm rotational capabilities, powerful grasping in the hands and feet, moderate hind limb mobility, and in a few taxa such as *Ekembo/Proconsul* and *Nacholapithecus*, the absence of a tail (Ward et al. 1991; Rose 1993, 1994, 1997; MacLatchy and Bossert 1996; Ward 1997a, 1998, 2015; Ruff 2002; Sherwood et al. 2002; Nakatsukasa et al. 2004; Nakatsukasa and Kunimatsu 2009; McNulty et al. 2015; Russo 2016).

These derived Miocene hominoid features, in combination with a more TP-body plan, are suggestive of a positional repertoire that included **symmetrical gait quadrupedal walking**, perhaps with an increased use of **irregular gait quadrupedal walking** (e.g., **TP-clambering/scrambling**), a submode of **quadrupedal walking**, **bridging**, and most certainly some **TO-vertical climbing**. However, such a mosaic trait combination in fossil forms is not entirely surprising, because all of these locomotor modes (save **quadrupedal walking** in *Hylobates lar*) are also present in extant cercopithecoid and hominoid taxa of our comparative sample (Table 3.3). Compared to the extant cercopithecoid pattern, however, differences in the locomotor repertoire of early stem hominoids may have included a reduced emphasis on **quadrupedal bounding/running** and **leaping** (sensu Hunt et al. 1996), with an emphasized use of the locomotor modes discussed above.

Despite known variation, e.g., *Equatorius* exhibits a greater terrestrial component, *Nacholapithecus* may emphasize greater amounts of **TO-vertical climb**, as compared to other proconsuloids, and both may display a more advanced hominoid-like distal humerus relative to *Afropithecus* and *Ekembo/Proconsul* (McCrossin and Benefit 1997; Sherwood et al. 2002; Begun 2007; Patel et al. 2009; Nakatsukasa and Kunimatsu 2009; Alba et al. 2011; Kikuchi et al. 2015), this basal hominoid positional repertoire is present in the African Miocene fossil record from ∼20–13 mya (Harrison 2010a), and as is the case with the TO-positional behaviors of extant hominoids, would have similarly been adaptive to circumvent issues associated with a small substrate to body size ratio (e.g., maximizing the size of the base of support and/or enhancing the spread of body mass among multiple substrates; Napier 1967; Cartmill 1974, 1985; Preuschoft et al. 1995, 2016; Dunbar and Badam 2000; Preuschoft 2002). Most importantly, the positional repertoire of early stem hominoids may be interpreted to represent an initial phase in the development of hominoid TO-positional behaviors (Gebo 2004; Nakatsukasa and Kunimatsu 2009).

Morotopithecus bishopi is an interesting and enigmatic fossil specimen, which with an estimated age of between ∼21 and 15 mya (Pickford et al. 1999; Gebo et al. 1997) is well within the hypothesized time frame of ∼25–12 mya for the initial emergence of the crown hominoid TO-body plan. Consistent with notions of an early emergence of a generalized TO-behavioral pattern and in contrast to the dearth of proconsuloids that display a largely TP-locomotor pattern, the known postcranial remains of *Morotopithecus* are suggestive of a greater emphasis on TO-locomotion (Sanders and Bodenbender 1994; MacLatchy et al. 2000; MacLatchy 2004; Nakatsukasa 2008). For instance, the lumbar vertebrae have robust pedicles, a caudally inclined spinous process, a transverse process with a more dorsal origin from the pedicle, and a lack of anapophyses, whereas a prospective glenoid scapular fragment displays a dorsoventrally broad and uniformly curved glenoid fossa (Sanders and Bodenbender 1994; Pickford et al. 1999; MacLatchy et al. 2000; MacLatchy 2004; Nakatsukasa 2008; Harrison 2010a). The morphology of the *Morotopithecus* proximal femur, appears to display a femoral head surface area relative to midshaft anteroposterior bending strength that is intermediate between that of extant cercopithecoids and hominoids, whereas the diaphyseal cortical area and the distal articular surface indicate an emphasis on muscular contraction and mobility, respectively (MacLatchy and Bossert 1996; MacLatchy et al. 2000; Ruff 2002; MacLatchy 2004).

Overall, the combination of postcranial morphological features is indicative of a unique form of locomotion in early stem hominoids that may have included some forelimb-dominated suspensory locomotor modes and/or locomotor modes requiring greater lumbar stability, and the use of any of the locomotor modes outlined above for other stem hominoids is not precluded. A reduced femoral head articular surface area might be indicative of a reduced ability to abduct the hind limb (MacLatchy and Bossert 1996; Ruff 2002), which may imply that the use of locomotor modes such as **TO-clamber/transfer**, if present, was more limited and/or biomechanically divergent in *Morotopithecus* compared to the extreme form utilized by extant taxa such as orangutans (Table 3.3). Significant to this discussion, Remis (1995) notes that western lowland gorillas utilize a locomotor mode she designated as 'scramble' and suggested to be most similar to **TO-clamber**. However, her description of the 'scramble' category differs from **TO-clamber** in that the hind limbs are never loaded in tension. A similar distinction may have applied to the TO-locomotor modes utilized by *Morotopithecus*. Aside from this potential difference, the available postcranial evidence from *Morotopithecus* cannot be used to conclusively suggest that *Morotopithecus* did or did not utilize any of the other TO-positional behaviors that are shared by extant hominoids (Tables 3.2 and 3.3).

More problematic for a discussion of the TO-behavior of stem hominoids, is the suggestion that based on craniodental features, *Afropithecus*, a proconsuloid in the subfamily Afropithecinae, and *Morotopithecus* may be synonymous (Pickford 2002; Patel and Grossman 2006). If this is the case, then the combined postcranial material for *Afropithecus* (commonly described as *Ekembo/Proconsul*-like; Rose 1993, 1994, 1997; Ward 1997a, 1998, 2015) and *Morotopithecus* suggests a locomotor repertoire that is not readily comparable to any single living primate taxon. However, more recently, MacLatchy et al. (2010) and Rossie and MacLatchy (2013) questioned integrating *Morotopithecues* with *Afropithecus* material and instead have suggested that despite craniodental similarities, sufficient dental evidence exists for a generic differentiation of the two taxa. Because *Morotopithecus* seems to display close affinities with the Afropithecinae (Harrison 2010a), it is possible that the derived postcranial features that *Morotopithecus* shares with extant hominoids have evolved in parallel. Nevertheless, the presence of hominoid-like postcranial specializations in a stem hominoid at ~21–15 mya suggests that the selective pressures promoting TO-postcranial specializations were present at this early time in some parts of the African Miocene hominoid habitat range.

As the earliest potential crown hominoids, *Griphopithecus* and *Kenyapithecus* are two important taxa for our understanding of hominoid anatomical and behavioral evolution. Unfortunately, *Griphopithecus* is currently only represented postcranially by a humeral shaft fragment and proximal ulna from Klein Hadersdorf, dated to ∼ 12–11 mya, while *Kenyapithecus* postcranial material is currently only represented by a distal humeral fragment from Fort Ternan, dated to ∼ 14 mya (Begun 1992, 2002b; Benefit and McCrossin 1995; Harrison, 2010a; Casanovas-Vilar et al. 2011). Postcranial material has been published from the Turkish site of Paşalar dated to ∼ 15–14 mya (Casanovas-Vilar et al. 2011); however, it is still unclear if this material belongs to *Griphopithecus alpani* or *Kenyapithecus kizili*, although the phalanges are most commonly attributed to *Griphopithecus alpani* (Ersoy et al. 2008).

The humeral shaft of *Griphopithecus* displays a straight/slightly anteroflexed shaft with prominent muscle marking that is also anteroposteriorly broad, whereas the proximal ulna fragment has a relatively narrow trochlear notch and a proximally projecting olecranon process (Begun 1992, 2002b; Alba et al. 2011). The Paşalar phalanges display features intermediate between those of cercopithecoids and hominoids, including an arboreal and a terrestrial signal (Ersoy et al. 2008). This morphological pattern is more consistent with proconsuloids and unlike that of modern hominoids (Begun 1992, 2002b; Alba et al. 2011), and thus like the majority of proconsuloids, the positional repertoire of *Griphopithecus* probably included some **symmetrical gait quadrupedal walking, irregular gait quadrupedal walking** (i.e., **TP-clambering/scrambling**), **bridging**, and **TO-vertical climbing**. Unfortunately, the available fossil material does not permit a more detailed behavioral reconstruction, though the Paşalar phalanges suggest perhaps a partial use of the terrestrial setting.

Compared with the incipient hominoid morphology of many proconsuloids (e.g., *Afropithecus* and *Ekembo/Proconsul*), the distal humerus of *Kenyapithecus* more closely approaches the morphological pattern of modern hominoids. It displays a deep olecranon fossa, a broad trochlea, a moderately developed lateral trochlear keel, and a globular moderately developed capitulum (Benefit and McCrossin 1995; Alba et al. 2011). Like *Griphopithecus*, Alba et al. (2011) characterize the humeral shaft as being more similar to the pattern observed for known proconsuloids. Benefit and McCrossin (1995) have also noted that the medial epicondyle of the distal humerus displays posterior inclination, which they suggest is indicative of some terrestriality. Given the paucity of postcranial fossil remains and that the humeral shaft is relatively similar to proconsuloids, while the distal humerus appears more advanced than that of the early proconsuloid condition (e.g., Rose 1993, 1994, 1997; Ward 1998, 2015; Alba et al. 2011), we can only infer that the locomotor repertoire of *Kenyapithecus* may have been slightly more derived than the general proconsuloid pattern, in that it may have further emphasized locomotor modes requiring elbow stabilization and forearm rotation (e.g., **irregular gait quadrupedal walking** [i.e., **TP-clambering/scrambling**], **bridging**, and **TO-vertical climbing**), and may also have been capable of exploiting both the arboreal and terrestrial settings.

In summary, the available evidence for both *Griphopithecus* and *Kenyapithecus* suggests that early crown hominoids likely utilized an ancestral hominoid positional repertoire, but that their lifestyle may have incorporated slightly greater amounts of terrestriality. Most importantly, there is a general lack of evidence for an advanced, modern hominoid TO-body plan. If *Griphopithecus* and *Kenyapithecus* represent crown hominoids that diverged from the ancestral line after hylobatids, and further fossil discoveries prove that they were in fact postcranially most similar to the proconsuloids, then either (1) a substantial amount of postcranial parallel evolution took place between at least the last common ancestor of hylobatids and the hominid lineages, (2) the few craniodental similarities *Griphopithecus* and *Kenyapithecus* share with hominids developed in parallel, or (3) *Griphopithecus* and *Kenyapithecus* have re-evolved a more primitive postcranial pattern. If, however, *Griphopithecus* and *Kenyapithecus* are interpreted as either crown or stem hominoid lineages that diverged from the hominoid ancestral lineage prior to the divergence of the hylobatid lineage, then similar to proconsuloids, they become informative regarding morphological patterns present in ancestral hominoids, but do not contribute to our understanding of the modern hominoid radiation (Benefit and McCrossin 1995). Evidently, more fossil samples are needed to clarify the morphological patterns and taxonomic affiliations of these key Miocene taxa.

The younger *Dryopithecus* and *Pierolapithecus,* along with some of their European middle-late Miocene[7] counterparts (e.g., *Hispanopithecus, Oreopithecus,*[8] and *Rudapithecus*), more clearly display morphological features associated with the modern hominoid TO-body plan, relative to that of proposed stem (e.g., proconsuloids) or crown hominoids (e.g., *Griphopithecus* and *Kenyapithecus*). While not known/present for all of the aforementioned taxa, key features that appear in this group include a broad/shallow thorax, an emphasis on lumbar stability, a reduced ulnar olecranon process, the lack of an ulnar styloid-triquetrum articulation in the wrist, relatively curved phalanges, and moderate iliac flare, in addition to further enhancements of the relatively derived humeral features found in early stem hominoids (Moyà-Solà and Köhler 1996; Harrison and Rook 1997; Köhler and Moyà-Solà 1997; Moyà-Solà et al. 1999, 2004, 2005a, b, 2009; Deane and Begun 2008; Almécija et al. 2007, 2009, 2012; Ward 2015; Alba et al. 2010, 2011, 2012; Pina et al. 2012, 2014; Hammond et al. 2013; Tallman et al. 2013; Susanna et al. 2014). While the interpretation of the positional behavior of each individual stem hominid is no doubt debatable, and most taxa do not display the entire suite of extant hominoid-like postcranial specializations, each of these taxa appear to be more derived than proconsuloids, and each taxon is more similar to extant hominoids with regards to their locomotor

---

[7]The fossil locations for *Dryopithecus, Hispanopithecus, Oreopithecus, Pierolapithecus*, and *Rudapithecus* collectively span from ∼13 to 7 Ma (Casanovas-Vilar et al. 2011).

[8]The taxonomic placement of *Oreopithecus* is highly enigmatic. Currently, many authors consider *Oreopithecus* to be a stem hominid (Harrison and Rook 1997; Begun 2007, 2015; Harrison 2010b; Begun et al. 2012). Some analyses suggest, however, that *Oreopithecus* shares derived dental traits with one or more of the nyanzapithecines (McCrossin 1992; Harrison 2002, 2010a). If this later scenario is proven to be correct, the postcranial specializations that are shared between *Oreopithecus* and extant hominoids are likely homoplastic.

osteology. Therefore, it would not be surprising if these taxa had been capable of utilizing most if not all of the positional behaviors that are shared by extant apes (Tables 3.2 and 3.3). At the same time, it would equally not be surprising if individual taxa emphasized certain positional modes over others, similar to the variation in positional modes seen among extant apes. What is most important from a comparative perspective is that by the middle-late Miocene at ~13–7 mya, a TO-body plan is present in Europe in some of the branches of the early hominid lineage.

While the taxonomic placement of *Griphopithecus* and *Kenyapithecus* represents a relatively minor challenge to the hypothesis of the evolution of a homologous hominoid postcranial skeleton, interpreting the position of *Sivapithecus* which is dated to ~13–7 mya (Begun 2015) is a greater challenge (Pilbeam and Young 2001; Young 2003). In particular, *Sivapithecus* has been shown to possess derived craniofacial features that are shared with extant orangutans (Ward 1997b; Kelley 2002; Begun 2015), whereas two humeral specimens indicate that these Asian hominoids may not have possessed an extant ape-like shoulder joint complex, and a recently described innominate specimen suggests that *Sivapithecus* had relatively narrow hips, and as a corollary, a narrow torso (Pilbeam et al. 1990; Richmond and Whalen 2001; Madar et al. 2002; Morgan et al. 2015), both features indicating that *Sivapithecus* may not have had an extant apelike TO-body plan. Nevertheless, similar to the aforementioned crown hominoids and stem hominids, *Sivapithecus* does share with extant hominoids an advanced distal humerus with a broad/spool-shaped trochlea, a deep olecranon fossa, well-developed medial and lateral trochlear keels, a globular and well-developed capitulum, and a deep *zona conoidea*, traits associated with elbow stability and the potential to rotate the forearm while extending or flexing the forelimb (Pilbeam et al. 1990; Richmond and Whalen 2001; Madar et al. 2002; Alba et al. 2011). *Sivapithecus* also displays a femur that was capable of multidirectional loading, and relatively mobile hands and feet that were capable of powerful grasping (Madar et al. 2002; DeSilva et al. 2010). Most recently, Begun and Kivell (2011) have suggested that there is evidence from the capitate and hamate that *Sivapithecus* may also have utilized knuckle-walking, but additional fossil material are needed to substantiate these claims.

Due to the mosaic nature of the available postcranial skeletal material, it is difficult to definitively interpret the positional repertoire of *Sivapithecus*. However, as with all Miocene hominoid fossils, a locomotor repertoire that included **quadrupedal walking** (i.e., **TP-clambering/scrambling**), **bridging**, and **TO-vertical climbing** seems very likely. Given the non-hominoid-like configuration of the proximal humerus, the use of suspensory positional behaviors (i.e., **TO-suspend**; **TO-brachiation/forelimb swing**, and **TO-clamber/transfer**), as they are utilized by modern hominoids, seems more unlikely. Furthermore, the potential knuckle-walking features of the *Sivapithecus* hand may suggest that it was also capable of exploiting the terrestrial setting.

What is most troublesome is that if the craniofacial remains of *Sivapithecus* do in fact signify a close phylogenetic relationship between *Sivapithecus* and extant orangutans, then the possibility remains that many of the hominoid postcranial specializations commonly associated with TO-locomotion and/or posture may have developed in parallel in lineages leading to *Morotopithecus*, hylobatids, *Oreopithecus*, any number of the aforementioned stem hominids, *Lufengpithecus*,[9] *Pongo*, and/or the extant hominines. As discussed above for *Griphopithecus* and *Kenyapithecus*, two other evolutionary scenarios are also possible, both of which require a reduced number of evolutionary steps relative to the aforementioned scenario, including the parallel development of craniofacial similarities between *Sivapithecus* and extant *Pongo*, or that *Sivapithecus* has re-evolved a more primitive postcranial pattern. Under either of these alternative scenarios, the TO-specializations of crown hominoids could still be homologous (Pilbeam and Young 2001, 2004; Andrews and Harrison 2005).

Conclusions

Our comparative approach to the evolution of hominoid locomotion and posture provides strong support for the notion that a collective use of TO-positional behaviors is shared by all living hominoids, as the entire hominoid sample investigated here consistently differs from the cercopithecoid sample by the total frequency of use of TO-positional behavior and the total number of TO-positional modes utilized. Given these new insights, in addition to the numerous postcranial anatomical similarities that are shared by extant apes (Keith 1923; Schultz 1930; Harrison 1987, 2016; Hunt 1991b, 2016; Gebo 1996; Ward 2015; Larson 1998; Gibbs et al. 2002; Young 2003; Kagaya et al. 2008, 2009, 2010; Diogo and Wood 2011; Chan 2014; Williams and Russo 2015), the most parsimonious explanation is that the last common ancestor of hominoids possessed TO-behavioral and morphological specializations similar to that of extant apes (Harrison 1987, 2016).

Such an interpretation, however, is not entirely consistent with the available fossil data, because so far no definitive Miocene crown hominoid displays clear TO-postcranial specializations within our conservative upper and lower temporal bounds of 25–12 mya. *Morotopithecus* seems to be the most plausible candidate to display TO-adaptations in an early Miocene hominoid, but its controversial affinities with the Afropithecinae may push it outside of the modern hominoid radiation, which implies that the postcranial similarities *Morotopithecus* shares with extant hominoids have been the result of parallel evolution under similar selective pressures. Furthermore, in some reconstructions of hominoid evolution, *Griphopithecus*, *Kenyapithecus*, and/or *Sivapithecus* represent challenges to the hypothesis of hominoid TO-behavioral and postcranial homology. As such, either

---

[9]While not very many of their postcranial remains have been published, Begun (2015) and Deane and Begun (2008) suggest that *Lufengpithecus*, which is dated to ~9–7 mya, (Xijun and Zhuding 2002; Qi et al. 2006) displays curved phalanges that are strongly indicative of suspensory positional behaviors.

the majority of taxa that currently comprise the Miocene fossil record have had little
to do with the modern hominoid radiation (Benefit and McCrossin 1995; Pilbeam
1996; Pilbeam and Young 2001, 2004), or conversely, they are telling a story of
recurring homoplasy (Rose 1997; Larson 1998; Ward 1997a, 2015; Begun 2007;
Alba 2012; Hunt 2016).

What then can be said about the evolution of hominoid TO-specializations?
Provided that our broad characterizations of hominoid taxonomy are somewhat
correct, the relatively large body size of both stem hominoids, i.e., the primates that
range from 9 to 50 kg (Fleagle 2013), and stem hominids, i.e., primates ranging
from 17 to 48 kg (Fleagle 2013), suggests that a large body size of >10 kg was
likely a key component to the initial hominoid radiation. Under this evolutionary
scenario, hylobatids are the result of phyletic dwarfing (Groves 1972; Wheatley
1982, 1987; Tyler 1991; Pilbeam 1996; Young 2003; Gebo 2004; MacLatchy 2004;
Pilbeam and Young 2004; Young and MacLatchy 2004; Jablonski and Chaplin
2009). If, however, the majority of proconsuloids are in fact stem hominoids, then
in contrast to hypotheses predicting a congruent relationship between body size and
TO-positional behaviors, the initial radiation of relatively large-bodied apes may
not have had a TO-body plan characteristic of extant apes (Rose 1994, 1997; Rae
1999; Ward 1997a, 2015; Gebo 2004).

Nevertheless, the available Miocene fossil material suggests that early stem
hominoids were likely utilizing a relatively generalized torso-pronograde positional
repertoire; however, this repertoire was more derived compared to cercopithecoids
and already highlights the presence of selection pressures associated with balance
and branch deformation/fracture (Napier 1967; Grand 1972, 1984; Cartmill 1974,
1985; Cant 1992; Preuschoft et al. 1995, 2016; Dunbar and Badam 2000;
Preuschoft 2002), the same selection pressures which inevitably helped to shape the
TO-behavioral and morphological traits characteristic of the hominoid lineage.
Thus, in general, the positional repertoire of stem hominoids does not demean a
hypothesized causal relationship between body size and TO-positional behaviors,
but instead, may be indicative of an initial evolutionary solution to problems
associated with increasing body size in the arboreal setting (Gebo 2004; MacLatchy
2004; Nakatsukasa and Kunimatsu 2009). It therefore seems likely that the
TO-specializations of crown hominoids developed from an ancestor who was
similar to one of the proconsuloid lineages. However, the notoriously fragmentary
state of the Miocene fossil record does not conclusively support a scenario where
the last common ancestor of crown hominoids possessed the complete suite of
extant ape TO-specializations, or a scenario where the TO-specializations that are
so prevalent in extant hominoids developed independently in multiple hominoid
lineages.

Even so, perhaps current models are too narrow in searching for evidence to
support either of two extremes and are thereby overlooking the very real possibility
that the TO-postcranial specializations that characterize modern hominoids may not
have evolved as a single functional package, but may have appeared in mosaic
fashion, implying that extant ape specializations are perhaps a mix of homologous
and homoplastic features, and TO-locomotor and postural traits were not as extreme

in their initial evolutionary development, with more advanced TO-specializations only appearing later in long-lived terminal taxa (Larson 1998; Rae 1999; Begun 2007). While such an intermediate position is not the most parsimonious evolutionary scenario, because it requires additional evolutionary steps within multiple hominoid lineages, such a scenario is also not entirely implausible, as this is precisely what one could predict for closely related evolving lineages and complex trait-packages that share similar, if not, homologous developmental mechanisms and ecological pressures (Hall 2003, 2007; Begun 2007; Shubin et al. 2009; Scotland 2011; Wake et al. 2011; Reno 2014).

In conclusion, a better understanding of present and past hominoid diversity, including how and why our ape ancestors developed TO-locomotion and posture are key factors to the understanding of our own lineage. As the hominoid fossil record develops and we undertake further study of living forms, we will be able to further address some of the inconsistencies inherent to our current comparisons. Most importantly, only with a detailed and comprehensive view of past and present ape species, both large and small, as is attempted here, will we be able to truly understand the vectors of selection that have resulted in the evolution of our own species.

Acknowledgements This chapter greatly benefited from field studies on Sumatran *Hylobates agilis* and *Symphalangus syndactylus*, which would not have been possible without the assistance of an NSF Doctoral Dissertation Improvement Grant (BCS 1061477 awarded to MGN).

References

Abourachid A, Höfling E (2012) The legs: a key to bird evolutionary success. J Ornithol 153:193–198

Alba DM (2012) Fossil apes from the Vallés-Penedés Basin. Evol Anthropol 21:254–269

Alba DM, Moyà-Solà S (2009) The origin of the great-ape-and-human clade (Primates: Hominidae) reconsidered in the light of recent hominoid findings from the middle Miocene of the Vallés-Penedés Basin (Catalonia, Spain). Paleolustina 1:75–83

Alba DM, Almécija S, Moyà-Solà S (2010) Locomotor inferences in *Pierolapithecus* and *Hispanopithecus*: reply to Deane and Begun (2008). J Hum Evol 59:143–148

Alba DM, Moyà-Solà S, Almécija S (2011) A partial hominoid humerus from the middle Miocene of Castell de Barberà (Vallés-Penedés Basin, Catalonia, Spain). Am J Phys Anthropol 144:365–381

Alba DM, Almécija S, Casanovas-Vilar I, Méndez JM, Moyà-Solà S (2012) A partial skeleton of the fossil great ape *Hispanopithecus laietanus* from Can Feu and the mosaic evolution of crown-hominoid positional behaviors. PLoS ONE 7(6):e39617

Alba DM, Almécija S, DeMiguel D, Fortuny J, Pérez de los Ríos M, Pina M, Robles JM, Moyà-Solà S (2015) Miocene small-bodied ape from Eurasia sheds light on hominoid evolution. Science 350:aab2625

Almécija S, Alba DM, Moyà-Solà S, Köhler M (2007) Orang-like manual adaptations in the fossil hominoid *Hispanopithecus laietanus*: first steps towards great ape suspensory behaviours. Proc R Soc Biol Sci B 274:2375–2384

Almécija S, Alba DM, Moyà-Solà S (2009) *Pierolapithecus* and the functional morphology of Miocene ape hand phalanges: paleobiological and evolutionary implications. J Hum Evol 57:284–297

Almécija S, Alba DM, Moyà-Solà S (2012) The thumb of Miocene apes: new insights from Castell de Barberà (Catalonia, Spain). Am J Phys Anthropol 148:436–450

Andrews P, Groves CE (1976) Gibbon and brachiation. In: Rumbaugh DM (ed) Gibbon and siamang: a series of volumes on the lesser apes. Suspensory behavior, locomotion, and other behaviors of captive gibbons: cognition, vol 4. Basel, Karger, pp 167–218

Andrews P, Harrison T (2005) The last common ancestor of apes and humans. In: Lieberman DE, Smith RJ, Kelley J (eds) Interpreting the past: essays on human, primate, and mammal evolution. Brill Academic Publishers Inc., Boston, pp 104–121

Arias-Martorell J, Tallman M, Maria Potau J, Bello-Hellegouarch G, Pérez-Pérez A (2015) Shape analysis of the proximal humerus in orthograde andsemi-orthograde primates: Correlates of suspensory behavior. Am J Primatol 77:1–19

Avis V (1962) Brachiation: the crucial issue for man's ancestry. Southwest J Anthropol 18:119–148

Begun DR (1992) Phyletic diversity and locomotion in primitive European hominids. Am J Phys Anthropol 87:311–340

Begun D (2002a) The Pliopithecoidea. In: Hartwig WC (ed) The primate fossil record. Cambridge University Press, Cambridge, pp 221–240

Begun D (2002b) European hominoids. In: Hartwig WC (ed) The primate fossil record. Cambridge University Press, Cambridge, pp 339–368

Begun DR (2007) How to identify (as opposed to define) a homoplasy: examples from fossil and living great apes. J Hum Evol 52:559–572

Begun DR (2015) Fossil record of Miocene hominoids. In: Henke W, Tattersall I (eds) Handbook of paleoanthropology, vol 2. Springer, New York, pp 1261–1332

Begun DR, Kivell TL (2011) Knuckle-walking in *Sivapithecus*? The combined effects of homology and homoplasy with possible implications for pongine dispersals. J Hum Evol 60:158–170

Begun DR, Ward CV, Rose MD (1997) Events in hominoid evolution. In: Begun DR, Ward CV, Rose MD (eds) Function, phylogeny, and fossils: Miocene hominoid evolution and adaptations. Plenum Press, New York, pp 389–415

Begun DR, Nargolwalla MC, Kordos L (2012) European Miocene hominids and the origin of the African ape and human clade. Evol Anthropol 21:10–23

Bell AD, Bryan A (2008) Plant form: an illustrated guide to flowering plant morphology. Oxford University Press, Oxford

Benefit BR, McCrossin ML (1995) Miocene hominoids and hominid origins. Annu Rev Anthropol 24:237–256

Benefit BR, McCrossin ML (2015) A window into ape evolution. Science 350:515

Cannon CH, Leighton M (1994) Comparative locomotor ecology of gibbons and macaques: selection of canopy elements for crossing gaps. Am J Phys Anthropol 93:505–524

Cant JGH (1987a) Positional behavior of female Bornean orangutans (*Pongo pygmaeus*). Am J Primatol 12:71–90

Cant JGH (1987b) Effects of sexual dimorphism in body size on feeding postural behavior of Sumatran orangutans (*Pongo pygmaeus*). Am J Phys Anthropol 74:143–148

Cant JG (1988) Positional behavior of long-tailed macaques (*Macaca fascicularis*) in northern Sumatra. Am J Phys Anthropol 76:29–37

Cant JGH (1992) Positional behavior and body size of arboreal primates: a theoretical framework for field studies and an illustration of its application. Am J Phys Anthropol 88:273–283

Cant JGH, Youlatos D, Rose MD (2001) Locomotor behavior of *Lagothrix lagothrica* and *Ateles belzebuth* in Yasuni National Park, Ecuador: general patterns and non suspensory modes. J Hum Evol 41:141–166

Carbone L, Harris RA, Gnerre S, Veeramah KR, Lorente-Galdos B, Huddleston J et al (2014) The gibbon genome provides a novel perspective on the accelerated karyotype evolution of small apes. Nature 513:195–201

Carlson KJ (2005) Investigating the form-function interface in African apes: relationships between principal moments of area and positional behaviors in femoral and humeral diaphyses. Am J Phys Anthropol 127:312–334

Cartmill M (1974) Pads and claws in arboreal locomotion. In: Jenkins FA (ed) Primate locomotion. Academic Press, New York, pp 45–83

Cartmill M (1985) Climbing. In: Hildebrand M, Bramble DM, Liem KF, Wake BD (eds) Functional vertebrate morphology. Harvard University Press, Cambridge, pp 73–88

Cartmill M, Milton K (1977) The lorisiform wrist joint and the evolution of "brachiating" adaptations in the Hominoidea. Am J Phys Anthropol 47:249–272

Casanovas-Vilar I, Alba DM, Garcés M, Robles JM, Moyà-Solà S (2011) Updated chronology for the Miocene hominoid radiation in western Eurasia. Proc Nat Acad Sci 108:5554–5590

Chan LK (2007a) Scapular position in primates. Folia Primatol 78:19–35

Chan LK (2007b) Glenohumeral mobility in primates. Folia Primatol 78:1–18

Chan LK (2008) The range of passive arm circumduction in primates: do hominoids really have more mobile shoulders? Am J Phys Anthropol 136:265–277

Chan LK (2014) The thoracic shape of hominoids. Anat Res Int Article ID 324850, 8 p. doi:10.1155/2014/324850

Chan Y-C, Roos C, Inoue-Murayama M, Inoue E, Shih C-C, Pei KJ-C, Vigilant L (2010) Mitochondrial genome sequences effectively reveal the phylogeny of Hylobates gibbons. PLoS ONE 5(12):e14419

Chatterjee HJ, Ho SYW, Barnes I, Groves C (2009) Estimating the phylogeny and divergence times of primates using a supertree approach. BMC Evol Biol 9:259

Chivers DJ (1972) The siamang and the gibbon in the Malay peninsula. In: Rumbaugh DM (ed) The gibbon and siamang, vol 1. Karger, Basel, pp 103–135

Chivers DJ (1974) The siamang in Malaya: a field study of a primate in tropical rain forest. Contrib Primatol 4:1–335

Chivers D, Anandam M, Groves C, Molur S, Rawson BM, Richardson MC, Roos C, Whittaker D (2013) Family Hylobatidae (gibbons). In: Mittermeier RA, Rylands AB, Wilson DE (eds) Handbook of the mammals of the world: primates. Lynx Edicions, Barcelona, pp 754–791

Crompton RH, Vereecke EE, Thorpe SKS (2008) Locomotion and posture from the common hominoid ancestor to fully modern hominins, with special reference to the last common panin/hominin ancestor. J Anat 212:501–543

Deane AS, Begun DR (2008) Broken fingers: retesting locomotor hypotheses for fossil hominoids using fragmentary proximal phalanges and high-resolution polynomial curve fitting (HR-PCF). J Hum Evol 55:691–701

DeSilva JM, Morgan ME, Barry JC, Pilbeam D (2010) A hominoid distal tibia from the Miocene of Pakistan. J Hum Evol 58:147–154

Diogo R, Wood B (2011) Soft-tissue anatomy of the primates: Phylogenetic analyses based on the muscles of the head, neck, pectoral region and upper limb, with notes on the evolution of these muscles. J Anat 219:273–359

Diogo R, Potau JM, Pastor JF, de Paz FJ, Ferrero EM, Bello G, Barbosa M, Aziz MA, Burrows AM, Arias-Martorell J, Wood BA (2012) Photographic and descriptive musculoskeletal atlas of gibbons and siamangs (Hylobates). CRC Press, Florida

Donoghue PCJ (2005) Saving the stem group: a contradiction in terms? Paleobiology 31:553–558

Donoghue MJ, Doyle JA, Gauthier J, Kluge AG, Rowe T (1989) The importance of fossils in phylogeny reconstruction. Annu Rev Ecol Syst 20:431–460

Doran DM (1992) Comparison of instantaneous and locomotor bout sampling methods: a case study of adult male chimpanzee locomotor behavior and substrate use. Am J Phys Anthropol 89:85–99

Doran DM (1993a) Comparative locomotor behavior of chimpanzees and bonobos: the influence of morphology on locomotion. Am J Phys Anthropol 91:83–98

Doran DM (1993b) Sex differences in adult chimpanzee positional behavior: the influence of body size on locomotion and posture. Am J Phys Anthropol 91:99–115

Doran DM (1996) Comparative positional behavior of the African ape. In: McGrew WC, Marchant LF, Nishida T (eds) Great ape societies. Cambridge University Press, Cambridge, pp 213–224

Druelle F, Berillon G (2014) Bipedalism in non-human primates: a comparative review of behavioural and experimental explorations on catarrhines. BMSAP 26(3–4):111–120

Dunbar DC, Badam GL (2000) Locomotion and posture during terminal branch feeding. Int J Primatol 21:649–669

Erikson GE (1963) Brachiation in New World monkeys and anthropoid apes. Symp Zool Soc Lond 10:135–164

Ersoy A, Kelley J, Andrews P, Alpagut B (2008) Hominoid phalanges from the middle Miocene site of Paşalar, Turkey. J Hum Evol 54:518–529

Fan P, Scott MB, Hanlan FEI, Changyong MA (2013) Locomotion behavior of cao vit gibbon (*Nomascus nasutus*) living in karst forest in Bangliang Nature Reserve, Guangxi, China. Int Zool 8:356–364

Farlow JO, Gatesy SM, Holtz TR Jr, Hutchinson JR, Robinson JM (2000) Theropod locomotion. Am Zool 40:640–663

Fei H, Ma C, Bartlett TQ, Dai R, Xiao W, Fan P (2015) Feeding postures of Cao Vit gibbons (*Nomascus nasutus*) living in a low-canopy karst forest. Int J Primatol 36:1036–1054

Finstermeier K, Zinner D, Brameier M, Meyer M, Kreuz E, Hofreiter Roos C (2013) A mitogenomic phylogeny of living primates. PLoS ONE 8(7):e69504

Fleagle JG (1976) Locomotion and posture of the Malayan siamang and implications for hominid evolution. Folia Primatol 26:245–269

Fleagle JG (1980) Locomotion and posture. In: Chivers DJ (ed) Malayan forest primates: ten years' study in tropical rain forest. Plenum Press, New York, pp 191–207

Fleagle JG (2013) Primate adaptation and evolution, 3rd edn. Academic Press, New York

Fleagle JG, Stern JT, Jungers WL, Susman RL, Vangor AK, Wells JP (1981) Climbing: a biomechanical link with brachiation and with bipedalism. Symp Zool Soc Lond 48:359–375

Gebo DL (1996) Climbing, brachiation, and terrestrial quadrupedalism: historical precursors of hominid bipedalism. Am J Phys Anthropol 101:55–92

Gebo DL (2004) Paleontology, terrestriality, and the intelligence of great apes. In: Russon AE, Begun DR (eds) The evolution of thought: evolutionary origins of great ape intelligence. Cambridge University Press, Cambridge, pp 320–334

Gebo DL, Chapman CA (1995) Positional behavior in five sympatric old world monkeys. Am J Phys Anthropol 91:49–76

Gebo DL, MacLatchy L, Kityo R, Deino A, Kingston J, Pilbeam D (1997) A hominoid genus from the early Miocene of Uganda. Science 276:401–404

Gibbs S, Collard M, Wood B (2002) Soft-tissue anatomy of the extant hominoids: a review and phylogenetic analysis. J Anat 200:3–49

Gittins SP (1983) Use of the forest canopy by the agile gibbon. Folia Primatol 40:134–144

Godfrey LR, Jungers WL (2003) The extinct sloth lemurs of Madagascar. Evol Anthropol 12:252–263

Godinot M (2015) Fossil record of the primates from the Paleocene to the Oligocene. In: Henke W, Tattersall I (eds) Handbook of paleoanthropology, vol 2. Springer, New York, pp 1137–1261

Grand TI (1972) A mechanical interpretation of terminal branch feeding. J Mammal 53:198–201

Grand TI (1984) Motion economy within the canopy: four strategies for mobility. In: Rodman PS, Cant JGH (eds) Adaptations for foraging in nonhuman primates. Columbia University Press, New York, pp 54–72

Gregory WK (1916) Studies on the evolution of the primates. Part II. Phylogeny of recent and extinct anthropoids with special reference to the origin of man. Bull Am Mus Nat Hist 35:258–355

Gregory WK (1928) Were the ancestors of man primitive brachiators? Proc Am Phil Soc 67:129–150

Gregory WK (1934) Man's place among the anthropoids. Clarendon, Benjamin/Cummings, Oxford

Groves CP (1972) Systematics and phylogeny of gibbons. In: Rumbaugh D (ed) Gibbon and siamang, vol 1. Karger, Basel, pp 2–89

Hall BK (2003) Descent with modification: the unity underlying homology and homoplasy as seen through an analysis of development and evolution. Biol Rev 78:409–433

Hall BK (2007) Homoplasy and homology: dichotomy or continuum? J Hum Evol 52:473–479

Hammond AS, Alba DM, Almécija S, Moyà-Solà S (2013) Middle Miocene *Pierolapithecus* provides a first glimpse into early hominid pelvic morphology. J Hum Evol 64:658–666

Hanna JB (2006) Kinematics of vertical climbing in lorises and *Cheirogaleus medius*. J Hum Evol 50:469–478

Harrison T (1987) The phylogenetic relationships of the early catarrhine primates: a review of the current evidence. J Hum Evol 16:41–80

Harrison T (1991) The implications of *Oreopithecus bambolii* for the origins of bipedalism. In: Coppens Y, Senut B (eds) Origine(s) de la Bipédie chez les Hominidés. Fodation Singer-Polignac, Paris, pp 235–244

Harrison T (2002) Late Oligocene to middle Miocene catarrhines from Afro-Arabia. In: Hartwig WL (ed) The primate fossil record. Cambridge University Press, Cambridge, pp 311–338

Harrison T (2010a) Dendropithecoidea, Proconsuloidea, and Hominoidea. In: Werdelin L, Sanders WJ (eds) Cenozoic mammals of Africa. University of California Press, Berkeley, pp 429–469

Harrison T (2010b) Apes among the tangled branches of human origins. Science 327:532–534

Harrison T (2013) Catarrhine origins. In: Begun DR (ed) A companion to paleoanthropology. Blackwell Publishing, New York, pp 376–396

Harrison T, Andrews P (2009) The anatomy and systematic position of the early Miocene proconsulids from Meswa Bridge, Kenya. J Hum Evol 56:479–496

Harrison T, Rook L (1997) Enigmatic anthropoid or misunderstood ape? The phylogenetic status of *Oreopithecus bambolii* reconsidered. In: Begun DR, Ward CV, Rose MD (eds) Function, phylogeny, and fossils: Miocene hominoid evolution and adaptations. Plenum Press, New York, pp 327–362

Harrison T, Ji X, Zheng L (2008) Renewed investigations at the late Miocene hominoid locality of Leilao, Yunnan, China. Am J Phys Anthropol 135:113

Harrison T (2016) The fossil record and evolutionary history of hylobatids. In: Reichard UH, Hirohisa H, Barelli C (eds) Evolution of gibbons and siamang. Springer, New York, pp 91–110

Houle A, Chapman CA, Vickery WL (2007) Intratree variation in fruit production and implications for primate foraging. Int J Primatol 28:1197–1217

Hunt KD (1991a) Positional behavior in the Hominoidea. Int J Primatol 12:95–118

Hunt KD (1991b) Mechanical implications of chimpanzee positional behavior. Am J Phys Anthropol 86:521–536

Hunt KD (1992) Positional behavior of *Pan troglodytes* in the Mahale Mountains and Gombe Stream National Parks, Tanzania. Am J Phys Anthropol 87:83–105

Hunt KD (2004) The special demands of great ape locomotion and posture. In: Russon AE, Begun DR (eds) The evolution of thought: evolutionary origins of great ape intelligence. Cambridge University Press, Cambridge, pp 172–189

Hunt KD (2016) Why are there apes? Evidence for the co-evolution of ape and monkey ecomorphology. J Anat 228:630–685

Hunt KD, Cant JGH, Gebo DL, Rose MD, Walker SE, Youlatos D (1996) Standardized descriptions of primate locomotor and postural modes. Primates 37:363–387

Israfil H, Zehr SM, Mootnick AR, Ruvolo M, Steiper ME (2011) Unresolved molecular phylogenies of gibbons and siamangs (Family: Hylobatidae) based on mitochondrial, Y-linked, and X-linked loci indicate a rapid Miocene radiation or sudden vicariance event. Mol Phylogenet Evol 58:447–455

Jablonski NG, Chaplin G (2009) The fossil record of gibbons. In: Lappan S, Whittaker DJ (eds) The gibbons: new perspectives on small ape socioecology and population biology. Springer, New York, pp 111–130

Ji X, Jablonski NG, Su DF, Deng C, Flynn LJ, You Y, Kelley J (2013) Juvenile hominoid cranium from the terminal Miocene of Yunnan, China. Chin Sci Bull 58:3771–3779

Jungers WL (1984) Scaling of the hominoid locomotor skeleton with special reference to lesser apes. In: Preuschoft H, Chivers DJ, Brockelman WY, Creel N (eds) The lesser apes: evolutionary and behavioral biology. Edinburgh University Press, Edinburgh, pp 146–169

Jungers WL (1985) Body size and scaling of limb proportions in primates. In: Jungers WL (ed) Size and scaling in primate biology. Plenum Press, New York, pp 345–382

Jungers WL, Stern JT (1980) Telemetered electromyography of forelimb muscle chains in gibbons (*Hylobates lar*). Science 208(4444):617–619

Kagaya M, Ogihara N, Nakatsukasa M (2008) Morphological study of the anthropoid thoracic cage: scaling of thoracic width and an analysis of rib curvature. Primates 49:89–99

Kagaya M, Ogihara N, Nakatsukasa M (2009) Rib orientation and implications for orthograde positional behavior in nonhuman anthropoids. Primates 50:305–310

Kagaya M, Ogihara N, Nakatsukasa M (2010) Is the clavicle of apes long? An investigation of clavicular length in relation to body mass and upper thoracic width. In J Primatol 31:209–217

Kano T, Mulavwa M (1983) Feeding ecology of the pygmy chimpanzees (*Pan paniscus*) of Wamba. In: Susman RL (ed) The pygmy chimpanzee: evolutionary biology and behavior. Plenum Press, New York, pp 275–300

Keith A (1891) Anatomical noted on Malay apes. J Straits Br R Asiat Soc 23:77–94

Keith A (1899) On the chimpanzees and their relation to the gorilla. Proc Zool Soc Lond 67:296–312

Keith A (1903) The extent to which the posterior segments of the body have been transmuted and suppressed in the evolution of man and allied primates. J Anat Physiol 37:18–40

Keith A (1923) Man's posture: its evolution and disorder. Br Med J 1:451–454, 499–502, 545–548, 587–590, 624–626, 669–672

Kelley J (2002) The hominoid radiation in Asia. In: Hartwig WL (ed) The primate fossil record. Cambridge University Press, Cambridge, pp 369–384

Kikuchi K, Nakatsukasa M, Nakano Y, Kunimatsu Y, Shimizu D, Ogihara N, Tsujikawa H, Takano T, Ishida H (2015) Morphology of thethoracolumbar spine of the middle Miocene hominoid Nacholapithecus kerioi from northern Kenya. J Hum Evol 88:25–42

Köhler M, Moyà-Solà S (1997) Ape-like or hominid-like? The positional behavior of Oreopithecus bambolii reconsidered. Proc Nat Acad Sci 94:11747–11750

Lambert J (2015) Evolutionary biology of ape and monkey feeding and nutrition. In: Henke W, Tattersall I (eds) Handbook of paleoanthropology, vol 2. Springer, New York, pp 1631–1660

Larson S (1988) Subscapularis function in gibbons and chimpanzees: implications for interpretation of humeral head torsion in hominoids. Am J Phys Anthropol 76:449–462

Larson SG (1998) Parallel evolution in the hominoid trunk and forelimb. Evol Anthropol 6:87–99

Lovejoy CO, Suwa G, Simpson SW, Matternes JH, White TD (2009) The great divides: *Ardipithecus ramidus* reveals the postcrania of our last common ancestors with African apes. Science 326:73–106

MacLatchy L (2004) The oldest ape. Evol Anthropol 13:90–103

MacLatchy LM, Bossert WH (1996) An analysis of the articular surface distribution of the femoral head and acetabulum in anthropoids, with implications for hip function in Miocene hominoids. J Hum Evol 31:425–453

MacLatchy LM, Gebo D, Kityo R, Pilbeam D (2000) Postcranial functional morphology of *Morotopithecus bishopi*, with implications for the evolution of modern ape locomotion. Hum Evol 39:159–183

MacLatchy LM, Rossie JB, Smith TM, Tafforeau P (2010) Evidence for dietary niche separation in the Miocene hominoids *Morotopithecus* and *Afropithecus*. Am J Phys Anthropol S 50:160

Madar SI, Rose MD, Kelley J, MacLatchy L, Pilbeam D (2002) New *Sivapithecus* postcranial specimens from the Siwaliks of Pakistan. J Hum Evol 42:705–752

Manduell KL, Morrogh-Bernard HC, Thorpe SKS (2011) Locomotor behavior of wild orangutans (*Pongo pygmaeus wurmbii*) in disturbed peat swamp forest, Sabangau, Central Kalimantan, Indonesia. Am J Phys Anthropol 145:348–359

Manduell KL, Harrison ME, Thorpe SKS (2012) Forest structure and support availability influence orangutan locomotion. Am J Primatol 74:1128–1142

Matsui A, Rakotondraparany F, Munechika I, Hasegawa M, Horai S (2009) Molecular phylogeny and evolution of prosimians based on complete sequences of mitochondrial DNAs. Gene 441:53–66

McCrossin ML (1992) An oreopithecid proximal humerus from the middle Miocene of Maboko Island, Kenya. Int J Primatol 13:659–677

McCrossin ML, Benefit BR (1997) On the relationship and adaptations of *Kenyapithecus*, a large-bodied hominoid from the middle Miocene of eastern Africa. In: Begun DR, Ward CV, Rose MD (eds) Function, phylogeny and fossils: Miocene hominoid origins and adaptations. Plenum Press, New York, pp 241–267

McNulty KP, Begun DR, Kelley J, Manthi FK, Mbua EN (2015) A systematic revision of Proconsul with the description of a new genus of early Miocene hominoid. J Hum Evol 84:42–61

Michilsens F, Vereecke EE, D'Août K, Aerts P (2009) Functional anatomy of the gibbon forelimb: adaptations to a brachiating lifestyle. J Anat 215:335–354

Miller ER, Benefit BR, McCrossin ML, Plavcan JM, Leakey MG, El-Barkooky AN, Hamdan MA, Abdel Gawad MK, Hassan SM, Simons EL (2009) Systematics of early and middle Miocene Old World monkeys. J Hum Evol 57:195–211

Morgan ME, Lewton KL, Kelley J, Otárola-Castillo E, Barry JC, Flynn LJ, Plbeam D (2015) A partial hominoid innominate from the Miocene of Pakistan: description and preliminary analyses. PNAS 112:82–87

Morton DJ (1926) Evolution of man's erect posture (preliminary report). J Morph Physiol 43:147–179

Moyà-Solà S, Köhler M (1996) A *Dryopithecus* skeleton and the origins of great-ape locomotion. Nature 379:156–169

Moyà-Solà S, Köhler M, Rook L (1999) Evidence of hominid-like precision grip capability in the hand of the Miocene ape *Oreopithecus*. Proc Nat Acad Sci 96:313–317

Moyà-Solà S, Köhler M, Alba DM, Casanovas-Vilar I, Galindo J (2004) *Pierolapithecus catalaunicus*, a new middle Miocene great ape from Spain. Science 306:1339–1344

Moyà-Solà S, Köhler M, Alba DM, Casanovas-Vilar I, Galindo J (2005a) Response to comment on *Pierolapithecus catalaunicus* a new middle Miocene great ape from Spain. Science 308:203d

Moyà-Solà S, Köhler M, Rook L (2005b) The *Oreopithecus* thumb: a strange case in hominoid evolution. J Hum Evol 49:395–404

Moyà-Solà S, Köhler M, Alba DM, Casanovas-Vilar I, Galindo J, Robles JM, Cabrera L, Garcés M, Almécija S, Beamud E (2009) First partial face and upper dentition of the middle Miocene hominoid *Dryopithecus fontani* from Abocador de Can Mata (Vallès-Penedès Basin, Catalonia, NE Spain): taxonomic and phylogenetic implications. Am J Phys Anthropol 139:126–145

Myatt JP, Thorpe SKS (2012) Postural strategies employed by orangutans (*Pongo abelii*) during feeding in the terminal branch niche. Am J Phys Anthropol 46:73–82

Nakatsukasa M (2008) Comparative study of Moroto vertebral specimens. J Hum Evol 56:581–588

Nakatsukasa N, Kunimatsu Y (2009) *Nacholapithecus* and its importance for understanding hominoid evolution. Evol Anthropol 18:103–119

Nakatsukasa N, Ward CV, Walker A, Teaford MF, Kunimatsu Y, Ogihara N (2004) Tail loss in *Proconsul heseloni*. J Hum Evol 46:777–784

Napier JR (1967) Evolutionary aspects of primate locomotion. Am J Phys Anthropol 27:333–342

Niemitz C (2010) The evolution of the upright posture and gait-a review and synthesis. Naturwissenschaften 97:241–263

Nowak MG, Reichard UH (2016) The torso-orthograde positional behavior of wild white-handed gibbons (hylobates lar). In: Reichard UH, Hirohisa H, Barelli C (eds) Evolution of gibbons and siamang. Springer, New York, pp 203–225

Patel BA, Grossman A (2006) Dental metric comparisons of *Morotopithecus* and *Afropithecus*: implications for the validity of the genus *Morotopithecus*. J Hum Evol 51:506–512

Patel BA, Susman RL, Rossie JB, Hill A (2009) Terrestrial adaptations in the hands of *Equatorius africanus* revisited. J Hum Evol 57:763–772

Pawłowski B, Nowaczewska W (2015) Origins of Hominiae and putative slection pressures acting on the early Hominins. In: Henke W, Tattersall I (eds) Handbook of paleoanthropology, vol 2. Springer, New York, pp 1887–1918

Perelman P, Johnson WE, Roos C, Seuánez HN, Horvath JE, Moreira MAM, Kessing B, Pontius J, Roelke M, Rumpler Y, Schneider MPC, Silva A, O'Brien SJ, Pecon-Slattery J (2011) A molecular phylogeny of living primates. PLoS Genet 7(3):e1001342

Pickford M (2002) New reconstruction of the Moroto hominoid snout and a reassessment of its affinities to *Afropithecus turkanensis*. Hum Evol 17:1–19

Pickford M, Senut B, Gommery D (1999) Sexual dimorphism in *Morotopithecus bishopi*, an early middle Miocene hominoid from Uganda, and a reassessment of its geological and biological contexts. In: Andrews P, Banham P (eds) Late Cenozoic environments and hominid evolution: a tribute to Bill Bishop. Geological Society, London, pp 27–38

Pilbeam D (1996) Genetic and morphological records of the Hominoidea and hominid origins: a synthesis. Mol Phylogenetic Evol 5:155–168

Pilbeam D, Young N (2001) *Sivapithecus* and hominoid evolution: some brief comments. In: De Bonis L, Koufos GD, Andrews P (eds) Hominoid evolution and climatic change in Europe, vol 2., Phylogeny of the neogene hominoid primates of EurasiaCambridge University Press, Cambridge, pp 349–364

Pilbeam D, Young N (2004) Hominoid evolution: synthesizing disparate data. C R Palevol 3:305–321

Pilbeam D, Rose MD, Barry JC, Shah SMI (1990) New *Sivapithecus* humeri from Pakistan and the relationship of *Sivapithecus* and *Pongo*. Nature 348:237–239

Pina M, Alba DM, Fortuny J, Almécija S, Moyà-Solà S (2012) Brief communication: paleobiological inferences on the locomotor repertoire of extinct hominoids based on femoral neck cortical thickness: the fossil great ape *Hispanopithecus laietanus* as a test-case study. Am J Phys Anthropol 149:42–48

Pina M, Almécija S, Alba DM, O'Neill M, Moyà-Solà S (2014) The middle Miocene ape *Pierolapithecus catalaunicus* exhibits extant great ape-like morphometric affinities on its patella: inferences on knee function and evolution. PLoS ONE 9:1–10

Preuschoft H (2002) What does "arboreal locomotion" mean exactly and what are the relationships between "climbing", environment and morphology? Z Morph Anthrop 83:171–188

Preuschoft H, Demes H (1984) Biomechanics of brachiation. In: Preuschoft H, Chivers D, Brockelman W, Creel N (eds) The lesser apes: evolutionary and behavioural biology. Edinburgh University Press, Edinburgh, pp 96–118

Preuschoft H, Demes B (1985) Influence of size and proportions on the biomechanics of brachiation. In: Jungers WL (ed) Size and scaling in primate biology. Springer, New York, pp 383–399

Preuschoft H, Witte H, Fischer M (1995) Locomotion in nocturnal prosimians. In: Alterman L, Doyle GA, Izard MK (eds) Creatures of the dark: the nocturnal prosimians. Plenum Press, New York, pp 453–472

Preuschoft H, Schönwasser K-H, Witzel U (2016) Selective value of characteristic size parametersin hylobatids. A biomechanical approach to smallape size and morphology. In: Reichard UH, Hirohisa H, Barelli C (eds) Evolution of gibbons and siamang. Springer, New York, pp 227–263

Qi G, Dong W, Zheng L, Zhao L, Gao F, Yue L, Zhang Y (2006) Taxonomy, age and environment status of the Yuanmou hominoids. Chin Sci Bull 51:704–712

Raaum RL, Sterner KN, Noviello CM, Stewart C-B, Disotell TR (2005) Catarrhine primate divergence dates estimated from complete mitochondrial genomes: concordance with fossil and nuclear DNA evidence. J Hum Evol 48:237–257

Rae TC (1999) Mosaic evolution in the origin of the Hominoidea. Folia Primatol 70:125–136

Rasmussen DT (2002) Early catarrhines of the African Eocene and Oligocene. In: Hartwig WC (ed) The primate fossil record. Cambridge University Press, Cambridge, pp 203–220

Remis M (1995) Effects of body size and social context on the arboreal activities of lowland gorillas in the Central African Republic. Am J Phys Anthropol 97:413–433

Remis M (1998) The gorilla paradox: the effects of body size and habitat on the positional behavior of lowland and mountain gorillas. In: Strasser E, Fleagle J, Rosenberger A, McHenry H (eds) Primate locomotion. Plenum Press, New York, pp 95–108

Reno PL (2014) Genetic and developmental basis for parallel evolution and its significance for hominoid evolution. Evol Anthropol 23:188–200

Richmond BG, Whalen M (2001) Forelimb function, bone curvature and phylogeny of *Sivapithecus*. In: De Bonis L, Koufos GD, Andrews P (eds) Hominoid evolution and climatic change in Europe, vol 2., Phylogeny of the neogene hominoid primates of EurasiaCambridge University Press, Cambridge, pp 326–348

Ripley S (1979) Environmental grain, niche diversification, and positional behavior in Neogene primates: an evolutionary hypothesis. In: Morbeck ME, Preuschoft H, Gomberg N (eds) Environment, behavior, and morphology: dynamic interactions in primates. Gustav Fischer, New York, pp 37–74

Rose MD (1991) The process of bipedalization in hominoids. In: Coppens Y, Senut B (eds) Origine(s) de la Bipédie chez les Hominidés. Fodation Singer-Polignac, Paris, pp 37–48

Rose MD (1993) Locomotor anatomy of Miocene hominoids. In: Gebo DL (ed) Postcranial adaptations in nonhuman primates. Northern Illinois University Press, DeKalb, pp 252–272

Rose MD (1994) Quadrupedalism in some Miocene catarrhines. J Hum Evol 26:387–411

Rose MD (1997) Functional and phylogenetic features of the forelimb in Miocene hominoids. In: Begun DR, Ward CV, Rose MD (eds) Function, phylogeny, and fossils: Miocene hominoid evolution and adaptations. Plenum Press, New York, pp 79–100

Rossie JB, MacLatchy L (2013) Dentognathic remains of an *Afropithecus* individual from Kalodirr, Kenya. J Hum Evol 65:199–208

Ruff CB (2002) Long bone articular and diaphyseal structure in Old World monkeys and apes. I: locomotor effects. Am J Phys Anthropol 119:305–342

Russo GA (2016) Comparative sacral morphology and the reconstructed tail lengths of five extinct primates: Proconsul heseloni, Epipliopithecusvindobonensis, Archaeolemur edwardsi, Megaladapis grandidieri, and Palaeopropithecus kelyus. J Hum Evol 90:135–162

Sanders WJ, Bodenbender BE (1994) Morphometric analysis of lumbar vertebra UMP 67-28: implications for spinal function and phylogeny of the Miocene Moroto hominoid. J Hum Evol 26:203–237

Sarringhaus LA, MacLatchy LM, Mitani JC (2014) Locomotor and postural development of wild chimpanzees. J Hum Evol 66:29–38

Sati JP, Alfred JRB (2002) Locomotion and posture in hoolock gibbon. Ann For 10:298–306

Schultz AH (1930) The skeleton of the trunk and limbs of higher primates. Hum Biol 2:303–438

Schultz AH (1973) The skeleton of the Hylobatidae and other observations on their morphology. In: Rumbaugh DM (ed) Gibbon and siamang: a series of volumes on the lesser apes. Anatomy, dentition, taxonomy, molecular evolution, and behavior, vol 2. Karger, Basel, pp 1–54

Scotland RW (2011) Deep homology: a view from systematics. BioEssays 32:438–449

Sherwood RJ, Ward S, Hill A, Duren DL, Brown B, Downs W (2002) Preliminary description of the *Equatorius africanus* partial skeleton (KNM-TH 28860) from Kipsaramon, Tugen Hills, Baringo District, Kenya. J Hum Evol 42:63–73

Shubin N, Tabin C, Carroll S (2009) Deep homology and the origins of evolutionary novelty. Nature 457:818–823

Smith RJ, Jungers WL (1997) Body mass in comparative primatology. J Hum Evol 32:523–559

Springer MS, Meredith RW, Gatesy J, Emerling CA, Park J, Rabosky DL, Sadler T, Stainer C, Ryder OA, Janečka Fischer CA, Murphy WJ (2013) Macroevolutionary dynamics and historical biogeography of primate diversification inferred from a species supermatrix. PLoS ONE 7(11):e49521

Srikosamatara S (1984) Notes in the ecology and behavior of the hoolock gibbon. In: Preuschoft H, Chivers DJ, Brockelman WY, Creel N (eds) The lesser apes. Edinburgh University Press, Edinburgh, pp 242–257

Steiper ME, Seiffert ER (2012) Evidence for a convergent slowdown in primate molecular rates and its implications for the timing of early primate evolution. Proc Nat Acad Sci 109:6006–6011

Steiper ME, Young NM (2006) Primate molecular divergence dates. Mol Phylogenetic Evol 41:384–394

Steiper ME, Young NM (2008) Timing primate evolution: lessons from the discordance between molecular and paleontological estimates. Evol Anthropol 17:179–188

Stern JT (1971) Functional myology of the hip and thigh of cebid monkeys and its implications for the evolution of erect posture. Biblio Primatol 14:1–319

Stern JT (1975) Before bipedality. Yrbk Phys Anthropol 19:59–68

Stern JT Jr, Larson SG (2001) Telemetered electromyography of the supinators and pronators of the forearm in gibbons and chimpanzees: implications for the fundamental positional adaptation of hominoids. Am J Phys Anthropol 115:253–268

Stern JT Jr, Wells JP, Vangor AK, Fleagle JG (1977) Electromyography of some muscles of the upper limb in *Ateles* and *Lagothrix*. Yrbk Phys Anthropol 20:498–507

Stevens NJ, Seiffert E, O'Connor PM, Roberts EM, Schmitz MD, Krause C, Gorscak E, Ngasala S, Hieronymus TL, Temu J (2013) Paleontological evidence for an Oligocene divergence between Old World monkeys and apes. Nature 497:611–614

Su DF, Jablonski NG (2009) Locomotor behavior and skeletal morphology of the odd-nosed monkeys. Folia Primatol 80:189–219

Sugardjito J, van Hooff JARAM (1986) Age-sex class differences in the positional behaviour of the Sumatran orang-utan (*Pongo pygmaeus abelii*) in the Gunung Leuser National Park, Indonesia. Folia Primatol 47:14–25

Susanna I, Alba DM, Almécija S, Moyà-Solà S (2014) The vertebral remains of the late Miocene great ape *Hispanopithecus laietanus* from Can Llobateres 2 (Vallés-Penedés Basin, NE Iberian Peninsula). J Hum Evol 73:15–34

Sussman RW (1991) Primate origins and the evolution of angiosperms. Am J Primatol 23:209–223

Takahashi LK (1990) Morphological basis of arm-swinging: multivariate analysis of the forelimbs of *Hylobates* and *Ateles*. Folia Primatol 54:70–85

Takahashi LK (1991) Forearm elongation in gibbons: hypothesis and preliminary results. Int J Primatol 12:599–614

Tallman M, Almécija S, Reber SL, Alba DM, Moyà-Solà S (2013) The distal tibia of *Hispanopithecus laietanus*: more evidence for mosaic evolution in Miocene apes. J Hum Evol 64:319–327

Temerin A, Cant JGH (1983) The evolutionary divergence of old world monkeys and apes. Am Nat 122:335–351

Thinh VN, Mootnick AR, Geissmann T, Li M, Ziegler T, Agil M, Moisson P, Nadler T, Walter L, Roos C (2010) Mitochondrial evidence for multiple radiations in the evolutionary history of small apes. BMC Evol Biol 10:74–87

Thorpe SKS, Crompton RH (2005) Locomotor ecology of wild orangutans (*Pongo pygmaeus abelii*) in the Gunung Leuser Ecosystem, Sumatra, Indonesia: a multivariate analysis using log-linear modeling. Am J Phys Anthropol 127:58–78

Thorpe SKS, Crompton RH (2006) Orangutan positional behavior and the nature of arboreal locomotion in Hominoidea. Am J Phys Anthropol 131:384–401

Thorpe SKS, Crompton RH (2009) Orangutan positional behavior. In: Wich SA, Utami Atmoko SS, Mitra Setia T, van Schaik CP (eds) Orangutans: geographic variation in behavioral ecology and conservation. Oxford University Press, Oxford, pp 33–48

Tuttle RH (1974) Darwin's apes, dental apes, and the descent of man: normal science in evolutionary anthropology. Curr Anthropol 15:389–426

Tuttle RH, Watts DP (1985) The positional behavior and adaptive complexes of *Pan gorilla*. In: Kondo S (ed) Primate morphophysiology, locomotor analyses and human bipedalism. University of Tokyo Press, Tokyo, pp 261–288

Tyler DE (1991) The problems of the Pliopithecidae as a hylobatid ancestor. Hum Evol 6:73–80

van Casteren A, Sellers WL, Thorpe SKS, Coward S, Crompton RH, Roland Ennos A (2013) Factors affecting the compliance and sway properties of tree branches used by the Sumatran orangutan (*Pongo abelii*). PLoS ONE 8(7):e67877

Wake DB, Wake MH, Specht CD (2011) Homoplasy: from detecting pattern to determining process and mechanism of evolution. Science 331:1032–1035

Ward CV (1997a) Functional anatomy and phyletic implications of the hominoid trunk and hindlimb. In: Begun DR, Ward CV, Rose MD (eds) Function, phylogeny, and fossils: Miocene hominoid evolution and adaptations. Plenum Press, New York, pp 101–130

Ward S (1997b) The taxonomy and phylogenetic relationships of *Sivapithecus* revisited. In: Begun DR, Ward CV, Rose MD (eds) Function, phylogeny, and fossils: miocene hominoid evolution and adaptations. Plenum Press, New York, pp 269–290

Ward CV (1998) *Afrpithecus, Pronconsul*, and the primitive hominoid skeleton. In: Strasser E, Fleagle J, Rosenberger A, McHenry H (eds) Primate locomotion: recent advances. Plenum Press, New York, pp 337–352

Ward CV (2002) Interpreting the posture and locomotion of *Australopithecus afarensis*: where do we stand? Yearb Phys Anthropol 45:185–215

Ward CV (2015) Postcranial and locomotor adaptations of hominoids. In: Henke W, Tattersall I (eds) Handbook of paleoanthropology, vol 2. Springer, New York, pp 1363–1387

Ward CV, Walker A, Teaford MF (1991) *Proconsul* did not have a tail. J Hum Evol 21:215–220

Washburn SL (1950) The analysis of primate evolution with particular reference to the origin of man. Cold Spring Harb Symp Quant Biol 15:67–78

Washburn SL, Avis V (1958) Evolution and human behavior. In: Simpson GG, Roe A (eds) Behavior and evolution. Yale University Press, New Haven, pp 421–436

Wheatley BP (1982) Energetics of foraging in *Macaca fascicularis* and *Pongo pygmaeus* and a selective advantage of large body size in the orang-utan. Primates 23:348–363

Wheatley BP (1987) The evolution of large body size in orangutans: a model for hominoid divergence. Am J Primatol 13:313–324

Wilkinson RD, Steiper ME, Soligo C, Martin RD, Yang Z, Tavare S (2011) Dating primate divergences through an integrated analysis of paleontological and molecular data. Syst Biol 60:16–31

Williams SA, Russo GA (2015) Evolution of the hominoid vertebral column: the long and the short of it. Evol Anthropol 24:15–32

Wood B, Harrison T (2011) The evolutionary context of the first hominins. Nature 470:347–352

Xijun N, Zhuding Q (2002) The micromammalian fauna from the Leilao, Yuanmou hominoid locality: implications for biochronology and paleoecology. J Hum Evol 42:535–546

Youlatos D (2008) Locomotion and positional behavior of spider monkeys. In: Campbell CJ (ed) Spider monkeys: behavior, ecology, and evolution of the genus Ateles. Cambridge University Press, Cambridge, pp 185–219

Young NM (2003) A reassessment of living hominoid postcranial variability: implications for ape evolution. J Hum Evol 45:441–464

Young NM (2008) A comparison of the ontogeny of shape variance in the anthropoid scapula: functional and phylogenetic signal. Am J Phys Anthropol 136:247–264

Young NM, MacLatchy L (2004) The phylogenetic position of *Morotopithecus*. J Hum Evol 46:163–184

Zalmout IS, Sanders WJ, MacLatchy LM, Gunnell GF, Al-Mufarreh YA, Ali MA, Nasser AAH, Al-Masari AM, Al-Sobhi SA, Nadhra AO, Matari AH, Wilson JA, Gingerich PD (2010) New Oligocene primate from Saudi Arabia and the divergence of apes and Old World monkeys. Nature 466:360–365

Chapter 4
The Fossil Record and Evolutionary History of Hylobatids

Terry Harrison

White-handed gibbon female foraging on fruit, Khao Yai National Park, Thailand. Photo credit: Ulrich H. Reichard

T. Harrison (✉)
Department of Anthropology, Center for the Study of Human Origins,
New York University, New York, NY 10003, USA
e-mail: terry.harrison@nyu.edu

© Springer Science+Business Media New York 2016
U.H. Reichard et al. (eds.), *Evolution of Gibbons and Siamang*,
Developments in Primatology: Progress and Prospects,
DOI 10.1007/978-1-4939-5614-2_4

Introduction

The fossil record of hylobatids is extremely meager, and little is known about the details of their evolutionary history. Earlier claims of fossil hylobatids from the Oligocene and Miocene of Africa and Eurasia have not been substantiated, and these taxa are now best interpreted as primitive stem catarrhines (Harrison 1987, 2002, 2005, 2010a, b, 2013; Andrews et al. 1996; Harrison and Gu 1999; Begun 2002; Fleagle 2013). The only definitive representatives of the family are from sites in Asia dating from the late Miocene and Pleistocene (Gu 1986, 1989; Harrison et al. 2008; Harrison 2010b). As a consequence, the nature and timing of the origins of the family and the subsequent radiation and diversification of the crown members of the clade are poorly known. The best evidence, in this regard, comes from molecular phylogenetic studies of extant taxa. These data confirm that hylobatids are monophyletic, as are the clusters of species included in the four currently recognized genera—*Hylobates*, *Hoolock*, *Symphalangus* and *Nomascus* (Mittermeier et al. 2013). The relationships between the genera are still not fully resolved, but there is a reasonable consensus that they are related in the following manner—(*Nomascus* (*Symphalangus* (*Hoolock*, *Hylobates*))) (Hayashi et al. 1995; Roos and Geissmann 2001; Chatterjee 2006, 2009; Fabre et al. 2009; Thinh et al. 2010; Chan et al. 2010, 2012; Matsudaira and Ishida 2010; Perelman et al. 2011; Finstermeier et al. 2013). This set of inferred relationships is consistent with evidence from morphology and behavior (see Groves 1972; Haimoff et al. 1982). Nevertheless, a number of alternative phylogenetic schemes have been proposed (Takacs et al. 2005; Whittaker et al. 2007; Chatterjee et al. 2009; Israfil et al. 2011; Meyer et al. 2012; Springer et al. 2012; Wall et al. 2013; Carbone et al. 2014; Chatterjee 2016), and the lack of agreement probably stems from the rapid radiation of the four main subclades and the potentially confounding influences of introgression (Meyer et al. 2012; Wall et al. 2013; Carbone et al. 2014). Using molecular clock models, the time of divergence of the hylobatids from the other hominoids is estimated at ~ 19.0 Ma (with mean estimates ranging from 16.3 to 21.8 Ma), during the early Miocene (Fig. 3.1). Crown hylobatids diverged much more recently, during the late Miocene, with an estimated divergence date of *Nomascus* from the other hylobatids at ~ 8.0 Ma and the *Symphalangus* and *Hoolock* lineages diverging around ~ 6.7 and ~ 6.2 Ma, respectively (Chatterjee 2006, 2009, 2016; Chatterjee et al. 2009; Chan et al. 2010, 2012; Thinh et al. 2010; Matsudaira and Ishida 2010; Perelman et al. 2011; Israfil et al. 2011; Finstermeier et al. 2013; Carbone et al. 2014) (Fig. 3.1). From an evolutionary perspective, and potentially profoundly important for interpreting the fossil record, the molecular data imply that stem hylobatids existed, largely as a ghost lineage, for about 10 myrs before the initial divergence of crown members of the clade.

The aim of this contribution is to document what is known about the evolutionary history of the hylobatids based on a review of the fossil evidence. Since there is much that we do not know and cannot deduce about the phylogeny of hylobatids from the incomplete fossil record, a fuller appreciation of the evolutionary history necessarily relies on what can also be learned from comparative anatomy and molecular data.

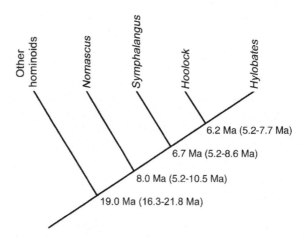

Fig. 3.1 Phylogenetic relationships and estimated divergence dates of extant hylobatids (Hayashi et al. 1995; Purvis 1995; Roos and Geissmann 2001; Raaum et al. 2005; Chatterjee 2006, 2009; Bininda-Emonds et al. 2007; Fabre et al. 2009; Thinh et al. 2010; Chan et al. 2010, 2012; Matsudaira and Ishida 2010; Israfil et al. 2011; Perelman et al. 2011; Finstermeier et al. 2013; Carbone et al. 2014). The first date is the average of the published estimates. Those in parentheses are the range of the average estimated divergence dates

Ancestral Morphotypes and Hylobatid Specializations

There has been far too much emphasis placed in the past on generalized and superficial phenetic characteristics to identify possible hylobatids in the fossil record, rather than on a critical assessment and application of potential synapomorphies. These generalized features include small size, a relatively short face, a lightly built and globular neurocranium, simple molars with a prominent hypoconulid on the lowers and a crista obliqua on the uppers, and relatively long and slender limbs. However, these features are not exclusive to gibbons. They can be inferred to be part of the ancestral catarrhine morphotype, and are features that are typically found in many of the smaller stem catarrhines from the Oligocene and Miocene of Africa and Eurasia (Groves 1968, 1972, 1974; Delson and Andrews 1975; Delson 1977; Ciochon and Corruccini 1977; Szalay and Delson 1979; Fleagle 1984, 2013; Harrison 1982, 1987, 1988, 2002, 2005, 2010a, 2013; Harrison et al. 1991; Andrews et al. 1996; Harrison and Gu 1999). Fortunately, extant hylobatids do share a suite of craniodental and postcranial specializations that can be used to distinguish the clade from other hominoids, and these represent an important aid in identifying hylobatids in the fossil record (and in ruling out small primitive catarrhine that superficially resemble hylobatids). These unique features include: very low sexual dimorphism in body size, with males only slightly larger on average than females; incisors relatively small and low-crowned; canines high-crowned in both sexes, with limited sexual dimorphism (canines of males are slightly larger); canines with strong lateral splay; p3 has a short honing face; upper premolars narrow; upper and lower premolars with a high incidence of fused roots; simple molars with well-rounded corners, large occlusal basins, weak crest development, and low conical cusps; lower molars with poorly defined distal foveae; M3 and m3 subequal in size or smaller than M1 and m1, respectively; mandibular corpus shallow, with steeply inclined symphysis and lack of a simian shelf; lower face short with a small premaxilla (possibly a primitive catarrhine feature); orbits relatively large with protruding circumorbital rims (except *Nomascus*); nasals short and trapezoidal; enlarged sphenoidal sinus and no frontal sinus; neurocranium globular and lightly built, lacking sagittal and nuchal crests (except *Symphalangus*); slender body; high intermembral index, exceeding 120 (also in *Pongo*); limbs long and gracile, with weakly developed muscle markings; radius longer than the humerus (as in some *Pongo* individuals); carpometacarpal joint of ball and socket type; narrow hook-like hands with very long and curved nonpollicial metacarpals and phalanges; manual digit II long in relation to digit IV; and reduced number of coccygeal vertebrae (Schultz 1930, 1963; Midlo 1934; Cave and Haines 1940; Day and Napier 1963; Frisch 1963, 1965; Tuttle 1969; Groves 1972; Harrison 1982; Jenkins 1990). Given this extensive suite of unique features, crown members of the clade should be relatively easy to identify in the fossil record, but it may not be such a simple task to recognize stem hylobatids, especially craniodentally, since inferences about what they may have looked like are exceedingly sketchy (Harrison 1982). Nevertheless, hylobatids share with other hominoids an extensive range of specializations (compared with other anthropoid primates) of the postcranium that are functionally and behaviorally associated with orthograde postures and a strongly differentiated usage of the forelimbs and hindlimbs, possibly related to enhanced vertical climbing abilities and forelimb suspension (Harrison and Rook 1997; Young 2003). These features include: high intermembral index; broad and shallow thorax with dorsally positioned scapula; long clavicle; increased potential to raise the forelimb above the head; greater ranges of extension and rotation at the elbow, circumduction at the shoulder joint, and pronation–supination and abduction–adduction at the wrist; long and curved metacarpals and manual phalanges; relatively short pollex; short and broad thoracolumbar vertebrae, with a reduced number of lumbar elements; increased number of sacral vertebrae; absence of external tail; greater ranges of rotation of the hip and knee joints, and inversion–eversion at the ankle joint; increased ability of the foot to grasp and to provide powerful push off from large-diameter vertical supports (Harrison 1991; Harrison and Rook 1997). Given how extensive and pervasive these specializations are it seems very unlikely, as some authors have suggested (e.g., Le Gros Clark and Thomas 1951; Napier and Davis 1959; Napier 1963; Simons 1967; Pilbeam 1969, 1996; Begun et al. 1997; Larson 1998; Pilbeam and Young 2004), that hylobatids and hominids (i.e., great apes and humans) acquired these features in parallel. It is much more likely that this suite of specializations characterized the postcranial skeleton of the last common ancestor of hominoids, and that we should expect to identify these features in stem hylobatids (Harrison 1982; Fleagle 1984).

Given the available evidence, it is reasonable to deduce that the earliest stem hylobatids would have had a relatively primitive skull (similar to stem catarrhines), with a short face and relatively large neurocranium, and most of the derived features of the postcranium shared by extant hominoids associated with their specialized mode of locomotion. There has been considerable speculation about whether ancestral gibbons were originally small, like extant hylobatids (4–13 kg), or whether their small size is the result of phyletic dwarfism (e.g., Groves 1972, 1974; Andrews and Groves 1975; Tyler 1991, 1993; Nisbett and Ciochon 1993; Jablonski and Chaplin 2009). While the idea of an evolutionary trend for reduced body size in hylobatids from a large-bodied ancestral hominoid is an attractive scenario, especially given that most of the known fossil hominoids are relatively large (estimated 20–200 kg), there appears to be little empirical support for such an hypothesis. By contrast, there is one compelling piece of evidence that supports the inference that ancestral hylobatids (and stem hominoids) were relatively small, and comparable in size to extant hylobatids. Old World monkeys and hylobatids primitively share ischial callosities as part of the ancestral catarrhine morphotype. These are roughened hairless sitting pads attached to an expanded bony tuberosity on the ischium of the pelvis (Schultz 1936; Miller 1945; Rose 1974). Ischial callosities represent an important adaptation among catarrhines for sitting on relatively slender arboreal supports during feeding, resting and sleeping (Washburn 1957; Rose 1974; Vilensky 1978; Harrison and Sanders 1999; McGraw and Sciulli 2011). At night, this allows catarrhines to sleep in the periphery of the tree crown, and offers an enhanced ability to detect the approach of large arboreal predators, such as felids. For great apes, the large diameter of the support needed to accommodate their mass would confer little or no effective advantage as a predator avoidance strategy. Great apes, which entirely lack or retain only vestigial ischial callosites, are able to sleep in terminal branch settings by constructing nests, which effectively spreads their large body mass over many slender supports (Harrison and Sanders 1999). Data on chimpanzees (*Pan troglodytes*) support the inference that the location of nests is influenced, at least in part, by predator pressure (Pruetz et al. 2008; Stewart and Pruetz 2013; Ogawa et al. 2014). If this scenario is correct, and I believe that it is the most parsimonious explanation of the anatomical and behavioral evidence, then it implies that ancestral hominoids and hylobatids were relatively small-bodied apes that primitively retained ischial callosities, and that the last common ancestor of great apes and humans was a specialized large-bodied form that constructed nests and developed highly reduced ischial callosities.

Purported Hylobatids from the Oligocene and Miocene

In the past, a number of small catarrhine primates from the Oligocene and Miocene of Africa and Eurasia have been linked phylogenetically to the hylobatids, and these extinct forms implied that the evolutionary history of the clade extended back to almost 30 Ma. These putative early hylobatids included *Propliopithecus* (=*Aeolopithecus*) *chirobates* from the early Oligocene of the Fayum, Egypt (Simons

Table 3.1 Family-group classification of the Catarrhini (after Harrison 2002, 2005, 2010a, 2013; Andrews and Harrison 2005)

Infra order	Superfamily	Family	Subfamily	Genera mentioned in the text
Catarrhini	Propliopithecoidea	Propliopithecidae[a]		*Propliopithecus*
	Pliopithecoidea	Pliopithecidae[a]	Pliopithecinae	*Pliopithecus*
			Crouzeliinae	*Anapithecus, Barberapithecus, Egarapithecus, Laccopithecus, Plesiopliopithecus*
		Dionysopithecidae[a]		*Dionysopithecus*
	Saadanioidea	Saadaniidae[a]		
	Dendropithecoidea	Dendropithecidae[a]		*Dendropithecus, Micropithecus*
	Proconsuloidea	Proconsulidae[a]		
	Cercopithecoidea	Cercopithecidae	Cercopithecinae	
			Colobinae	
	Hominoidea	Hylobatidae		*Bunopithecus, Hoolock, Hylobates, Nomascus, Symphalangus, Yuanmoupithecus*
		Hominidae	Kenyapithecinae	
			Ponginae	*Pongo*
			Homininae	*Pan*
	Incertae sedis			*Limnopithecus, Krishnapithecus,* 'Kansupithecus'

[a]Denotes extinct families

1965, 1972), the dendropithecids and smaller proconsulids from the early Miocene of East Africa, including *Dendropithecus*, *Limnopithecus* and *Micropithecus* (Hopwood 1933; Le Gros Clark and Leakey 1951; Le Gros Clark and Thomas 1951; Leakey 1963; Frisch 1965; Simons 1972; Simons and Fleagle 1973; Fleagle 1975; Andrews and Simons 1977; Andrews 1978; Fleagle and Simons 1978), and the pliopithecids from the Miocene of Eurasia (Hürzeler 1954; Zapfe 1958, 1961; Simons 1972; Simons and Fleagle 1973; Wu and Pan 1985; Meldrum and Pan 1988) (Table 3.1 for a current classification of catarrhine primates). However, these purported relationships were primarily based on a combination of small size and primitive catarrhine craniodental traits, all of which bear only a superficial similarity to the morphology seen in extant hylobatids. Subsequently, a more critical reappraisal of the phylogenetic status of these fossil taxa demonstrated conclusively that they are all stem catarrhines (i.e., the primitive sister taxa to extant cercopithecoids and hominoids) and much too primitive to be closely allied with extant hylobatids (Remane 1965; Groves 1968, 1972, 1974; Delson and Andrews 1975; Delson 1977; Ciochon and Corruccini 1977; Szalay and Delson 1979; Fleagle 1984; Ginsburg and Mein 1980; Harrison 1982, 1987, 1991, 1993, 2002, 2010a; Harrison et al. 1991; Andrews et al. 1996; Harrison and Gu 1999; Moyà-Solà et al. 2001; Begun 2002; Fleagle 2013).

However, one stem catarrhine, *Laccopithecus robustus*, still persists in the literature as a potential hylobatid contender (Tyler 1991, 1993; Nisbett and Ciochon 1993; Jablonski and Chaplin 2009). *Laccopithecus*, from the late Miocene locality of Shihuiba, Lufeng County, Yunnan in southern China, is known from a partial cranium, a large collection of teeth and crushed jaw fragments, and an isolated phalanx (Wu and Pan 1984, 1985; Meldrum and Pan 1988; Pan et al. 1989). The notion that it might represent a fossil hylobatid is based primarily on its small size, and the possession of a relatively short face, a primitive dentition, and a long, slender and curved manual phalanx (Pan 1988; Meldrum and Pan 1988). It is also tantalizingly in the right place at the right time (i.e., southern China during the late Miocene) (Pan 1988). However, comparisons clearly demonstrate that *Laccopithecus* is not an early hylobatid, but rather a late surviving pliopithecoid, a group of primitive catarrhines with a wide geographical distribution throughout much of Eurasia during the Miocene (Ginsburg and Mein 1980; Ginsburg 1986; Harrison 1987, 1991, 2005; Harrison et al. 1991; Andrews et al. 1996; Begun 2002; Moyà-Solà et al. 2001; Alba and Moyà-Solà 2012). Cranially and postcranially pliopithecoids conform closely to the inferred ancestral catarrhine morphotype, with the retention of a number of primitive characters not found in crown catarrhines (i.e., a partially enclosed tubular ectotympanic, an entepicondylar foramen in the distal humerus, relatively broad upper molars, and detailed features of the occlusal morphology of the lower molars) (Andrews 1985; Harrison 1982, 1987; Harrison et al. 1991; Andrews et al. 1996; Fleagle 2013). In addition, pliopithecoids are distinguished from other catarrhines by a number of dental specializations, which include: upper and lower incisors mesiodistally waisted toward the base of the crown; p3 mesiodistally short and relatively high-crowned, with a steeply inclined mesiobuccal face; and lower molars narrow with well-developed occlusal crests and a pliopithecine triangle (Andrews et al. 1996). *Laccopithecus* shares many of these specialized features with other pliopithecoids. It is also important to note that the cranium of *Laccopithecus* is remarkably similar to those of *Pliopithecus vindobonensis* and *Anapithecus hernyaki* from Central Europe and *Pliopithecus zhanxiangi* from China, the only other pliopithecoids for which partial crania are known (Zapfe 1961; Wu and Pan 1985; Harrison et al. 1991; Begun 2002). They share a relatively short and narrow muzzle, extensive incisive fenestra, shallow lower face in which the orbits overlap extensively with the nasal aperture in the dorsoventral plane, very short sub-nasal clivus, anterior root of zygomatic arch located close to the alveolar margin of the cheek teeth, relatively broad interorbital region, subcircular orbits with a slightly projecting inferior rim, slender lateral orbital rim, and strongly marked temporal lines. Moreover, *Laccopithecus* exhibits a suite of dental specializations that link it uniquely with the crouzeliine pliopithecoids, a clade that also includes *Plesiopliopithecus*, *Anapithecus*, *Egarapithecus*, and *Barberapithecus* from the Miocene of Europe (Andrews et al. 1996; Moyà-Solà et al. 2001; Begun 2002; Alba and Moyà-Solà 2012).

These features include: relatively small and low-crowned incisors; lower lateral incisors broader than central incisors; p4 with well-developed hypoconid and entoconid, and a mesial fovea that opens mesiolingually; lower molars relatively long and narrow, with buccolingual waisting, elongated mesial fovea, trigonid distinctly elevated above the talonid basin, cristid obliqua long and obliquely directed, hypoconulid reduced in size and situated in the midline of the crown, and small distal fovea; and lower molars increasing markedly in size from m1 to m3. Little is known about the postcranial morphology of crouzeliines, but what we do know indicates that they were specialized for suspensory behaviors (Begun 1993), which is consistent with the morphology of the isolated manual phalanx attributed to *Laccopithecus*. The evidence definitively establishes *Laccopithecus* as a crouzeliine pliopithecid, and, like other pliopithecoids from Eurasia, it represents a primitive stem catarrhine only distantly related to extant hylobatids and other hominoids.

In addition to these well-known fossil catarrhines from the Miocene of Eurasia and Africa, there are a number of isolated teeth and jaw fragments of small catarrhine primates from fossil localities in Asia that might bear on the origins and evolutionary history of hylobatids. These include *Pliopithecus posthumus* from Ertemte, China; *Kansupithecus* from Tabenbuluk, China; *Krishnapithecus krishnaii* from Haritalyangar, India; isolated teeth from the Kamlial and Manchar Formations, Pakistan, tentatively referred to *Dionysopithecus* sp.; and *Dendropithecus orientalis* from northern Thailand. These specimens provide tantalizing clues to the taxonomic diversity of small catarrhine primates in Asia during the Miocene (Harrison 2005), but unfortunately the specimens are too fragmentary or too poorly known to determine their precise relationships.

Schlosser (1924) described a heavily worn left M3 from the late Miocene ($\sim$5–7 Ma) locality of Ertemte, Nei Monggol (Inner Mongolia) northern China (Qiu and Qiu 1995; Qiu et al. 1999) as belonging to a new species, *Pliopithecus posthumus*. The tooth is heavily worn, so it is not possible to determine its taxonomic affinities, and several authors have questioned whether it belongs to a primate at all (Hürzeler 1954; Simons 1972; Simons and Fleagle 1973; Fleagle 1984; Harrison et al. 1991).

An edentulous mandibular symphysis of a small catarrhine primate was recovered during the 1930s from the early Miocene ($\sim$19–20 Ma) locality of Xishuigou (=Hsi-shui) Tabenbuluk, Gansu Province, northern China (Bohlin 1946; Wang et al. 2008). The symphysis is low and robust with a well-developed superior transverse torus. Comparisons show that the Tabenbuluk specimen is generally similar in morphology to dendropithecids and pliopithecoids (Harrison 1982; Harrison et al. 1991), but beyond that it is difficult to conclude anything meaningful about its relationships. Similarly, a molar fragment from the late Oligocene ($\sim$23–25 Ma) site of Yandantu (=Yindirte), 8 km to the northeast of Xishuigou (Wang et al. 2008), is too incomplete to determine its affinities (Harrison et al. 1991). Bohlin (1946) originally described these two specimens as belonging to a new genus, *Kansupithecus*, but the nomen remains unavailable because he failed to provide a species name (Harrison et al. 1991).

Chopra and Kaul (1979) described a worn M3 from the late Miocene (~8–10 Ma) locality of Haritalyangar in northern India as a new species, *Pliopithecus krishnaii*. Later, Ginsburg and Mein (1980) included the species, along with *P. posthumus*, in a separate genus, *Krishnapithecus*. The Haritalyangar molar is comparable to small catarrhine primates from the Miocene of East Africa in the shape of the crown, the distribution and size of the cusps, and the development of the lingual cingulum, and differs from pliopithecoids in being narrower and less rectangular in occlusal outline (Harrison et al. 1991; Harrison and Gu 1999). The possibility that the tooth belongs to a stem hylobatid cannot be entirely ruled out, but morphologically it is most similar to dendropithecids and small proconsulids from the early Miocene of East Africa.

The same is true for several isolated teeth of a small catarrhine primate recovered from the early middle Miocene Kamlial and Manchar Formations in Pakistan (~16–17 Ma) (Raza et al. 1984; Bernor et al. 1988). These specimens were initially considered to have their closest affinities with *Dionysopithecus*, a primitive pliopithecoid from the early Miocene of China, on the basis of their general similarity in size and molar morphology (Fleagle 1984; Barry et al. 1986; Bernor et al. 1988). However, subsequent discoveries and detailed comparisons showed that there are important differences from *Dionysopithecus* (Harrison and Gu 1999), with no indications that the small catarrhines from Pakistan are allied to the pliopithecoids. In addition, the teeth are readily distinguishable from those of extant hylobatids in having lower molars with tall cusps, elevated occlusal crests, and a well-developed buccal cingulum, as well as a low-crowned canine, at least in presumed female individuals. In general, the Kamlial and Manchar teeth appear to be most similar to dendropithecids from the Miocene of East Africa, and this is likely to be the group to which they are most closely related.

Finally, an isolated m1 of a small catarrhine primate has been recovered from the middle Miocene (~15–17 Ma) locality of Ban San Klang in northern Thailand (Ducrocq et al. 1994; Suteethorn et al. 1990). Suteethorn et al. (1990) recognized similarities to *Dendropithecus* from the early Miocene of East Africa, and referred it to a new species, *Dendropithecus orientalis*. However, the molar from Ban San Klang can be distinguished from dendropithecids by a suite of features that it shares with Eurasian pliopithecoids (Harrison and Gu 1999; Harrison 2005). On this basis, the Thai specimen probably has its closest affinities with the Eurasian pliopithecids, rather than with the East African dendropithecids. Further comparisons have shown that "*Dendropithecus*" *orientalis* is quite similar in size and morphology to *Dionysopithecus shuangouensis* from Sihong, China (Harrison and Gu 1999). Additional material will be needed to resolve the taxonomic relationships of the Ban San Klang catarrhine, but given its overall morphological similarity to *D. shuangouensis*, it can be provisionally referred to *Dionysopithecus orientalis* (Harrison and Gu 1999; Harrison 2005).

Yuanmoupithecus xiaoyuan from the late Miocene (~7–9 Ma) of Yunnan in southern China, originally considered to be closely related to East African dendropithecids and proconsulids (Pan 2006), can be shown to be a stem hylobatid based on a number of key synapomorphies (Harrison et al. 2008). These include:

lower canines tall and slender with a stout base, distally recurved crown, and slight degree of lateral splay relative to the long axis of the root; upper and lower premolars with high frequency of fused roots; p3 with mesiodistally elongated crown, low protoconid and short mesiobuccal honing face; low-crowned molars with rounded corners, low, rounded and peripherally arranged cusps, and weakly developed crests; lower molars with metaconid and entoconid relatively small, widely spaced and linked by an elevated crest, shallow and expansive talonid basin, and marked reduction or absence of the buccal cingulum; and upper molars with relatively small paracone and metacone linked by a well-developed crest. This suite of unique dental features provides strong evidence to support the inference that *Yuanmoupithecus* is closely related to extant hylobatids. However, *Yuanmoupithecus* differs from modern hylobatids in a number of primitive catarrhine features of the dentition (i.e., upper molars with well-developed lingual cingula, lower molars with better developed crests and well-defined mesial and distal foveae, and a relatively larger m3) that indicate that it represents the sister taxon of crown hylobatids. Based on this evidence, one can conclude that *Yuanmoupithecus* is the only known stem hylobatid and the only fossil representative of the clade that predates the Pleistocene.

Fossil Hylobatids from the Pleistocene of Asia

After *Yuanmoupithecus* there is a 5 million year gap in the fossil record of hylobatids, until specimens, morphologically very similar to extant taxa, occur at paleontological and archaeological localities throughout China and Southeast Asia, dating from the beginning of the Pleistocene onwards. Isolated teeth of gibbons have been reported from a number of sites in China (south of the Yangtze River), dating from the early Pleistocene to recent. Today, the distribution of gibbons in China is restricted to Yunnan Province, Guangxi Zhuang Autonomous Region and Tibet in southwestern China (i.e., *Hylobates lar yunnanensis*, *Hoolock leuconedys*, *Hoolock hoolock*, *Nomascus concolor*, *Nomascus leucogenys* and *Nomascus nasutus*) and on the island of Hainan (i.e., *Nomascus hainanensis*) (Ji and Jiang 2004; Geissmann 2007; Fan and Huo 2009; Mittermeier et al. 2013), but during the Pleistocene they were more widely distributed across central and southern China, with occurrences in Yunnan, Guangxi, Guangdong, Guizhou, Sichuan, Chongqing, Hunan and Hainan Provinces (von Koenigswald 1935; Gu 1989; Gu et al. 1996; Jablonski et al. 2000; Jablonski and Chaplin 2009; Ortiz et al. 2015). Moreover, historic records indicate that the range of gibbons extended as far north as the Yellow River and eastwards as far as Zhejiang Province (van Gulik 1967; Groves 1972; Delson 1977; Gao et al. 1981; Gu 1989; Geissmann 1995; Jablonski and Chaplin 2009). The oldest Plio-Pleistocene occurrence in China appears to be isolated teeth of *Nomascus* sp. from the early Pleistocene site of Baikong Cave, Guangxi, dated to ~2.2 Ma (Jin et al. 2008; Takai et al. 2014). Gu (1986, 1989) has published the most detailed account of the material recovered from cave sites and traditional drugstores in Guangxi Autonomous Region. The upper molars retain a prominent lingual cingulum, which occurs commonly in

Nomascus (Frisch 1965). Most of the specimens from southern China are likely attributable to *Nomascus* sp. (Gu 1986, 1989; Gu et al. 1996; Takai et al. 2014; Harrison et al. 2014), but a few occurrences from Yunnan and Guangxi may be attributable to *Hoolock* (Gu 1986, 1989; Jablonski and Chaplin 2009).

The best-known and most celebrated fossil gibbon from the Pleistocene of China is the specimen published by Matthew and Granger (1923). They described and illustrated a left mandibular fragment with m2 and m3 from Yanjinggou (=Yenchingkou), Chongqing (formerly in Sichuan), as a new genus and species, *Bunopithecus sericus*. Establishing the age of the specimen has proved problematic, especially since Granger's collections represent mixed faunas of different ages, but it is likely to be early or middle Pleistocene in age (Colbert and Hooijer 1953; Chen et al. 2013). Colbert and Hooijer (1953) and Hooijer (1960) noted that the specimen is comparable in size and morphologically indistinguishable from the modern hoolock gibbon, and questioned whether a generic distinction was justified. Frisch (1965) identified molar characteristics that linked *B. sericus* with the modern-day hoolock gibbon, and Prouty et al. (1983) provided additional support for this association, and included the fossil and extant hoolock gibbon in the genus *Bunopithecus*. Gu (1989) argued that the Yanjinggou specimen most closely resembles *Nomascus concolor,* and Jablonski and Chaplin (2009) suggested that it should be referred to *Nomascus sericus* (although the nomen *Bunopithecus* (Matthew and Granger 1923) has priority over *Nomascus* (Miller 1933)). However, Mootnick and Groves (2005) highlighted a number of features of the molars that distinguish *Bunopithecus* from all extant genera of hylobatids, and argued that it should be recognized as a distinct genus (see also Groves 2001). A recent study (Ortiz et al. 2015) has demonstrated that the lower molars of *Bunopithecus* exhibit a suite of unique features that fall outside the range of variation of extant hylobatids and provides support for the generic distinction of *Bunopithecus*. The results further indicate that *B. sericus* likely represents a crown hylobatid, and one that may possibly represent the extinct sister taxon of *Hoolock*.

Remains of fossil hylobatids are also reported from the Middle and Late Pleistocene cave sites in Vietnam, Thailand and Laos, but few details on their morphology or taxonomic attribution are available (Nisbett and Ciochon 1993; Schwartz et al. 1994; Tougard 2001; Bacon et al. 2008; Jablonski and Chaplin 2009; Zeitoun et al. 2010). The most interesting specimen from Vietnam is a relatively complete cranium, still awaiting description, from the middle Pleistocene site of Tham Khuyen that is probably referable to *Nomascus* (Ciochon and Olsen 1986).

Fossil and subfossil hylobatids are relatively rare in Pleistocene and Holocene sites on the Sunda Islands of Borneo, Sumatra and Java. A small collection of isolated teeth of hylobatids from the late Pleistocene (~ 120 ka) Punung Fissures of Java, excavated by von Koenigswald in the 1930s, was briefly described by Badoux (1959). Most of the teeth are similar in size and morphology to extant *Symphalangus syndactylus*, and can be attributed to this species (von Koenigswald 1940; Badoux 1959; Hooijer 1960). Additional remains of fossil siamangs from Java have been reported from the late Pleistocene site of Gunung Dawang (Storm and de Vos 2006; Westaway et al. 2007) and the early Pleistocene of Sangiran (von Koenigswald 1940). The evidence demonstrates that siamangs, which are restricted to Sumatra and the Malay Peninsula today, had a wider geographical distribution in the past and that their range extended to Java during the Pleistocene. A single upper molar from Punung, considerably smaller than those referred to *Symphalangus*, is probably best referred to *Hylobates moloch* (Badoux 1959; Hooijer 1960), even though it does fall outside the upper size limit of the extant species. An isolated tooth of *Hylobates moloch* from Sangiran in Java (von Koenigswald 1940) extends the temporal range of the species to the early Pleistocene. A fragmentary proximal femur from the early Pleistocene ($\sim$ 1 Ma) of Trinil, Java may represent the earliest record of a hylobatid from insular Southeast Asia (Ingicco et al. 2014).

In 1888–1890, Eugene Dubois excavated a large sample of subfossil siamangs, comprising more than five hundred isolated teeth, from three cave sites (i.e., Sibrambang, Djambu, and Lida Ayer) in the Padang Highlands of west central Sumatra (Hooijer 1960). Based on the associated mammalian fauna, the localities are Late Pleistocene in age (Drawhorn 1995). The teeth are slightly larger on average than those of extant *Symphalangus syndactylus*, and as a consequence they have been assigned to an extinct subspecies, *Symphalangus syndactylus subfossilis* (Hooijer 1960). In addition to their larger size, the teeth from Padang differ from those of extant siamang in molar proportions and in having a relatively larger p3 compared to p4. Associated upper cheek teeth (P4-M2) from Lida Ayer and three isolated teeth from Sibrambang document the rare occurrence of *Hylobates* sp. in the Padang sites (Hooijer 1960). These are similar in size or slightly larger than *Hylobates lar* and *H. agilis*; the two gibbon species that occur today on Sumatra.

Hylobatid remains are also known from several late Pleistocene and Holocene archaeological sites on Borneo, including Niah, Gua Sireh and Madai Caves (Hooijer 1960; Harrison 1996, 1998, 2000; Harrison et al. 2006). These are comparable in morphology and similar in size or slightly larger than extant *Hylobates muelleri* (Hooijer 1960, 1962; Harrison et al. 2006), and can be assigned with reasonable confidence to this species. It is worth noting that hylobatids are exceedingly rare elements of the fauna at these sites, and in the most extensively sampled collection from Niah Cave they represent less than 0.5 % of all primates recovered (Harrison et al. 2006).

Conclusions

The fossil record documenting the evolutionary history of hylobatids is extremely poor (Fig. 3.2). Based on molecular evidence, hylobatids can be inferred to have diverged from other hominoids during the early Miocene, at $\sim$ 19 Ma, and crown hylobatids are estimated to have originated at $\sim$ 8 Ma. The only pre-Pleistocene hylobatid known is *Yuanmoupithecus* from the late Miocene of China, dating to $\sim$ 7–9 Ma. It is a stem hylobatid that represents the sister taxon of all extant members of the clade. The dentition of *Yuanmoupithecus* is less specialized than modern hylobatids, but it does share a unique suite of derived dental features that support a close relationship. The discovery of a stem hylobatid in the late Miocene

Fig. 3.2 Phylogenetic tree illustrating the inferred evolutionary relationships of fossil and extant hylobatids. The *solid gray bars* represent the known temporal ranges of each genus based on the fossil record. The *broken lines* represent inferred relationships using a combination of fossil and molecular evidence (see Fig. 3.1 for estimated divergence dates)

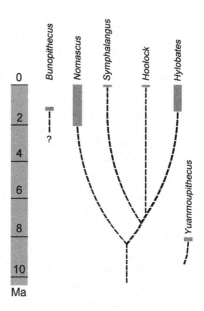

of southern China is consistent with the timing of the estimated divergence date of crown members of the clade.

The combination of molecular and paleontological evidence indicates that there is a ghost lineage for the initial 10 myrs of hylobatid evolutionary history, during which there is no trace of a fossil record (Fig. 3.2). The phylogenetic and biogeographic origins of the hylobatids are, therefore, obscure. The earliest appearance of hominoids in Eurasia was not until ∼ 16 Ma, so it is likely, given their estimated divergence date at ∼ 19 Ma, that hylobatids initially diverged in Africa where stem catarrhines and stem hominoids were restricted until the early Miocene (Harrison 2005, 2010a, 2013). Without a more complete fossil record it is impossible to ascertain when hylobatids arrived in Asia or any details of their early zoogeographic distribution. Since no recognizable precursors of *Yuanmoupithecus* have been found at earlier Chinese localities, it is reasonable to speculate that stem hylobatids originated in Southeast Asia, where the Neogene fossil record is poorly known, and later migrated northwards into China during the late Miocene.

No fossil hylobatids are known from the Pliocene, so there is another substantial gap in the record between 7 and 2 Ma. By the Pleistocene, hylobatids occur at a number of archaeological and paleontological sites throughout southern China and Southeast Asia, but tend to be relatively rare elements of the primate fauna. Hylobatids from Pleistocene sites are generally referable to extant lineages or species. The exception is *B. sericus* from the early or middle Pleistocene of China, which appears to be a distinct genus most closely related to *Hoolock*.

Acknowledgements I am grateful to the Editors for inviting me to prepare a contribution for this volume, and to Ulrich Reichard and an anonymous reviewer for helpful suggestions that improved the manuscript. I thank the following institutions and their staff for allowing me access to the fossil specimens and comparative material in their care: American Museum of Natural History, New York; National Museums of Kenya, Nairobi; The Natural History Museum, London; Muséum National D'Histoire Naturelle, Paris; Institut Català de Paleontologia 'Miquel Crusafont', Sabadell; Naturalis Biodiversity Center, Leiden; Naturhistorisches Museum, Basel; Naturhistorisches Museum, Vienna; Landesmuseum 'Joanneum', Graz; Magyar Állami Földtani Intézet, Budapest; and Institute of Vertebrate Paleontology and Paleoanthropology, Beijing; Beijing Museum of Natural History, Beijing; Yunnan Cultural Relics and Archaeology Institute, Kunming; and Geological Survey, Bangkok. Numerous colleagues have contributed to the research and ideas presented here, but the following deserve special mention: Peter Andrews, David Begun, Ray Bernor, Eric Delson, John Fleagle, Ji Xueping, Jin Changzhu, Pan Yuerong, Martin Pickford, Bill Sanders, and Zhang Yingqi.

References

Alba DM, Moyà-Solà S (2012) A new pliopithecid genus (Primates, Pliopithecoidea) from Castell de Barberà (Vallès-Penedès Basin, Catalonia, Spain). Am J Phys Anthropol 147:88–112

Andrews P (1978) A revision of the Miocene Hominoidea of East Africa. Bull Br Mus Nat Hist (Geol) 30:85–224

Andrews P (1985) Family group systematics and evolution among catarrhine primates. In: Delson E (ed) Ancestors: the hard evidence. Alan R Liss, New York, pp 14–22

Andrews P, Groves C (1975) Gibbons and brachiation. Gibbon Siamang 4:167–218

Andrews PJ, Simons EL (1977) A new Miocene gibbon-like genus, *Dendropithecus* (Hominoidea, Primates) with distinctive postcranial adaptations: its significance to origin of Hylobatidae. Folia Primatol 28:161–169

Andrews P, Harrison T (2005) The last common ancestor of apes and humans. In: Lieberman DE, Smith RJ, Kelley J (eds) Interpreting the past: essays on human, primate, and mammal evolution. Brill, Boston, pp 103–121

Andrews P, Harrison T, Delson E, Bernor RL, Martin L (1996) Distribution and biochronology of European and Southwestern Asian Miocene catarrhines. In: Bernor RL, Fahlbusch V, Mittmann H-W (eds) The evolution of western Eurasian Neogene mammal faunas. Columbia University Press, New York, pp 168–207

Bacon A-M, Demeter F, Tougard C, de Vos J, Sayavongkhamdy T, Antoine P-O, Bouasisengpaseuth B, Sichanthongtip P (2008) Redécouverte d'une faune pléistocène dans les remplissages karstiques de Tan Hang au Laos: Premiers résultats. C R Palevol 7:277–288

Badoux DM (1959) Fossil mammals from two fissure deposits at Punung (Java). Kemink en Zoon, Utrecht

Barry JC, Jacobs LL, Kelley J (1986) An early middle Miocene catarrhine from Pakistan with comments on the dispersal of catarrhines into Eurasia. J Hum Evol 15:501–508

Begun DR (1993) New catarrhine phalanges from Rudabánya (northeastern Hungary) and the problem of parallelism and convergence in hominoid postcranial morphology. J Hum Evol 24:373–402

Begun DR (2002) The Pliopithecoidea. In: Hartwig WC (ed) The primate fossil record. Cambridge University Press, Cambridge, pp 221–240

Begun DR, Ward CV, Rose MD (1997) Events in hominoid evolution. In: Begun DR, Ward CV, Rose MD (eds) Function, phylogeny, and fossils: Miocene hominoid evolution and adaptation. Plenum Press, New York, pp 389–415

Bernor R, Flynn L, Harrison T, Hussain ST, Kelley J (1988) *Dionysopithecus* from southern Pakistan and the biochronology and biogeography of early Eurasian catarrhines. J Hum Evol 17:339–358

Bininda-Emonds ORP, Cardillo M, Jones KE, MacPhee RDE, Beck RMD, Grenyer R, Price SA, Vos RA, Gittelman JL, Purvis A (2007) The delayed rise of present-day mammals. Nature 446:507–512

Bohlin B (1946) The fossil mammals from the Tertiary deposit of Taben-Buluk, western Kansu. II. Simplicidentata, Carnivora, Artiodactyla, Perissodactyla and Primates. Palaeontol Sinica N.S. C 8b:1–256

Carbone L, Harris RA, Gnerre S, Veeramah KR, Lorente-Galdos B, Huddleston J, Meyer TJ Herrero J, Roos C, Aken B, Anaclerio F, Archidiacono N, Baker C, Barrell D, Batzer MA, Beal K, Blancher A, Bohrson CL, Brameier M, Campbell MS, Capozzi O, Casola C, Chiatante G, Cree A, Damert A, de Jong PJ, Dumas L, Fernandez-Callejo M, Flicek P, Fuchs NV, Gut I, Gut M, Hahn MW, Hernandez-Rodriguez J, Hillier LW, Hubley R, Ianc B, Izsvák Z, Jablonski NG, Johnstone LM, Karimpour-Fard A, Konkel MK, Kostka D, Lazar NH, Lee SL, Lewis LR, Liu Y, Locke DP, Mallick S, Mendez FL, Muffato M, Nazareth LV, Nevonen KA, O'Bleness M, Ochis C, Odorn DT, Pollard KS, Quilez J, Reich D, Rocchi M, Schumann GG, Searle S, Sikela JM, Skollar G, Smit A, Sonmez K, ten Hallers B, Terhune E, Thomas GWC, Ullmer B, Ventura M, Walker JA, Wall JD, Walter L, Ward MC, Wheelan SJ, Whelan CW, White S, Wilhelm LJ, Woerner AE, Yandell M, Zhu B, Hammer MF, Marques-Bonet T, Eichler EE, Fulton L, Fronick C, Muzny DM, Warren WC, Worley KC, Rogers J, Wilson RK, Gibbs RA (2014) Gibbon genome and the fast karyotype evolution of small apes. Nature 513:195–201

Cave AJ, Haines RW (1940) The paranasal sinuses of the anthropoid primates. J Anat 74:493–523

Chan Y-C, Roos C, Inoue-Murayama M, Inoue E, Shih C-C, Pei KJ-C, Vigilant L (2010) Mitochondrial genome sequences effectively reveal the phylogeny of *Hylobates* gibbons. PLoS ONE 5(12):e124419

Chan Y-C, Roos C, Inoue-Murayama M, Inoue E, Shih C-C, Vigilant L (2012) A comparative analysis of Y chromosome and mtDNA phylogenies of the *Hylobates* gibbons. BMC Evol Biol 12:150

Chatterjee HJ (2006) Phylogeny and biogeography of gibbons: a dispersal-vicariance analysis. Int J Primatol 27:699–712

Chatterjee HJ (2009) Evolutionary relationships among the gibbons: a biogeographic perspective. In: Lappan S, Whittaker DJ (eds) The gibbons. Springer, Dordrecht, pp 13–36

Chatterjee, HJ (2016) The role of historical and fossil records in predicting changes in the spatial distribution of hylobatids. In: Reichard UH, Hirohisa H, Barelli C (eds) Evolution of gibbons and siamang. Springer, New York, pp 43–54

Chatterjee HJ, Ho SYW, Barnes I, Groves C (2009) Estimating the phylogeny and divergence times of primates using a supermatrix approach. BMC Evol Bio 9:259

Chen SK, Pang LB, He CD, Wei GB, Huang WB, Yue ZY, Zhang XH, Zhang H, Qin L (2013) New discoveries from the classic quaternary mammalian fossil area of Yanjinggou, Chongqing, and their chronological explanations. Chinese Sci Bull 58:3780–3787

Chopra SRK, Kaul S (1979) A new species of *Pliopithecus* from the Indian Sivaliks. J Hum Evol 8:475–477

Ciochon RL, Corruccini R (1977) The phenetic position of *Pliopithecus* and its phylogenetic relationship to the Hominoidea. Syst Zool 26:290–299

Ciochon RL, Olsen JW (1986) Paleoanthropological and archaeological research in the Socialist Republic of Vietnam. J Hum Evol 15:623–633

Colbert EH, Hooijer DA (1953) Pleistocene mammals from the limestone fissures of Szechwan, China. Bull Am Mus Nat Hist 102:1–134

Day MH, Napier J (1963) The functional significance of the deep head of flexor pollicis brevis in primates. Folia Primatol 1:122–134

Delson E (1977) Vertebrate paleontology, especially of nonhuman primates, in China. In: Howells WW, Tsuchitani PJ (eds) Paleoanthropology in the People's Republic of China. National Academy of Sciences, Washington, DC, pp 40–64

Delson E, Andrews P (1975) Evolution and interrelationships of the catarrhine primates. In: Luckett W, Szalay FS (eds) Phylogeny of the primates: a multidisciplinary approach. Plenum Press, New York, pp 405–446

Drawhorn GM (1995) The systematics and paleodemography of fossil orangutans (Genus *Pongo*). Dissertation, University of California, Davis

Ducrocq S, Chaimanee Y, Suteethorn V, Jaeger J-J (1994) Ages and paleoenvironment of Miocene mammalian faunas from Thailand. Palaeogeog Palaeoclimatol Palaeoecol 108:149–163

Fabre P-H, Rodrigues A, Douzery EJP (2009) Patterns of macroevolution among primates inferred from a supermatrix of mitochondrial and nuclear DNA. Mol Phylogenet Evol 53:808–825

Fan PF, Huo S (2009) The northern white-cheeked gibbon (Nomascus leugogenys) is on the edge of extinction in China. Gibbon J 5:44–52

Finstermeier K, Zinner D, Brameier M, Meyer M, Kreuz E, Hofreiter M, Roos C (2013) A mitogenomic phylogeny of primates. PLoS ONE 8(7):e69504

Fleagle JG (1975) A small gibbon-like hominid from the Miocene of Uganda. Folia Primatol 24:1–15

Fleagle JG (1984) Are there any fossil gibbons? In: Preuschoft H, Chivers DJ, Brockelman WY, Creel N (eds) The lesser apes. Edinburgh University Press, Edinburgh, pp 431–447

Fleagle JG (2013) Primate adaptation and evolution, 3rd edn. Academic Press, New York

Fleagle JG, Simons EL (1978) Micropithecus clarki, a small ape from the Miocene of Uganda. Am J Phys Anthropol 49:427–440

Frisch JE (1963) Sex-differences in the canines of the gibbon (Hylobates lar). Primates 4:1–10

Frisch JE (1965) Trends in the evolution of the hominoid dentition. Bibliotheca Primatologica Fasc. 3. Karger, Basel

Gao Y, Wen H, He Y (1981) The change of historical distribution of Chinese gibbons (Hylobates). Zool Res 2:1–8

Geissmann T (1995) Gibbon systematics and species identification. Int Zoo News 42:467–501

Geissmann T (2007) Status reassessment of the gibbons: results of the Asian primate red list workshop 2006. Gibbon J 3:5–15

Ginsburg L (1986) Chronology of the European pliopithecoids. In: Else JG, Lee PC (eds) Primate evolution. Cambridge University Press, Cambridge, pp 47–57

Ginsburg L, Mein P (1980) Crouzelia rhodanica, nouvelle espèce de primate catarhinien, et essai sur la position systématique des Pliopithecidae. Bull Mus Natn Hist Nat, Paris, Sér 4 2:57–85

Groves CP (1968) The classification of the gibbons (Primates, Pongidae). Z Saugetierkde 33:239–246

Groves CP (1972) Systematics and phylogeny of the gibbons. Gibbon Siamang 1:1–89

Groves CP (1974) New evidence on the evolution of apes and man. Vest ústred Úst geol 49:53–56

Groves CP (2001) Primate taxonomy. Smithsonian Institution Press, Washington, DC

Gu Y (1986) Preliminary research on the fossil gibbon of Pleistocene China. Acta Anthropol Sinica 5:208–219

Gu Y (1989) Preliminary research on the fossil gibbons of the Chinese Pleistocene and recent. Human Evol 4:509–514

Gu Y, Huang W, Chen D, Guo X, Jablonski NG (1996) Pleistocene fossil primates from Luoding, Guangdong. Vert PalAs 34:235–250

Haimoff EH, Chivers DJ, Gittins SP, Whitten T (1982) A phylogeny of gibbons (Hylobates spp.) based on morphological and behavioural characters. Folia Primatol 39:213–237

Harrison T (1982) Small-bodied apes from the Miocene of East Africa. Dissertation, University of London, London

Harrison T (1987) The phylogenetic relationships of the early catarrhine primates: a review of the current evidence. J Hum Evol 16:41–80

Harrison T (1988) A taxonomic revision of the small catarrhine primates from the early Miocene of East Africa. Folia Primatol 50:59–108

Harrison T (1991) The implications of Oreopithecus for the origins of bipedalism. In: Coppens Y, Senut B (eds) Origine(s) de la Bipédie Chez les Hominidés. Cahiers de Paléoanthropologie. CNRS, Paris, pp 235–244

Harrison T (1993) Cladistic concepts and the species problem in hominoid evolution. In: Kimbel W, Martin L (eds) Species, species concepts, and primate evolution. Plenum Press, New York, pp 345–371

Harrison T (1996) The paleoecological context at Niah Cave Sarawak: evidence from the primate fauna. Bull Indo-Pacific Prehist Assoc 14:90–100

Harrison T (1998) Vertebrate faunal remains from Madai Cave (MAD 1/28), Sabah, East Malaysia. Bull Indo-Pacific Prehist Assoc 17:85–92

Harrison T (2000) Archaeological and ecological implications of the primate fauna from prehistoric sites in Borneo. Bull Indo-Pacific Prehist Assoc 20:133–146

Harrison T (2002) Late Oligocene to middle Miocene catarrhines from Afro-Arabia. In: Hartwig WC (ed) Primate fossil record. Cambridge University Press, Cambridge, pp 311–338

Harrison T (2005) The zoogeographic and phylogenetic relationships of early catarrhine primates in Asia. Anthropol Sci 113:43–51

Harrison T (2010a) Dendropithecoidea, Proconsuloidea and Hominoidea. In: Werdelin L, Sanders WJ (eds) Cenozoic mammals of Africa. University of California Press, Berkeley, pp 429–469

Harrison T (2010b) Apes among the tangled branches of human origins. Science 327:532–534

Harrison T (2013) Catarrhine origins. In: Begun DR (ed) A companion to paleoanthropology. Wiley-Blackwell, Oxford, pp 376–396

Harrison T, Delson E, Guan J (1991) A new species of *Pliopithecus* from the middle Miocene of China and its implications for early catarrhine zoogeography. J Hum Evol 21:329–361

Harrison T, Gu Y (1999) Taxonomy and phylogenetic relationships of early Miocene catarrhines from Sihong, China. J Hum Evol 37:225–277

Harrison T, Krigbaum J, Manser J (2006) Primate biogeography and ecology on the Sunda Shelf Islands: a paleontological and zooarchaeological perspective. In: Lehman SM, Fleagle JG (eds) Primate biogeography. Springer, New York, pp 331–372

Harrison T, Ji X, Zheng L (2008) Renewed investigations at the late Miocene hominoid locality of Leilao, Yunnan, China. Am J Phys Anthropol 135(S46):113

Harrison T, Jin C, Zhang Y, Wang Y (2014) Fossil *Pongo* from the early Pleistocene *Gigantopithecus* fauna of Chongzuo, Guangxi, southern China. Quatern Int 354:59–67

Harrison T, Rook L (1997) Enigmatic anthropoid or misunderstood ape? The phylogenetic status of *Oreopithecus bambolii* reconsidered. In: Begun DR, Ward CV, Rose MD (eds) Function, phylogeny, and fossils: Miocene hominid evolution and adaptations. Plenum, New York, pp 327–362

Harrison T, Sanders WJ (1999) Scaling of lumbar vertebrae in anthropoid primates: its implications for the positional behavior and phylogenetic affinities of *Proconsul*. Am J Phys Anthropol S 28:146

Hayashi S, Hayasaka K, Takenaka O, Horai S (1995) Molecular phylogeny of gibbons inferred from mitochondrial DNA sequences: preliminary report. J Mol Evol 41:359–365

Hooijer DA (1960) Quaternary gibbons from the Malay Archipelago. Zool Verhand Leiden 46:1–42

Hooijer DA (1962) Prehistoric bone: the gibbons and monkeys of Niah Great Cave. Sarawak Mus J 10:428–449

Hopwood AT (1933) Miocene primates from Kenya. J Linn Soc London Zool 38:431–464

Hürzeler J (1954) Contribution à l'odontologie et à la phylogénèse du genre *Pliopithecus* Gervais. Ann Paléontol 40:1–63

Ingicco T, de Vos J, Huffman OF (2014) The oldest gibbon fossil (Hylobatidae) from insular Southeast Asia: evidence from Trinil (East Java, Indonesia), lower/middle Pleistocene. PLoS ONE 9(6):e99531

Israfil H, Zehr SM, Mootnick AR, Ruvolo M, Steiper ME (2011) Unresolved molecular phylogenies of gibbons and siamangs (Family: Hylobatidae) based on mitochondrial, Y-linked, and X-linked loci indicate a rapid Miocene radiation or sudden vicariance event. Mol Phylogenet Evol 58:447–455

Jablonski NG, Chaplin G (2009) The fossil record of gibbons. In: Lappan S, Whittaker DJ (eds) The gibbons. Springer, Dordrecht, pp 111–130

Jablonski NG, Whitworth MJ, Roberts-Smith N, Qingqi X (2000) The influence of life history and diet on the distribution of catarrhine primates during the Pleistocene in eastern Asia. J Hum Evol 39:131–157

Jenkins PD (1990) Catalogue of primates in the British Museum (Natural History) and elsewhere in the British Isles. Part V: The apes, superfamily Hominoidea. Natural History Museum, London

Ji W, Jiang X (2004) Primatology in China. Int J Primatol 5:1077–1092

Jin CZ, Qin DG, Pan WS, Tang ZL, Liu JY, Wang Y, Deng CL, Zhang YQ, Dong W, Tong HW (2008) A newly discovered *Gigantopithecus* fauna from Sanhe Cave, Chongzuo, Guangxi, South China. Chinese Sci Bull 54:788–797

Larson SG (1998) Parallel evolution in the hominoid trunk and forelimb. Evol Anthropol 6:87–99

Le Gros Clark WE, Leakey LSB (1951) The Miocene Hominoidea of East Africa. Fossil mammals of Africa, No. 1. British Museum (Natural History), London

Le Gros Clark WE, Thomas DP (1951) Associated jaws and limb bones of *Limnopithecus macinnesi*. Fossil mammals of Africa, No. 3. British Museum (Natural History), London

Leakey LSB (1963) East African fossil Hominoidea and the classification of the superfamily. In: Washburn SL (ed) Classification and Human Evolution. Aldine, Chicago, pp 32–49

Matsudaira K, Ishida T (2010) Phylogenetic relationships and divergence dates of the whole mitochondrial genome sequences among three gibbon genera. Mol Phylogenet Evol 55:454–459

Matthew WD, Granger W (1923) New fossil mammals from the Pliocene of Szechuan, China. Bull Am Mus Nat Hist 48:568–598

McGraw WS, Sciulli PW (2011) Posture, ischial tuberosities, and tree zone use in West African cercopithecids. In: D'Août K, Vereecke EE (eds) Primate locomotion: linking field and laboratory research. Springer, Dordrecht, pp 215–245

Meldrum J, Pan Y (1988) Manual proximal phalanx of *Laccopithecus robustus* from the latest Miocene site of Lufeng. J Hum Evol 17:719–732

Meyer TJ, McLain AT, Oldenburg M, Faulk C, Bourgeois MG, Conlin EM, Mootnick AR, de Jong PJ, Roos C, Carbone L, Batzer MA (2012) An *Alu*-based phylogeny of gibbons (Hylobatidae). Mol Biol Evol 29:3441–3450

Midlo C (1934) Form of hand and foot in primates. Am J Phys Anthropol 19:337–389

Miller GS (1933) The classification of the gibbons. J Mammal 14:158–159

Miller RA (1945) The ischial callosities of primates. Am J Anat 76:67–91

Mittermeier RA, Rylands AB, Wilson DE (eds) (2013) Handbook of mammals of the world. Primates, vol 3. Lynx Edicions, Barcelona

Mootnick A, Groves C (2005) A new generic name for the hoolock gibbon (Hylobatidae). Int J Primatol 26:971–976

Moyà-Solà S, Köhler M, Alba DM (2001) *Egarapithecus narcisoi*, a new genus of Pliopithecidae (Primates, Catarrhini) from the late Miocene of Spain. Am J Phys Anthropol 114:312–324

Napier JR (1963) Brachiation and brachiators. Symp Zool Soc Lond 10:183–195

Napier JR, Davis PR (1959) The forelimb skeleton and associated remains of *Proconsul africanus*. Fossil mammals of Africa, No. 16. British Museum (Natural History), London

Nisbett RA, Ciochon RL (1993) Primates in northern Viet Nam: a review of the ecology and conservation status of extant species, with notes on Pleistocene localities. Int J Primatol 14:765–795

Ogawa H, Yoshikawa M, Idani G (2014) Sleeping site selection by savanna chimpanzees in Ugalla, Tanzania. Primates 55:269–282

Ortiz A, Pilbrow V, Villamil CI, Korsgaard JG, Bailey SE, Harrison T (2015) The taxonomic and phylogenetic affinities of *Bunopithecus sericus*, a fossil hylobatid from the Pleistocene of China. PLoS ONE 10(7):e0131206

Pan Y (1988) Small fossil primates from Lufeng, a latest Miocene site in Yunnan Province, China. J Hum Evol 17:359–366

Pan Y (2006) Primates Linnaeus, 1758. In: Qi G, Dong W (eds) *Lufengpithecus hudienensis* site. Science Press, Beijing, pp 131–148

Pan Y, Waddle DM, Fleagle JG (1989) Sexual dimorphism in *Laccopithecus robustus*, a late Miocene hominoid from China. Am J Phys Anthropol 79:137–158

Perelman P, Johnson WE, Roos C, Seuánez HN, Horvath JE, Moreira MAM, Kessing B, Pontius J, Roelke M, Rumpler Y, Scheider MPC, Silva A, O'Brien SJ, Pecon-Slattery J (2011) A molecular phylogeny of living primates. PLoS Genet 7(3):e1001342

Pilbeam DR (1969) Tertiary Pongidae of East Africa: evolutionary relationships and taxonomy. Yale Peabody Mus Bull 31:1–185

Pilbeam D (1996) Genetic and morphological records of the Hominoidea and hominid origins: a synthesis. Mol Phylogenet Evol 5:155–168

Pilbeam DR, Young NM (2004) Hominoid evolution: synthesizing disparate data. C R Palevol 3:303–319

Pruetz JD, Fulton SJ, Marchant LF, McGrew WC, Schiel M, Waller M (2008) Arboreal nesting as anti-predator adaptation by savanna chimpanzees (Pan troglodytes verus) in southeastern Senegal. Am J Primatol 70:393–401

Prouty LA, Buchanan PD, Pollitzer WS, Mootnick AR (1983) Bunopithecus: a genus-level taxon for the hoolock gibbon (Hylobates hoolock). Am J Primatol 5:83–87

Purvis A (1995) A composite estimate of primate phylogeny. Phil Trans R Soc Lond B 348:405–421

Qiu Z, Qiu Z (1995) Chronological sequence and subdivision of Chinese Neogene mammalian faunas. Palaeogeog Palaeoclimatol Palaeoecol 116:41–70

Qiu Z, Wu W, Qiu Z (1999) Miocene mammal faunal sequence of China: palaeozoogeography and Eurasian relationships. In: Rössner GE, Heissig K (eds) The Miocene land mammals of Europe. Verlag Dr. Friedrich Pfeil, Munich, pp 443–455

Raaum RL, Sterner KN, Noviello CM, Stewart C-B, Disotell TR (2005) Catarrhine primate divergence dates estimated from complete mitochondrial genomes: concordance with fossil and nuclear DNA evidence. J Hum Evol 48:237–257

Raza SM, Barry JC, Meyer GE, Martin L (1984) Preliminary report on the geology and vertebrate fauna of the Miocene Manchar Formation, Sind, Pakistan. J Vert Paleontol 4:584–599

Remane A (1965) Die Geschichte der Menschenaffen. In: Heberer G (ed) Menschliche Abstammungslehre. Gustav Fischer, Göttingen, pp 249–309

Roos C, Geissmann T (2001) Molecular phylogeny of the major hylobatid divisions. Mol Phylogenet Evol 19:486–494

Rose MD (1974) Ischial tuberosities and ischial callosities. Am J Phys Anthropol 40:375–384

Schlosser M (1924) Fossil primates from China. Palaeont Sinica (c) 1, (2):1–16

Schultz AH (1930) The skeleton of the trunk and limbs of the higher primates. Hum Biol 2:303–438

Schultz AH (1936) Characters common to higher primates and characters specific for man. Quart Rev Biol 11:259–283

Schultz AH (1963) Relations between the lengths of the main parts of the foot skeleton in primates. Folia Primatol 1:150–171

Schwartz JH, Long VT, Cuong NL, Kha LT, Tattersall I (1994) A diverse hominoid fauna from the late Middle Pleistocene breccia cave of Tham Khuyen, Socialist Republic of Vietnam. Anthropol Papers Am Mus Nat Hist 73:1–11

Simons EL (1965) New fossil apes from Egypt and the initial differentiation of Hominoidea. Nature 205:135–139

Simons EL (1967) Fossil primates and the evolution of some primate locomotor systems. Am J Phys Anthropol 26:241–254

Simons EL (1972) Primate evolution: an introduction to man's place in nature. MacMillan, New York

Simons EL, Fleagle JG (1973) The history of extinct gibbon-like primates. Gibbon Siamang 2:121–148

Springer MS, Meredith RW, Gatesy J, Emerling CA, Park J, Rabosky DL, Stadler T, Steiner C, Ryder OA, Janeka JE, Fisher CA, Murphy WJ (2012) Macroevolutionary dynamics and historical biogeography of primate diversification inferred from a species supermatrix. PLoS ONE 7(11):e49521

Stewart FA, Pruetz JD (2013) Do chimpanzee nests serve an anti-predatory function? Am J Primatol 75:593–604

Storm P, de Vos J (2006) Rediscovery of the Late Pleistocene Punung hominin sites and the discovery of a new site Gunung Dawung in East Java. Senckenb Lethaea 86:271–281

Suteethorn V, Buffetaut E, Buffetaut-Tong H, Ducrocq S, Helmcke-Ingavat R, Jaeger J-J, Jongkanjanasoontorn Y (1990) A hominoid locality in the Middle Miocene of Thailand. CR Acad Sci Paris 311, Sér II:1449–1454

Szalay FS, Delson E (1979) Evolutionary history of the primates. Academic Press, New York

Takai M, Zhang Y, Kono RT, Jin C (2014) Changes in the composition of the Pleistocene primate fauna in southern China. Quatern Int 354:75–85

Takacs Z, Morales JC, Geissmann T, Melnick DJ (2005) A complete species-level phylogeny of the Hylobatidae based on mitochondrial ND3-ND4 gene sequences. Mol Phylogenet Evol 36:456–467

Thinh VN, Mootnick AR, Geissmann T, Li M, Ziegler T, Agil M, Moisson P, Nadler T, Walter L, Roos C (2010) Mitochondrial evidence for multiple radiations in the evolutionary history of small apes. BMC Evol Biol 10:74

Tougard (2001) Biogeography and migration routes of large mammal faunas in South-East Asia during the late Middle Pleistocene: focus on the fossil and extant faunas from Thailand. Palaeogeog Palaeoclimatol Palaeoecol 168:337–358

Tuttle RH (1969) Knuckle-walking and the problem of human origins. Science 166:953–961

Tyler DE (1991) The problems of the Pliopithecidae as a hylobatid ancestor. Human Evol 8:73–80

Tyler DE (1993) The evolutionary history of the gibbons. In: Jablonski NG (ed) Evolving landscapes and evolving biotas of East Asia since the mid-tertiary. Centre of Asian Studies, The University of Hong Kong, Hong Kong, pp 228–240

van Gulik RH (1967) The gibbon in China: an essay in Chinese animal lore. EJ Brill, Leiden

Vilensky JA (1978) The function of ischial callosities. Primates 19:363–369

von Koenigswald GHR (1935) Eine fossile Säugertierfauna mit Simia aus Südchina. Proc Koninkl Ned Akad Wetensch, Amsterdam 38:872–879

von Koenigswald GHR (1940) Neue Pithecanthropus-Funde 1936–1938. Wetensch Meded Mijnbouw Nederlands Indië (Batavia) 28:1–232

Wall JD, Kim SK, Luca F, Carbone L, Mootnick AR, de Jong PJ, Di Rienzo A (2013) Incomplete lineage sorting is common in extant gibbon genera. PLoS ONE 8(1):e53682

Wang XM, Wang BY, Qiu ZX (2008) Early explorations of Tabenbuluk region (western Gansu Province) by Birger Bohlin—reconciling classic vertebrate fossil localities with modern stratigraphy. Vert PalAs 46:1–19

Washburn SL (1957) Ischial callosities as sleeping adaptations. Am J Phys Anthropol 15:269–276

Westaway KE, Morwood MJ, Roberts RG, Rokus AD, Zhao JX, Storm P, Aziz F, van den Bergh G, Hadi P, Jatmiko, de Vos J (2007) Age and biostratigraphic significance of the Punung rainforest fauna, East Java, Indonesia, and implications for Pongo and Homo. J Hum Evol 53:709–717

Whittaker DJ, Morales JC, Melnick DJ (2007) Resolution of the Hylobates phylogeny: congruence of mitochondrial D-loop sequences with molecular, behavioral, and morphological data sets. Mol Phylogenet Evol 45:620–628

Wu R, Pan Y (1984) A late Miocene gibbon-like primate from Lufeng, Yunnan Province. Acta Anthropol Sinica 3:185–194

Wu R, Pan Y (1985) Preliminary observation on the cranium of Laccopithecus robustus from Lufeng, Yunnan with reference to its phylogenetic relationship. Acta Anthropol Sinica 4:7–12

Young NM (2003) A reassessment of living hominoid postcranial variability: implications for ape evolution. J Hum Evol 45:441–464

Zapfe H (1958) The skeleton of Pliopithecus (Epipliopithecus) vindobonensis Zapfe and Hürzeler. Am J Phys Anthropol 16:441–458

Zapfe H (1961) Die Primatenfunde aus der Miozänen Spaltenfüllung von Neudorf an der March (Devínská Nová Ves), Tschechoslowakei. Schweiz Palaeont Abh 78:4–293

Zeitoun V, Lenoble A, Laudet F, Thompson J, Rink WJ, Mallye J-B, Chinnawut W (2010) The cave of the Monk (Ban Fa Suai, Chiang Dao Wildlife Sanctuary, northern Thailand). Quatern Int 220:160–173

Chapter 5
Hylobatid Evolution in Paleogeographic and Paleoclimatic Context

Ulrich H. Reichard and Michelle M. Croissier

Adult male white-handed gibbon foraging on fruit, Khao Yai National Park, Thailand. Photo credit: Ulrich H. Reichard

Introduction

The land area that comprises East and Southeast Asia (E-SA) has one of Earth's most complex geologic and climatic histories (Metcalfe 1996, 2011, 2013a; Cannon et al. 2009). The region was assembled in a multi-step process beginning ~400 million years ago (mya) in the Paleozoic, when landmasses from the margins of the eastern parts of the southern supercontinent Gondwana broke off and began to rift and drift northward (Metcalfe 2013a; Corlett 2014). Successive ocean basins opened between

U.H. Reichard (✉)
Department of Anthropology and Center for Ecology, Southern Illinois University Carbondale, Carbondale, IL 62901, USA
e-mail: ureich@siu.edu

M.M. Croissier
Center for Ecology, Southern Illinois University Carbondale, Carbondale, IL 62901, USA
e-mail: mmcroiss@gmail.com

© Springer Science+Business Media New York 2016
U.H. Reichard et al. (eds.), *Evolution of Gibbons and Siamang*,
Developments in Primatology: Progress and Prospects,
DOI 10.1007/978-1-4939-5614-2_5

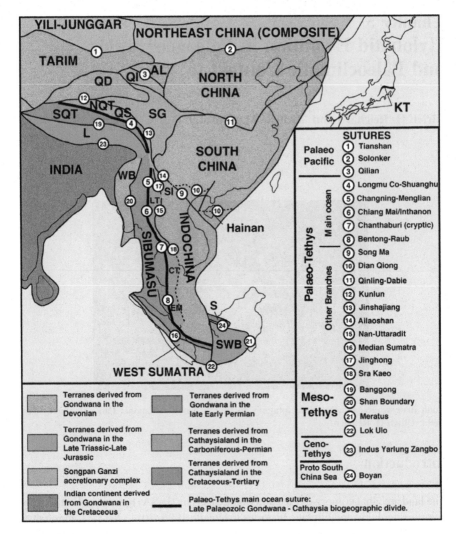

Fig. 5.1 Distribution of principal continental blocks, arc terranes, and sutures of eastern Asia. *WB* West Burma, *SWB* South West Borneo, *S* Semitau, *L* Lhasa, *SQT* South Qiangtang, *NQT* North Qiangtang, *QS* Qamdo–Simao, *SI* Simao, *SG* Songpan Ganzi accretionary complex, *QD* Qaidam, *QI* Qilian, *AL* Ala Shan, *KT* Kurosegawa Terrane, *LT* Lincang arc Terrane, *CT* Chanthaburi arc Terrane, *EM* East Malaya. After Metcalfe (2011, 2013a, b)

each fragment and Gondwana, closing again later through subduction and leading to the progressive joining of these fragments with Asian continental masses, primarily the North China, South China, and Indochina blocks. The process lasted until the end of the Cretaceous 65 mya, when the basic shape of modern E-SA was established (Metcalfe 2011; Fig. 5.1). Metcalfe (2013b: 16) concisely outlines the geologic processes of the Cenozoic, the following time period:

The Cenozoic evolution of East and Southeast Asia involved substantial movements along and rotations of strike-slip faults, rotations of continental blocks and oceanic plates, the development and spreading of 'marginal' seas, and the formation of important hydrocarbon-bearing sedimentary basins. This evolution was essentially due to the combined effects of the interactions of the Eurasian, Pacific, and Indo-Australian plates and the collisions of India and Eurasia and of Australia with Southeast Asia.

A profound tectonic event in the geologic history of E-SA, with significant consequences for the evolution of Asian primates, was the several million long process of the closing of the Tethys sea. The Tethys sea once connected the Mediterranean sea with the Indian ocean, but its closure was completed no later than ~20 mya (Okay et al. 2010; McQuarrie and van Hinsbergen 2013). Although the Afro-Arabian–Eurasian collision created a wide zone of diffuse deformation on the southern margin of Eurasia, it also allowed for the first time direct migration of non-avian fauna across the newly formed *Gomphotherium Landbridge* at ~19 mya (Fig. 5.2; Rögl 1998, 1999a, b; Harzhauser and Piller 2007). Besides the general, broad geo-biological significance of these newly joined landmasses, due to the creation of new biota through species exchanges (Zhou et al. 2012), we suggest that the Afro-Asian corridor across the *Gomphotherium Landbridge* became the primary migration route along which African primates, including hominoids, entered Eurasia. This interpretation finds support in the fact that some stem hominoids left Africa ~17–16 mya (Andrews and Kelley 2007) and, for example, *Sivapithecus* fossils dated to ~13.5 mya (Patnaik and Chauhan 2009) are present in the Siwalik foothills of present day Pakistan and North India.

Distinct topographic and ecological changes that resulted from climatic cycles and tectonic events (i.e., rifts, uplifts, sea-level changes, and volcanism [Vrba 2007]) once formed E-SA into viable primate habitat and these forces have contributed significantly to the evolutionary trajectories of Asia's diverse primate

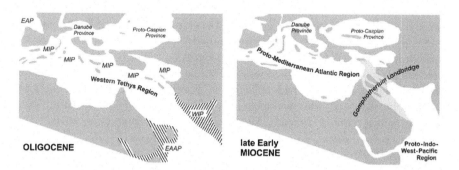

Fig. 5.2 Palaeogeography and mollusc-based biogeography of the Western Tethys during the Oligocene and Early Miocene (geography modified from Rögl (1998), Popov et al. (2004) and Harzhauser and Piller (2007); biogeography after Harzhauser et al. (2002) and Harzhauser (2007)). Note that the selected time slices are only snapshots from a much more complex palaeogeographic development as reflected by the numerous maps presented by Rögl (1998) and Popov et al. (2004). *Abbreviations: MIP* Mediterranean-Iranian Province, *EAP* Eastern Atlantic Province, *WIP* Western Indian Province, *EAAP* Eastern African-Arabian Province (after Harzhauser et al. 2007)

Table 5.1 Hylobatid evolution in geologic time

		Time (mya)[1]	Hylobatid evolution
Miocene	Step 1 - emergence of the family Hylobatidae	**16.26**(18.16-14.70)	split of stem hylobatids from stem hominoids
	Step 2 - emergence of four hylobatid genera	**8.34** (9.68-7.14)	emergence of *Nomascus*
		7.22 (8.44-5.99)	emergence of *Symphalangus*
		6.69 (7.88-5.56)	divergence of *Hoolock* and *Hylobates*
Pliocene	Wave 1 -speciation within *Nomascus* and *Hylobates*	**4.24** (5.06-3.46)	diversification within *Nomascus* and *Hylobates*
		2.83 (3.50-2.21)	divergence of *Nomascus concolor*
Pleistocene	Wave 2 -speciation within *Nomascus,Hoolock,* and *Hylobates*	**1.74** (2.22-1.78)	further diversifaction within *Nomascus* and *Hylobates*
		1.42 (1.90-0.75)	divergence of *H. hoolock* and *H. leuconedys*
		1.3 (1.68-0.95)	divergence of *H. agilis*
	Wave 3 -sub-speciation within *Hylobates* and *Nomascus*	**1.05** (1.35-0.75)	sub-speciation within *Nomascus* and *Hylobates*
		0.38 (0.55-0.18)	divergence of *Nomascus siki*

[1]Data from Thinh et al. (2010a)

fauna. A comprehensive understanding of the evolution of Asia's primates requires at least a basic knowledge of the area's paleogeographic and paleoclimatologic history. This chapter presents such background for a deeper understanding of hylobatid evolution focusing mostly on the post-Oligocene history of E-SA from ∼20 to 1 mya and, specifically, times preceding and overlapping with suggested radiation processes within hylobatids.

Thinh et al. (2010a) recently provided a comprehensive scenario for the evolution of the hylobatid family based on molecular divergence times, which we adopted to allow relating paleotopographic and paleoclimatic global and Asia-specific developments with significant moments in hylobatid evolution. Although details of the evolutionary history of hylobatids, in particular divergence times and the first 10 mya, after the family split from the stem hominoid lineage, are still unresolvable (Carbone et al. 2014), we believe Thinh et al. (2010a) offer an adequate and sufficient temporal guidance for a broad geo-climatic background to events preceding and accompanying hylobatid evolution.

Following Thinh et al. (2010a), hylobatid evolution can be broadly broken down into two diversification steps that occurred during the Miocene, followed by three waves of rapid speciation and radiation in Plio-Pleistocene times (Table 5.1). Briefly, the first step was marked by the split of stem hylobatids from stem hominoids at the end of the early Miocene, 16.26 mya (Thinh et al. 2010a). The second step was marked by swift diversification of the family into four genera *Nomascus, Hoolock, Symphalangus,* and *Hylobates.* This process began in the late middle Miocene ∼8.34 mya with the emergence of the genus *Nomascus* and was concluded only ∼1.65 my later with the split of the genera *Hoolock* and *Hylobates.* Diversification within genera is traceable as three speciation waves, the first occurring at 4.24 mya in the late Pliocene within *Nomascus* and *Hylobates* lasting until 2.83 mya and concluding with the emergence of *Nomascus concolor.* The second speciation wave began ∼1.74 mya in the Pleistocene, likewise within

Nomascus and *Hylobates*, and lasted until 1.3 mya with the appearance of *Hylobates agilis*. During this period, within the genus *Hoolock*, the two species *H. hoolock* and *H. leuconedys* also separated from each other. We considered the most recent diversification in genera *Nomascus* and *Hylobates*, between 1.05 and 0.38 mya, of somewhat lesser importance for an overall understanding of hylobatid evolution, because it primarily represents subspecies-level divergences.

East and Southeast Asia's Geologic History

East and Southeast Asia, which today includes the Malaysian Peninsula, Sumatra, Borneo, Java, and the Palawan islands (Corlett 2014), are characterized by relatively small continental land masses and numerous islands: The Indo-Malayan archipelago comprises around 17,000 islands, and includes two of the World's largest, Borneo and Sumatra, which together with the Philippines adds another 7100 islands to the region (Woodruff 2010). The history of being Sundaland, the principal landmass of E-SA, is a history of oscillations between E-SA's terrestrial backbone as a chain of islands to a solid subcontinent (Bird et al. 2005; Woodruff 2010; Metcalfe 2011; Fig. 5.3). As far back as the early Tertiary, even the central peninsula is were believed to have once been divided by a deep water channel and the land masses were so far apart that movement between them must have been nearly impossible for Indochinese land mammals (Woodruff and Turner 2009). At the other extreme, during the Last Glacial Maximum (LGM) $\sim$26.5–19 kya (Clark et al. 2009), E-SA was one large, continental landmass that included all islands where hylobatids live today, i.e., Hainan island, Sumatra, Java, Borneo, and the larger Mentawai Islands (Fig. 5.4).

Undoubtedly, significant events in E-SA's geologic history were the collision of the Indian subcontinent with landmasses of Eurasia at the southern tip of Tibet in the late Eocene $\sim$35 mya (Ali and Aitchison 2008), followed by the West-East emergence of the Himalayan mountain range and the stepwise rise of the Tibetan Plateau (Tapponnier et al. 2001; Nábelek et al. 2009). During the collision, pressures pushing East-northeast met with resistance from older formations of the strong South China block, leading to significant deformation in the contact region and the rise of the Tibetan Plateau, which since has been a major influence on E-SA's climate.

The Ancient Hengduan Mountains and Related Geographic Separations

The area that resisted the docking forces of India onto Tibet is referred to as the ancient Hengduan Mountains of Yunnan Province, China, an assemblage of peaks and ridges of the Eastern Himalayas bordering Tibet, India, and Myanmar whose

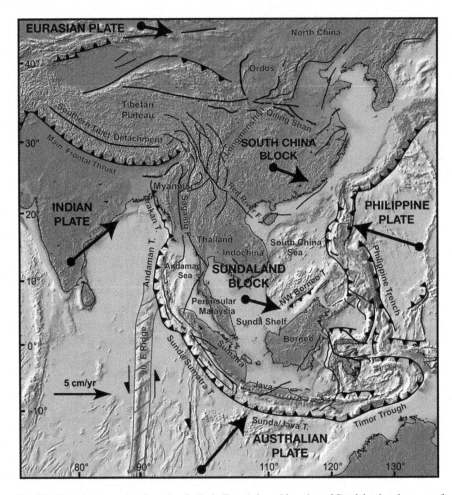

Fig. 5.3 Topography and main active faults in East Asia and location of Sundaland at the zone of convergence of the Eurasian, Philippine, and Indian–Australian plates. *Large arrows* represent absolute (International Terrestrial Reference Frame 2000) motions of plates (after Simons et al. 2007)

formation is suggested to have preceded the Tibetan Plateau uplift (Chaplin 2005). Geologically, the Hengduan Mountains are situated at the margins of the Eurasian Plate to the North, the Indochina Block to the South, and the Indian Plate to the West. These plates are constrained by the Philippine-Pacific Plates to the East and the Australasian Plate to the South (Hall 1997), which are all moving relative to the stable Eurasian Plate (Chaplin 2005). Past and present geophysical pressures have created and maintained steep rising peaks alongside with high altitude, contiguous

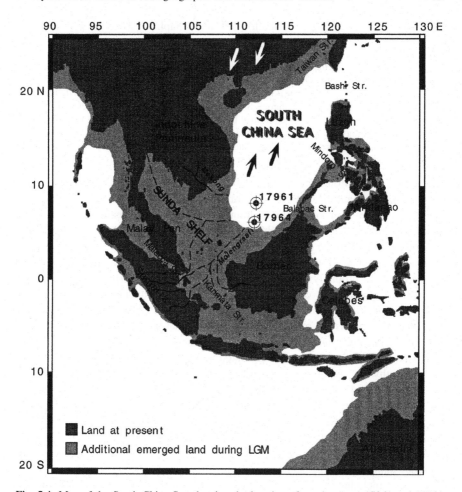

Fig. 5.4 Map of the South China Sea showing the location of gravity cores 17961 and 17964, collected during the R/V *Sonne* cruise in April–June 1994 (Sarnthein et al. 1994). The glacial drainage systems according to Molengraaff (1921) and the additional emerged land during the Last Glacial Maximum (LGM) are also plotted. *Thick arrows* represent typical monsoonal wind directions during winter (*white*) and summer (*black*). After Pelejero et al. 1999

mountain ranges such as the Gaoligong Shan[1], Tanian Tawen Shan, Nu Shan, Markam Shan, Ninchin Shan, and Yunling Shan separated by deep valleys running parallel along fault lines of the North China and South China Gondwanian

---

[1]We agree with Chaplin (2005: 528) to follow Zhao and Salter's (1986) suggestion that if Chinese names are translated they should not be contracted to less than two characters, for example, the "Tian Shan Mountains" is preferable to "Tian Mountains" even though the Chinese word "Shan" is best translated as Mountain.

fragments (Chaplin 2005; Metcalfe 2013a). Chaplin (2005: 531) concisely describes the situation:

As a consequence of the Indian Plate's collision into the Eurasian Plate, the eastern Himalaya syntaxis rotated clockwise, and crustal fragments of the Northern and Southern Terranes extruded southeastward. The extrusion was along the NW-trending Karakoram-Jiali, N-trending Gaoligong Shan, and Sagaing Faults (Lin et al. 2005). Ages of the faults indicate that deformation may have started from the south along the Sagaing Fault in Indochina and propagated toward the north along the Gaoligong Shan Fault. Subsequently, the deformation proceeded toward the northwest along the Jiali Fault and then the Karakoram Fault in southern Tibet.

The Hengduan Mountain ranges can be interpreted as tentacle-like extensions of the Tibetan Plateau through which grand Asian rivers, i.e., the Salween, the Mekong, and the Yangtze of China, drain waters from the plateau and surrounding high mountains into the Gulf of Bengal, the Gulf of Thailand, and the East China sea (Woodruff 2010). On its western side another major river, the Irrawaddy, drains the Gaoligong Shan Mountains through Myanmar into the Andaman Sea. Because hylobatids were probably always arboreal (Reichard et al. 2016), large bodies of water, including large rivers, have presumably always functioned as barriers for their dispersal (Marshall and Sugardjito 1986; Chatterjee 2009). A recent molecular phylogenetic analysis of crested gibbons suggests that the northernmost distributed species, *Nomascus concolar*, branched off first followed by *N. nasutus* and *N. hainanus*, implying the genus originated at a northern latitude and successively moved southward (Thinh et al. 2010b), perhaps along rivers such as the Mekong, the Black and Red rivers, which today separate crested gibbon species (Geissmann 2007). This scenario is consistent with a proposed northern latitude origin of hylobatids (Chatterjee 2006, 2009), such as the foothills of the Himalayas in what is today Yunnan province, China (Reichard et al. 2016).

Although evaluating the biological significance of historic biota is complicated, because many variables in concert influence such projections, the Eastern Himalayas, including the ancient Hengduan Mountains, have been of particular biological significance for the evolution of E-SA's flora and fauna since the Oligocene/Miocene. The Gaoligong Shan Mountains, for example, include three globally recognized biodiversity hotspots, i.e., 'Himalaya,' 'Indo-Burma,' and 'Mountains of Southwest China' (Wu 1988; Myers et al. 2000; Li et al. 2011). Chaplin (2005) has suggested that the Gaoligong Shan combine geophysical features that make the area ecologically unique. Generally, the mountains North-South orientation results in an absence of North or South facing slopes, and West facing slopes are larger than East-facing slopes, because most Eastern mountain sides are steeper. At lower latitude, where the Gaoligong Shan is trending more Northeast-Southwestward, slopes face West and also Northwest and then Northeast and East (Chaplin 2005). As a consequence of this general orientation, many slopes in these high mountains face away from the sun resulting in valleys that receive little sunlight, which is further dissipated across the steep slopes. The mountains' steepness compromises day break and sunset leading to relatively shorter day lengths in valleys compared to the lowlands, which curtails activity and productivity periods for plants and animals. As a result, more temperate climate-like conditions extend further South in the Gaoligong Shan valleys than in other places in the world, because the Gaoligong Shan extend further South than most of the Hengduan Mountain ranges (Chaplin 2005). The North-South and Northeast-Southwest orientated valleys form unique corridors along which species can and probably always have moved in response to dynamic North-South temperature gradients of Asia's monsoon climate, escaping temperature increases and decreases by migrating north- or southward, while East-West movements have been barricaded by the continuous high mountain ridges leading to isolation of flora and fauna deeper in the valleys. The regions prevailing southwesterly winds add to the particular climate with an influx of moist air and significant rainfall to western slopes while also forcing damp air over high ridges, which can lead to a peculiar weather pattern known as *Föhn*, which are strong, warm, and very dry winds descending from a high mountain into a valley or plain (Seluchi et al. 2003). As Chaplin (2005: 546) resumes:

> The unusual physical features combine to multiply the number of opportunities for microclimates. Furthermore, these physical features are not fixed in time but are dynamic due to the nature of the underlying geological processes. This dynamism provides ample opportunity for adaptation and vicariant events to further promote biological diversification.

We propose, in agreement with others (Jablonski and Chaplin), that the geophysical and climatic conditions detailed above for the Gaoligong Shan were representative, although perhaps along a graded scale, for the Southern parts of the ancient Hengduan Mountains as well as mountainous areas to the East that had a similar orientation and geophysical properties. Because of these features, the Southern Himalayan mountain valleys, which harbor significant hominoid fossil sites (Jablonski and Chaplin 2009; Ji et al. 2013; Jablonski et al. 2014), could function as refugia to which Asian hominoids retreaded or were compressed as climatic conditions deteriorated in the later Miocene and lowland rainforests became patchier. The middle Miocene climatic transition has broadly been recognized as a time of dramatic global climate change (Miller et al. 1991; Wright et al. 1992; Flower and Kennett 1994; Zachos et al. 2001; Shevenell et al. 2004). The deep Southern Himalaya valleys had the capacity to buffer and decelerate ongoing global climate perturbations of the time. While cooler temperatures, increasing seasonality and aridification resulted in tropical forests shrinking, becoming discontinuous and partially being replaced by grassland (Morley 2000; Kürschner et al. 2008; Eronen et al. 2009), the Hengduan Mountain valleys continued to provide a more humid and less seasonal environment to forest dwelling species such as stem hylobatids. Perhaps the temperature in valleys was slightly cooler than the surrounding lowlands, but nevertheless allowed stem hylobatids, which like crown hylobatids presumably had dense fur, to survive and adapt to the changing Miocene/Pliocene environment. It is likely that an evergreen forest flora covering the valley bottoms continued further northward in the Hengduan Mountains than elsewhere due to a more stable, favorable microclimate in the mountain valleys. Hence, it is plausible that different stem hylobatid populations may have become isolated in different Hengduan Mountain valley tentacles where they evolved into the four genera recognized today, and from where hylobatid species later radiated southward. Although no fossils except *Yuanmoupithecus* of the Yuanmou Basin (Harrison et al. 2008) are yet available to support this, the location of the Yuanmou Basin between the Mekong and Yangtze rivers and the fossil's tentative age at of 8 mya, are suggestive of a stem crested gibbon presence. This interpretation is consistent with the Asian hominoid fossil record, which shows that Asian hominoids were present in southern Himalayan refuge forests longer than elsewhere in E-SA (Jablonski 2005; Ji et al. 2013; Jablonski et al. 2014).

East and Southeast Asia's Paleoclimate and Consequences on Primate Evolution

Earth's orbital oscillations and changing distributions of landmasses, oceans, and oceanic gateways are primary forces that have shaped the world's climate (Zachos et al. 2001). Plate tectonics, which explains the distribution of Earth's land and sea areas, including subaerial and submarine mountains, plateaus, and basins, has played a pivotal role in reconstructing regional climatic histories (Hay 1996).

Climate significantly affects biodiversity; it can alter species' ranges and relative abundance, and change ecological dynamics (Heller and Zavaleta 2009), and can ultimately lead to species-level extinction (Thomas et al. 2004). Paleoclimates can be reconstructed by analyzing marine, freshwater, and terrestrial sediments, using a variety of geochemical indicators, including oxygen isotope ratios determined via mass spectrometry of planktonic foraminifera (Wei et al. 2007) as well as biomarkers, such as pollen and spores preserved in a swamp deposits—the latter useful in the reconstruction of past vegetation and plant communities (Ohlwein and Wahl 2012).

Paleoclimates histories are needed to understand the formation and composition of primate habitat relative to the dynamics of primate communities through time, because climate and weather patterns critically influence the distribution of plants, which are a fundamental resource for consumers such as primates.

Primate evolution has been linked to the rise of *angiosperms*, i.e., fruit-bearing trees (Sussman et al. 2013), which replaced ferns and gymnosperms and came to dominate terrestrial biota in the Cretaceous (Cronquist 1988; Benton 2010). Angiosperms were part of the *Cretaceous Terrestrial Revolution* (Lloyd et al. 2008) and underwent an explosive radiation 125–80 mya perhaps originating in E-SA (Buerki et al. 2014). The rise of angiosperms created new ecological opportunities for pollinating insects, leaf-eating flies, as well as butterflies and moths, all of which diversified rapidly in the Cretaceous (Grimaldi 1999; Lloyd et al. 2008). However, it was not until the Paleocene/Eocene that angiosperms evolved seeds in large stony and fleshy fruit, probably to attract birds and mammals as dispersers (Dilcher 2000). The angiosperms coevolution with birds and mammals was probably highly successful as all taxa underwent major diversification around the same time (Hedges et al. 1996; Bininda-Emonds et al. 2007). The highest diversification rates among stem angiosperm orders occurred 95–75 mya (Magallón and Castillo 2009), which is shortly before the evolution of primates in the later part of the Cretaceous or during the early Paleocene (Fleagle 2013).

It is still unclear whether the earliest primates conquered the arboreal ecological that emerged with the dominance of angiosperms in order to hunt insects feeding off nectar from angiosperm flowers (Cartmill et al. 2007; Dagosto 2007; Szalay 2007), or to consume the nectar and flowers themselves as well as harvest angiosperm fruits (Sussman et al. 2013). Less controversial is the assumption that primate evolution became inextricably linked to angiosperm-dominated tropical forests (Milton 1980; Lucas et al. 2007). Tropical rainforests are generally characterized by relatively low seasonality and high productivity (Morley 2000). Since their rise, angiosperms have quickly become the primary constituents of tropical rain forests providing important three-dimensional structural definition for terrestrial ecosystems at most latitudes (Crepet 2000). They produced multispecies, multilayered forest canopies that became most diverse at the end of the Paleocene (Eriksson et al. 2000) and have since been the stable, ecological backbone of primate existence. The climate during the Paleocene was favorable for angiosperm diversification as temperatures were higher than today and continued to rise toward the early Eocene (Zachos et al. 2001).

Since the Eocene climatic optimum ~50 mya, Earth's climate has become colder (Zachos et al. 2001). The general trend was punctuated by three pronounced cooling steps: (1) at the Eocene-Oligocene boundary, (2) in the late phase of the middle Miocene, and (3) during the late Pliocene.

1. Earth's temperature plunge at the beginning of the Oligocene ~34 mya is broadly associated with the onset of continental glaciation and the permanent establishment of Antarctic ice sheets (Zachos et al. 2001). For most of the Oligocene, Earth's climate remained at a lower, but stable temperature compared to the Eocene. Primate evolution during the Oligocene is complex, because the pronounced climatic changes led to equally dramatic changes among the primates (Fleagle 2013). Prosimians, which were abundant in the northern hemisphere during the Eocene, disappeared early in the Oligocene from Europe and became rare in North America while anthropoid primates began to dominate Africa (Fleagle 2013). The hominoid-cercopithecoid divergence likely fell into the early Oligocene period (Steiper et al. 2004). Overall, however, hominoid evolution during the Oligocene seems to have been slow, perhaps because the cool/dry Oligocene climate was not particularly ape friendly. Then, at the latest Oligocene/earliest Miocene, global temperature began to climb and humidity increased between 28 and 23.3 mya (Tang et al. 2011) leading to warm-wet conditions for a couple of million years, interspersed only by a few, insignificant short cooling spikes (Zachos et al. 2001). The warm Miocene conditions peaked around 17–15 mya at the late middle Miocene climatic optimum (Flower and Kennett 1994; Sun and Zhang 2008; Zhou et al.

2012). Perhaps these hominoid-friendly times favored hominoid diversification and perhaps prompted the divergence of stem hylobatids from stem hominoids around 16.26 mya (Thinh et al. 2010a). The warm-wet climate allowed for the expansion of moist megathermal forests poleward out of subtropical, high-pressure zones (Morley 2011). Climate alterations were most pronounced in the low latitudes of Southeast Asia, where a shift from a seasonally dry (monsoonal) to an ever-wet climate occurred at about 23.3 mya, evidenced by the sudden absence of *Gramineae* pollen and an increase in pollen from peat swamps coupled with an abrupt appearance of coals in the lithological record (Morley 2011). With an estimated emergence bracket of 18.2–14.7 mya (Thinh et al. 2010a), the onset of the independent hylobatid evolution seems closely linked to the warm-wet conditions of the early middle Miocene and predating the middle Miocene transition.

2. The middle Miocene climatic optimum was followed by the middle Miocene transition that began with a sharp global temperature drop (Zachos et al. 2001) and increasing aridification (Tang et al. 2011) at 14.2–13.8 mya. The climate change of the late Miocene led to major ice-sheet development on Antarctica by ∼10 mya (Shevenell et al. 2004, 2008; Lewis et al. 2008). Around the same time, the climate of E-SA was additionally impacted by Himalayan Plateau rises (Wang et al. 2008), which led to an even more pronounced increase in the strength of seasonality in the region. Although hylobatid evolution during the Miocene transition is undocumented due to a lack of fossils, it can be assumed that gibbons and siamang, like other hominoids, adapted to their changing environment. Recognizable diversification of the hylobatid family is suggested by molecular data, resulting in the separation of the current four genera over a short time span, between 8.34 and 6.69 mya (Thinh et al. 2010a), perhaps as a direct response to climatic and subsequent environmental changes. Increasing, extensive aridification followed in the later Miocene after 6 mya (Sun and Zhang 2008), but no corresponding divergences within hylobatids can be associated during the following 2 my until ∼4.24 mya when the first speciation wave within *Nomascus* and *Hylobates* led to significant changes in the hylobatid family (Thinh et al. 2010a).

3. During the middle Pliocene, Asia experienced a short warm period between ∼4.3 and 3.4 mya (Pollard and DeConto 2009), before Earth's climate took a final, sharp temperature drop in the late Pliocene at ∼3 mya that has broadly been related to the formation of permanent northern hemisphere ice sheets (Clark et al. 1999; Zachos et al. 2001). In Asia, these global climate developments were probably exacerbated by another rise of the Tibetan plateau also at ∼3 mya (Wang et al. 2008). The ongoing cooling found a direct expression in the onset of major northern hemispheric glaciations at 2.6 mya (Kennett and Hodell 1993). The first wave of hylobatid diversification within the genera *Nomascus* and *Hylobates*, which began within *Nomascus* at 4.24 mya and within *Hylobates* around 3.91 mya (Thinh et al. 2010a), can be broadly related to the relatively warmer and thus ecologically more favorable period of the middle Pliocene. This diversification wave lasted until ∼2.83 mya with the

separation of *Nomascus gabriellae* in the far south of current day Vietnam (Thinh et al. 2010a). During the harsher post-3 mya period, no speciation events within hylobatids have been suggested until ~1.7 mya (Table 5.1).

The Emergence of the Tibetan Plateau and Asia's Monsoons

The high Himalayan mountains and the Tibetan Plateau emerged as a consequence of the collision of the Indian subcontinent with Eurasia ~35 mya (Ali and Aitchison 2008). Since its rise, the Tibetan plateau has been a major influence on E-SA's climate (Flohn 1968; Zhisheng et al. 2001; Liu and Yin 2002). Although an early onset of the Asian monsoon ~24 mya is still debated (Guo et al. 2002; Garzione et al. 2005; Sun and Wang 2005; Clift et al. 2008), it appears less controversial that both the East and the South Asian monsoon systems together impacted Asia's climate with alternating but joint forces (Clift et al. 2008) from ~17 mya onwards.

The Tibetan Plateau's influence on E-SA's climate is well-established although not well-defined: consecutive episodes of tectonic-uplift correspond with an increase in monsoon rainfall, regional changes in monsoon weather patterns, and/or a decoupling between summer and winter monsoon intensities. Models favoring one or the other vary depending on the proxies used, but remain resolute on the significance of topographic growth on climate change in E-SA (Prell and Kutzbach 1992; Fluteau et al. 1999; Zhisheng et al. 2001; Steppuhn et al. 2006; Micheels et al. 2007; Jacques et al. 2011). Different scenarios for the uplift of the Tibetan Plateau have been put forth (Chung et al. 1998; Sun and Liu 2000), more recently a model of three punctuated height increases during the Miocene and Pliocene has gained support. It proposes an initial increase in elevation at ~15 mya, quickly followed by a second height increase at 13 mya, and a final increase during the Pliocene at 3 mya (Harris 2006; Wang et al. 2008). While the plateau rose to about half of its current elevation by the end of the Miocene, Lancang, in China's Yunnan province, underwent an uplift of 672–1,263 m since the Pliocene, and most notably since 3 mya (Jacques et al. 2011).

Interestingly, the Tibetan Plateau's first proposed height increase coincides neatly with climate changes comprehensively identified as the middle Miocene climatic transition (Flower and Kennett 1994; Zachos et al. 2001; Shevenell et al. 2004). The growth of the Tibetan plateau marked the onset of climate changes in E-SA Asia, and perhaps even more globally, which overall led to a rather abrupt high-latitude cooling during the middle Miocene. The subsequent height increase at 13 mya probably accelerated the ongoing global climatic changes of cooling and aridification, which promoted, for example, a global expansion of open habitats, grasslands, and consequently the emergence and spread of C_4 grasses (Kürschner et al. 2008; Eronen et al. 2009). At the same time, the region's increasingly seasonal climate led to fragmentation and contraction of tropical forests to much further southern latitudes, except perhaps in the sheltered, steep valleys of the southern Himalayas, where tropical forests may have persisted for much longer due to specific microclimatic conditions.

Asia's Miocene-Pliocene monsoon climate was characterized by warmer summers than today and by a clear seasonality in precipitation and temperature (Jacques et al. 2011). It seems unquestionable that the regions unique and complex monsoon climate has played a central role in the evolution of E-SA's flora and fauna due to a profoundly seasonal weather pattern characterized by a distinct summer and winter monsoon climate (Jacques et al. 2011). In summer, the weather is driven by two major convective heat sources: one over the Bay of Bengal from the Indian Ocean to the Arabian sea, and the other over the South China and the Philippines seas, respectively (Wang and Fan 1999). The two systems, which show poor inter-annual correlations (Wang and Fan 1999), produce the Southeast Asian summer monsoon and the East Asian winter monsoon, respectively (Jacques et al. 2011). Both systems result in high precipitation over parts of E-SA and China in summer, while arid conditions prevail in winter across all of Asia (Liu and Yin 2002).

China's Yunnan province, which harbors the southern Himalayas and Hengduan Mountains around where hylobatid evolution is hypothesized to have originated (Chatterjee 2006, 2009; Thinh et al. 2010a; Reichard et al. 2016), is impacted by the Southeast Asian summer and the East Asian winter monsoons (Jacques et al. 2011). Interestingly, the Ailao Mountains, where black crested gibbons (*Nomascus concolor*) are found today (Jiang et al. 2006), mark the boundary between East and West Yunnan (Yang et al. 2007). Eastern Yunnan is primarily under the influence of the East Asian monsoon, while Western Yunnan is strongly affected by the South Asian monsoon (Li and Li 1992). However, during the Miocene the two monsoon systems seem to have had an equally strong influence on Yunnan's weather as reflected in the flora of Lincang and Xialongtan (Jacques et al. 2011).

In conclusion, when considering Asia's climate it seems that from the beginning of their independent evolution (Thinh et al. 2010a), hylobatids were confronted with having to adapt to increasing marked seasonality. Despite these climatic and subsequent ecological challenges, the small Asian apes prospered and successfully radiated after 10 mya (Chatterjee 2006, 2009, 2016) and by ~6–7 mya had diversified into four genera (Thinh et al. 2010a; Table 5.1), during times when most Asian stem hominoids had disappeared. Reichard and colleagues (2016) have proposed three key hylobatid adaptations that allowed them to master the challenging climate of the middle to late Miocene: reduction in body size, reduction in group size, and perfection of forelimb suspensory locomotion. These adaptations characterize all extant hylobatids, thus they may have emerged already in stem hylobatids, before the emergence of the four distinct genera. Asia's monsoon pattern has not changed significantly since the Pliocene, and neither have hylobatids. Subsequent diversification is therefore probably less a direct response to monsoon rainfall patterns, but rather a product of global warming and cooling periods that led to changing sea levels with corresponding paleogeographic changes coast lines and rivers.

Sea-Level Changes and Sundaland's Paleorivers

Although the basic geography of E-SA continental outline and mountains were in place and relatively stable since the Miocene ~ 20 mya, rivers, shorelines, and the waxing and waning of continental islands, have changed repeatedly, sometimes dramatically, in response to climate and associated sea-level changes and the region's many shallow seas (Woodruff 2010; Corlett 2014).

After a late Cretaceous sea-level highstand, a gradual global sea-level decline began that has continued throughout the Cenozoic until today (Haq et al. 1987). During the early Eocene, major high-latitude ice sheets were still absent and the sea-level of E-SA was ~ 50–70 m higher than today (Kominz et al. 2008; Corlett 2014). While the geologic formation of E-SA was still incomplete during Eocene times, because the Indian and Australian plates had not yet joined together, it is likely that large parts of E-SA were inundated and islands were the regions defining geologic feature. During the middle Eocene a shallow seaway formed between Antarctica and South America, foreshadowing the formation of the deep Drake Passage that became established by 34–30 mya (Livermore et al. 2007). Through the Drake Passage, deep currents and oceanographic fronts started flowing eastward causing changes to the Southern Ocean circulations, which likely contributed to the Eocene cooling trend as well as growth of the Antarctic ice sheets. By the earliest Oligocene, when the Antarctic Circumpolar Current (ACC) had become established and Earth's ice volume had increased markedly (Miller et al. 1991, 2005; Zachos et al. 2001), global sea-level declined to 20–30 m above present (Corlett 2014). The growing Antarctic ice sheet remained a forcing agent of ocean circulation (Miller et al. 2005), and when Antarctica was fully covered in ice, absorbing the ocean's waters around ~ 23 mya, the global sea-level dropped to $-15/-25$ m of today's level, but only to spike back to 40 m above today's at 22 mya. These variations in sea-levels, however, all predate the independent evolution of hylobatids.

Following high sea-level stands during the early Miocene, sea-levels of the middle Miocene fell and oscillated around ± 10 m of today's sea-level (Kominz et al. 2008). At 15–14 mya, however, the data of Kominz et al. (2008) indicate a sea-level highstand of ~ 30 m above present, which coincides with the warm-moist conditions of the middle Miocene climatic optimum (Flower and Kennett 1994). It is unclear how significant a role changing sea-levels played on the earliest hylobatid evolution, following their proposed separation from stem hominoids at ~ 16.26 mya (Thinh et al. 2010a), other than its indirect effect on Asia's monsoon climate relative to global cooling and warming episodes. This is because the family's geographic origin has tentatively been placed at much Northern latitudes in non-coastal, mountainous areas (Reichard et al. 2016), and the first significant radiation within the hylobatid family has been hypothesized to have occurred not earlier than 10.5 mya (Chatterjee 2006).

During the middle Miocene transition, when temperatures decreased and more ocean water became trapped in ice sheets, sea-levels are suggested to have dropped to approximate that of present day (Holbourn et al. 2005; Kominz et al. 2008).

Shevenell et al. (2008) have suggested a stepwise orbitally paced cooling process for the middle Miocene between ∼15 and 13.8 mya. Interestingly, ice sheet growth (leading to the lower sea level), is hypothesized to begin during the warmest period of the Neogene, based on a poleward heat/moisture supply (Shevenell et al. 2008) and colder bottom water temperatures. This is indicative of generally cooler global temperature, which has been recognized at ∼16 mya (4.7 ± 1.0 °C) and 13.6 mya (4.6 ± 0.7 °C). These periods correspond with intervals of more positive $\delta^{18}O$ stages (Miller et al. 1991). Across the middle Miocene, Shevenell et al. (2008) observed $\delta^{18}O$ increases (14.2–13.8 mya) corresponding to a cooling of ∼2 ± 1.5 °C (∼6°–4 °C), which is similar to estimated cooling using indirect methods (Miller et al. 1991; Wright et al. 1992; Flower and Kennett 1994; John et al. 2004). However, within the general cooling of the middle Miocene, global temperatures also increased: at 16.4–16.2 mya (6.3 ± 0.3 °C), at 14.5–14.1 mya (5.6 ± 1.0 °C), and at 13.5–13.3 mya (5.7 ± 0.7 °C). Two of these warm intervals, which led to small sea-level rises of 15 and 22 m, respectively, above the present level (Miller et al. 2005; Kominz et al. 2008), fell within the Miocene climatic optimum of 17–14 mya (Flower and Kennett 1994). The third at ∼13.5 mya occurred just after a globally recognized climate cooling step during a time of inferred low atmospheric pCO_2 (Vincent and Berger 1985; Pagani et al. 1999; Holbourn et al. 2005) and coincided with the second uplifting of the Tibetan Plateau (Wang et al. 2008). At 12.6 mya, the effects of continuous cooling become evident with the formation of Northern Hemisphere ice sheets (Fronval and Jansen 1997). Generally, it seems the late Miocene period was a period of relative climatic stability despite minor cooling episodes between 11.7 and 8.8 mya (Westerhold et al. 2005) corresponded with a 10 m drop in sea-level (Westerhold et al. 2005).

During the late Miocene between 7 and 5 mya, ice sheets were building up on West Antarctica as evidenced by general increase in $\delta^{18}O$ (Miller and Fairbanks 1985; Kennett 1995). Around 6 mya, a cooling period had a significant impact on the Mediterranean basin, which dried up during the Messinian sea-level lowstand, and led to the salinity crisis of 5.8–5.3 mya (Haq et al. 1980; Hodell et al. 1994; Bernor and Lipscomb 1995; Aifa et al. 2003; Garcia et al. 2004). By this time, however, the four hylobatid genera had emerged with the last divergence of *Hoolock* at 6.69 mya (Thinh et al. 2010a), which predated the harsher climatic conditions of the latest Miocene.

The Pliocene of 5–4 mya showed a gradual warming (Haq et al. 1987) which resulted in 2–3 °C increase in temperature, and led to a loss of large areas of ice sheets and thus a sea-level increase to about 22 ± 10 m higher than today (Haywood et al. 2009; Naish and Wilson 2009; Woodruff 2010). The Pliocene warming and corresponding sea-level highstands coincide with the first wave of hylobatid speciation within the genera *Nomascus* and *Hylobates*. By and large, speciation within *Nomascus* was most likely confined to Northern latitudes of present day Yunnan province and parts of Vietnam. And the proposed emergence of *Nomascus hainanus*, the Hainan gibbon, at 3.25 mya (Thinh et al. 2010a) may be causally linked to a high sea level, because as sea levels climbed, Hainan became an island that isolated Hainan gibbons from mainland populations. Speciation also occurred within the genus *Hylobates*, specifically, the emergence of Kloss gibbons at 3.91 mya as well as the separation of Bornean gibbons at 3.65 mya and Javan gibbons by about 3.29 mya (Thinh et al. 2010a), which may be attributed to geographic isolation events caused by island formation. Although Sundaland was a continuous, large landmass during some of the Pleistocene, it is possible that the origin of Hainan, Bornean, Javan and Mentawai gibbons all fall into the warm, late Pliocene period of relatively higher sea-levels when these areas existed as islands.

The history of the Mentawai islands differs from other E-SA islands, because of a much longer separation time from other landmasses. The Mentawai islands are separated from Sumatra by a 1500 m deep trench (Wilting et al. 2012), which has existed at least since the middle Pleistocene (Batchelor 1979; Dring et al. 1990), but probably longer. Based on subsidences of Plio-Pleistocene origin, Izart et al. (1994) concluded that a land connection between the Mentawai Islands and Sumatra during the late Neogene was unlikely, including periods when sea-levels were much lower than today and when E-SA landmasses and large islands were joined together forming Sundaland (Voris 2000; Whitten et al. 2000). These observations are consistent with the hypothesized early isolation of Kloss gibbons on the Mentawai Islands during the Pliocene. Perhaps only the northernmost of the Mentawai Islands, Siberut, may subsequently have been linked to Sumatra through the Batu Islands (Batchelor 1979; Dring et al. 1990), given that the Siberut Strait is only ∼55 km wide today (Wilting et al. 2012). However, even during the Pleistocene glacial maxima, when sea-levels dropped up to 125 m relative to present sea-level (Rohling et al. 1998; Hanebuth et al. 2000), the sea passage between Siberut and Sumatra would have been approximately 10 km. Hypothetically, a discontinuous, indirect Siberut-Sumatra land connection may have existed (Meijaard 2003), but given hylobatids general aversion to open water and reliance on tropical forests, it seems unlikely that Kloss gibbon's could have crossed into Siberut at that time.

Although Earth's temperature continued to slightly decrease at the beginning of the Pleistocene, the incipit of strong glacial-interglacial cycles can be traced to a major cooling step at 2.75 mya followed by the first 'deep' glacial (i.e., a sea-level 70 m below present day) event at 2.15 mya (Foster and Rohling 2013). This chronology clearly postdates the suggested first hylobatid speciation wave, but only slightly predates the second speciation wave within the genera *Nomascus* and *Hylobates* beginning at 1.74 mya (Thinh et al. 2010a). This second speciation wave included the divergence of Western and Eastern Hoolock gibbons at 1.4 mya (Thinh et al. 2010a; Table 5.1). It also included divergences within Bornean gibbons. We suggest the second hylobatid diversification wave was unrelated to sea-level changes, and primarily represents speciation events in response to the formation of large rivers: the paleorivers of Sundaland. During sea-level lowstands the shallow Sunda shelf emerged forming a large, continuous landmass that became divided into geographic sections by paleorivers, which became barriers to hylobatids. Chatterjee (Chatterjee 2009) has proposed that western Hoolock (*Hoolock hoolock*) has been restricted in the North by the Brahmaputra and Salween rivers, to the South by the Brahmaputra, and to the East by the Dibang and Chidwin rivers, while the eastern Hoolock (*Hoolock leuconedys*) is found east of the Chidwin river. In the headwaters of the Irrawaddy river, outside the Gaoligong Shan mountains, separation of the two species by water ways is unclear. In fact, a population of eastern Hoolocks is found north of the western Hoolock distribution, perhaps because of the generally lower elevation where the mountain ranges begin to rise and where the Irrawaddy and Brahmaputra rivers are fed by smaller tributaries, Hoolocks had been able to cross. A second significant geographic separation is related to the southern margins of the Himalayan mountains, where western black crested gibbons (*Nomascus concolor*) are still found deeper in the steep Salween-Mekong river valleys and mountain slopes, perhaps because a favorable microclimate allows for such unusual northern latitudinal existence. Western black crested gibbons were probably prevented from radiating further westward by the far southward reaching Gaoligong Shan mountains, while at northern latitudes they are found to the East on both sides of the Mekong, Red and Black rivers (Thinh et al. 2010c). However, given that western black crested gibbons in northern latitudes occur on both sides of the Mekong and Black rivers, where mountain ranges are already lower and the rivers may not yet have presented an impassable barrier, it seems plausible that historically black crested gibbons continued much further south on both sides of the Mekong and Black rivers, because even today a small population of black crested gibbons persists in isolation on the eastern side of the Mekong where otherwise only northern white-cheeked gibbons (*Nomsascus leucogenys*) are found (Thinh et al. 2010c). It is also possible that the northern most distributed white-handed gibbon subspecies (*Hylobates lar yunnanensis*) once pushed crested gibbons more northward to their current location. Finally, the rivers of Borneo that separate agile gibbons from Müller's gibbons and white-bearded gibbons may have formed during the Pleistocene, and similarly some speciation events within the Laotian crested gibbons have been traced to the formation of paleorivers during this epoch (Thinh et al. 2010b). Today, river continue to be formidable barriers that separate many hylobatid species and subspecies.

Conclusion

Hylobatid evolution, as seen in paleogeographic and paleoclimatological contexts, can be summarized by highlighting a few salient features. The middle Miocene climatic optimum, a warm-humid, primate-friendly time when tropical forests had extended northward, coincides with the emergence of the hylobatid lineage in E-SA. However, subsequent climate changes forced stem hylobatids to quickly adapt to a more seasonal environment. The successful mastering of these early challenges is partially reflected in the emergence of the four current hylobatid genera, which can broadly be linked to a period of more favorable climatic conditions around 8 mya. The proposed geographic origin of hylobatids at northern latitudes, in the Southern Himalayas and deep mountain valleys of the ancient Hengduan may have sheltered hylobatids from the climatic perturbations of the Miocene and thus contributed significantly to the divergence of hylobatid genera into separate populations based on these geographic barriers. The first wave of speciation within the genera *Nomascus* and *Hylobates* can again be linked to a warmer period in the Pliocene, which may have lead to the isolation of several hylobatid populations on islands where they evolved into separate species, for example, the Mentawai, Bornean, and Hainan gibbons. The second wave of hylobatid speciation was most likely related to populations becoming separated by large paleorivers at a time when sea-level was relatively low and most of E-SA was a large, more or less connected landmass.

Although it is possible to contextualize some events in hylobatid evolution with climatic and geographic events, many open questions remain, and solid evidence for most propositions about causal relationships of the abiotic and biotic evolution of hylobatids is still absent. Since no early hylobatid fossils, other than *Yuanmoupithecus,* have been found so far, the time from the postulated emergence of the family at the beginning of the middle Miocene to the end of the late Miocene, or ~8 mya, is highly uncertain. This is the time when proposed adaptations of small body size, small group size, and the perfection of brachiation are postulated to have allowed hylobatids to flourish. However, no fossils yet exist to corroborate or challenge these assumptions. Similarly, it is unclear what happened within hylobatid genera between 7 mya and 4.5 mya, when a second rapid speciation and radiation wave began, because no large-scale speciation events coincide with the 7–4.5 mya time period. The hylobatids are the most diverse of today's ape lineages. They are distributed over a much wider geographic range than their large-bodied great ape cousins and have shown unparalleled flexibility in their ability adapt to climate change through time. Yet, hylobatids continue to be a largely misunderstood, underused primate model in studies of human evolution.

References

Aifa T, Feinberg H, Derder EM (2003) Magnetostratigraphic constraints on the duration of the marine communications interruption in the Western Mediterranean during the upper Messinian. Geodiversitas 25:617–631

Ali JR, Aitchison JC (2008) Gondwana to Asia: plate tectonics, paleogeography and the biological connectivity of the Indian sub-continent from the Middle Jurassic through latest Eocene (166–35 Ma). Earth-Sci Rev 88:145–166

Andrews P, Kelley J (2007) Middle Miocene dispersals of apes. Folia Primatol 78:328–343

Bard E (2001) Comparison of alkenone estimates with other paleotemperature proxies. Geochem Geophys Geosy 2:1002

Batchelor BC (1979) Geological characteristics of certain coastal and offshore placers as essential guides for tin exploration in Sundaland. SE Asia Geol Soc Malaysia Bull 11:283–313

Benton MJ (2010) The origins of modern biodiversity on land. Phil Trans R Soc 365:3667–3679

Bernor RL, Lipscomb D (1995) A consideration of old world Hipparionine horse phylogeny and global a biotic processes. In: Vrba ES, Denton GH, Partridge TC, Burckle LH (eds) Paleoclimate and evolution with emphasis on human origins. Yale University Press, New Haven, Connecticut, p 164

Bininda-Emonds OR, Cardillo M, Jones KE, MacPhee RD, Beck RM, Grenyer R et al (2007) The delayed rise of present-day mammals. Nature 446:507–512

Bird MI, Taylor D, Hunt C (2005) Palaeo environments of insular Southeast Asia during the Last Glacial Period: a savanna corridor in Sundaland? Quat Sci Rev 24:2228–2242

Buerki S, Forest F, Alvarez N (2014) Proto-South-East Asia as a trigger of early angiosperm diversification. Bot J Linn Soc 174:326–333

Cannon CH, Morley RJ, Bushe ABG (2009) The current refugial rainforests of Sundaland are unrepresentative of their biogeographic past and highly vulnerable to disturbance. Proc Natl Acad Sci 109:11188–11193

Carbone L, Harris RA, Gnerre S, Veeramah KR, Lorente-Galdos B, Huddleston J et al (2014) Gibbon genome and the fast karyotype evolution of small apes. Nature 513:195–201

Cartmill M, Lemelin P, Schmitt D (2007) Understanding the adaptive value of diagonal-sequence gaits in primates: a comment on Shapiro and Raichlen 2005. Am J Phys Anthropol 133:822–825

Chaplin G (2005) Physical geography of the Gaoligong Shan area of southwest China in relation to biodiversity. Proc Calif Acad Sci 56(27/37):527

Chatterjee HJ (2006) Phylogeny and biogeography of gibbons: a dispersal-vicariance analysis. Int J Primatol 27:699–712

Chatterjee HJ (2009) Evolutionary relationships among the hylobatids: a biogeographic perspective. In: Lappan S, Whittaker DJ (eds) The hylobatids: new perspectives on small ape socioecology and population biology. Springer, New York, pp 13–36

Chatterjee HJ (2016) The role of historical and fossil records in predictingchanges in the spatial distribution of hylobatids. In: Reichard UH, Hirohisa H, Barelli C (eds) Evolution of gibbons and siamang. Springer, New York, pp 43–54

Chung SL, Lo CH, Lee TY, Zhang Y, Xie Y, Li X et al (1998) Diachronous uplift of the Tibetan Plateau starting 40 Myr ago. Nature 394:769–773

Clark PU, Alley RB, Pollard D (1999) Northern Hemisphere ice-sheet influences on global climate change. Science 286:1104–1111

Clark PU, Dyke AS, Shakun JD, Carlson AE, Clark J, Wohlfarth B et al (2009) The last glacial maximum. Science 325:710–714

Clift PD, Hodges KV, Heslop D, Hannigan R, Van Long H, Calves G (2008) Correlation of Himalayan exhumation rates and Asian monsoon intensity. Nat Geosci 1:875–880

Corlett RT (ed) (2014) The ecology of tropical East Asia, 2nd edn. Oxford University Press, Oxford

Crepet WL (2000) Progress in understanding angiosperm history, success, and relationships: Darwin's abominably 'perplexing phenomenon'. Proc Natl Acad Sci 97:12939–12941

Cronquist A (ed) (1988) The evolution and classification of flowering plants. Columbia University Press, New York

Dagosto M (2007) The postcranial morphotype of primates. Primate origins: adaptations and evolution. Springer, New York, pp 489–534

Dilcher D (2000) Toward a new synthesis: major evolutionary trends in the angiosperm fossil record. Proc Natl Acad Sci 97:7030–7036

Ding Z, Liu T, Rutter NW, Yu Z, Guo Z, Zhu R (1995) Ice-volume forcing of East Asian winter monsoon variations in the past 800,000 years. Quat Res 44:149–159

Dring J, McCarthy C, Whitten A (1990) The terrestrial herpetofauna of the Mentawai Islands, Indonesia. Indo-Malayan Zool 6:119–132

Eriksson O, Friis EM, Löfgren P (2000) Seed size, fruit size, and dispersal systems in angiosperms from the Early Cretaceous to the Late Tertiary. Am Nat 156:47–58

Eronen JT, Ataabadi MM, Micheels A, Karme A, Bernor RL, Fortelius M (2009) Distribution history and climatic controls of the Late Miocene Pikermian chronofauna. Proc Natl Acad Sci 106:11867–11871

Fleagle JG (ed) (2013) Primate adaptation and evolution, 3rd edn. Academic Press Elsevier, Amsterdam

Flohn H (1968) Contributions to a meteorology of the Tibetan Highlands. J Atmos Sci 130

Flower BP, Kennett JP (1994) The middle Miocene climatic transition: east Antarctica ice sheet development, deep ocean circulation and global carbon cycling. Paleogeogr Paleoclimatol Paelaeocol 108:537–555

Fluteau F, Ramstein G, Besse J (1999) Simulating the evolution of the Asian and African monsoons during the past 30 Myr using an atmospheric general circulation model. J Geophys Res: Atmospheres 104:11995–12018

Foster GL, Rohling EJ (2013) Relationship between sea level and climate forcing by CO_2 on geological timescales. Proc Natl Acad Sci 110:1209–1214

Fronval T, Jansen E (1997) Eemian and early Weichselian (140–60 ka) paleoceanography and paleoclimate in the Nordic seas with comparisons to Holocene conditions. Paleoceanography 12:443–462

Garcia F, Conesa G, Munch P, Cornee JJ, Saint Martin JP, Andre JP (2004) Evolution of Melilla-Nador basin (NE Morocco) littoral environments during late Messinian between 6.0 and 5.77 Ma. Geobios 37:23–36

Garzione CN, Ikari MJ, Basu AR (2005) Source of Oligocene to Pliocene sedimentary rocks in the Linxia basin in northeastern Tibet from Nd isotopes: Implications for tectonic forcing of climate. Geol Soc Am Bull 117:1156–1166

Geissmann T (2007) Status reassessment of the gibbons: Results of the Asian Primate Red List Workshop 2006. Gibbon J 3:5–15

Grimaldi D (1999) The co-radiations of pollinating insects and angiosperms in the Cretaceous. Ann Miss Bot Gard 86:373–406

Guo ZT, Ruddiman WF, Hao QZ, Wu HB, Qiao YS, Zhu RX et al (2002) Onset of Asian desertification by 22 Myr ago inferred from loess deposits in China. Nature 416:159–163

Hall R (1997) Cenozoic plate tectonic reconstructions of SE Asia. Geol Soc 126:11–23

Hanebuth T, Stattegger K, Grootes PM (2000) Rapid flooding of the Sunda Shelf: a late-glacial sea-level record. Science 288:1033–1035

Haq BU, Worsley TR, Burckle LH, Douglas RG, Keigwin LD, Opdyke ND et al (1980) Late Miocene marine carbon-isotopic shift and synchroneity of some phytoplanktonic biostratigraphic events. Geology 8:427–431

Haq BU, Hardenbol J, Vail PR (1987) Chronology of fluctuating sea levels since the Triassic. Science 235:1156–1167

Harris N (2006) The elevation history of the Tibetan Plateau and its implication for the Asian monsoon. Palaeogeogr Palaeoclimatol Palaeoecol 241:4–15

Harrison T, Ji X, Zheng L (2008) Renewed investigations at the late Miocene hominoid locality of Leilao, Yunnan China. Am J Phys Anthropol 135(S46):113

Harzhauser M (2007) Oligocene and Aquitanian gastropod faunas from the Sultanate of Oman and their biogeographic implications for the early western Indo-Pacific. Palaeontographica 280:75–121

Harzhauser M, Piller WE (2007) Benchmark data of a changing sea—Palaeogeography, Palaeobiogeography and events in the Central Paratethys during the Miocene. Palaeogeogr Palaeoclimatol Palaeoecol 253:8–31

Harzhauser M, Piller WE, Steininger FF (2002) Circum-Mediterranean Oligo-Miocene biogeographic evolution—The gastropods' point of view. Palaeogeogr Palaeoclimatol Palaeoecol 183:103–133

Harzhauser M, Kroh A, Mandic O, Piller WE, Gröhlich U, Reuter M, Berning B (2007) Biogeographic responses to geodynamics: a key study all around the Oligo-Miocene Tethyan Seaway. Zool Anz 246:241–256

Hay WW (1996) Tectonics and climate. Geol Rundsch 85:409–437

Haywood AM, Dowsett HJ, Valdes PJ, Lunt DJ, Francis JE, Sellwood BW (2009) Introduction. Pliocene climate, processes and problems. Philos Trans R Soc A 367:3–17

Hedges SB, Parker PH, Sibley CG, Kumar S (1996) Continental breakup and the ordinal diversification of birds and mammals. Nature 381:226–229

Heller NE, Zavaleta ES (2009) Biodiversity management in the face of climate change: A review of 22 years of recommendations. Biol Conserv 142:14–23

Hodell DA, Benson RH, Kent DV, Boersma A, Bied RE (1994) Magnetostratigraphic, biostratigraphic, and stable isotope stratigraphy of an Upper Miocene drill core from the Salé Briqueterie (northwestern Morocco): a high-resolution chronology for the Messinian stage. Paleoceanography 9:835–855

Holbourn A, Kuhnt W, Schulz M, Erlenkeuser H (2005) Impacts of orbital forcing and atmospheric carbon dioxide on Miocene ice-sheet expansion. Nature 438:483–487

Izart A, Kemal BM, Malod JA (1994) Seismic stratigraphy and subsidence evolution of the northwest Sumatra fore-arc basin. Mar Geol 122:109–124

Jablonski NG (2005) Primate homeland: forests and the evolution of primates during the Tertiary and Quaternary in Asia. Anthropol Sci 113:117–122

Jablonski NG, Chaplin G (2009) The fossil record of gibbons. In: Lappan SM, Whittaker D (eds) The gibbons: new perspectives on small ape socioecology and population biology. Springer, Berlin, pp 111–130

Jablonski NG, Su DF, Flynn LJ, Ji X, Deng C, Kelley J et al (2014) The site of Shuitangba (Yunnan, China) preserves a unique terminal Miocene fauna. J Vertebr Paleontol 34:1251–1257

Jacques FM, Guo SX, Su T, Xing YW, Huang YJ, Liu YS et al (2011) Quantitative reconstruction of the Late Miocene monsoon climates of southwest China: a case study of the Lincang flora from Yunnan Province. Palaeogeogr Palaeoclimatol Palaeoecol 304:318–327

Ji XP, Jablonski NG, Su DF, Deng CL, Flynn LJ, You YS, Kelley J (2013) Juvenile hominoid cranium from the terminal Miocene of Yunnan, China. Chin Sci Bull 58:3771–3779

Jiang X, Luo Z, Zhao S, Li R, Liu C (2006) Status and distribution pattern of black crested gibbon (Nomascus concolor jingdongensis) in Wuliang Mountains, Yunnan, China: implication for conservation. Primates 47:264–271

John CM, Karner GD, Mutti M (2004) $\delta^{18}O$ and Marion Plateau backstripping: combining two approaches to constrain late middle Miocene eustatic amplitude. Geology 32:829–832

Kennett J (1995) A review of polar climatic evolution during the Neogene based on the marine sediment record. In: Vrba ES, Denton GH, Partridge TC, Burckle LH (eds) Paleoclimate and evolution with emphasis on human origins. Yale University Press, New Haven Connecticut, pp 49–64

Kennett JP, Hodell DA (1993) Evidence for relative climatic stability of Antarctica during the early Pliocene: a marine perspective. Geogr Ann 75A:202–222

Kominz MA, Browning JV, Miller KG, Sugarman PJ, Mizintseva S, Scotese CR (2008) Late Cretaceous to Miocene sea-level estimates from the New Jersey and Delaware coastal plain coreholes: an error analysis. Basin Res 20:211–226

Kürschner WM, Kvaček Z, Dilcher DL (2008) The impact of Miocene atmospheric carbon dioxide fluctuations on climate and the evolution of terrestrial ecosystems. Proc Natl Acad Sci 105:449–453

Lewis AR, Marchant DR, Ashworth AC, Hedenäs L, Hemming SR, Johnson JV et al (2008) Mid-Miocene cooling and the extinction of tundra in continental Antarctica. Proc Natl Acad Sci 105:10676–10680

Li Y, Zhai S-N, Qiu Y-X, Guo Y-P, Ge X-J, Comes HP (2011) Glacial survival east and west of the 'Mekong–Salween Divide' in the Himalaya-Hengduan Mountains region as revealed by AFLPs and cpDNA sequence variation in Sinopodophyllum hexandrum (Berberidaceae). Mol Phylogenet Evol 59:412–424

Li XW, Li J (1992) On the validity of Tanaka Line and its significance viewed from the distribution of Eastern Asiatic genera in Yunnan. Acta Bot Yunnanica 14:1–12

Lin T, Lo C, Chung S, Lee T, Lee H, Hsu F (2005) New geochronological constraints on the movement of Jiali and Gaoligong shear zones in SE Tibet, and its tectonic implication. Am Geophys Union, Western Pacific Geophysical Meeting, Abstracts 85:28

Liu X, Yin ZY (2002) Sensitivity of East Asian monsoon climate to the uplift of the Tibetan Plateau. Palaeogeogr Palaeoclimatol Palaeoecol 183:223–245

Livermore R, Hillenbrand CD, Meredith M, Eagles G (2007) Drake Passage and Cenozoic climate: An open and shut case? Geochem Geophys Geosy 8:1–11

Lloyd GT, Davis KE, Pisani D, Tarver JE, Ruta M, Sakamoto M et al (2008) Dinosaurs and the Cretaceous terrestrial revolution. Proc R Soc Lon B 275:2483–2490

Lucas PW, Dominy NJ, Osorio D, Peterson-Pereira W, Riba-Hernandez P, Solis-Madrigal S, Stoner KE, Yamashita N (2007) Perspectives on primate color vision. In: Ravosa MJ, Dagosto M (eds) Primate origins: adaptations and evolution. Springer, New York, pp 805–819

Magallón S, Castillo A (2009) Angiosperm diversification through time. Am J Bot 96:349–365

Marshall JT, Sugardjito J (1986) Gibbon systematics. In: Swindler DR, Erwin J (eds) Comparative primate biology, vol 1, systematics. Alan R Liss, New York, pp 137–185

McQuarrie N, van Hinsbergen DJJ (2013) Retrodeforming the Arabia-Eurasia collision zone: Age of collision versus magnitude of continental subduction. Geology 41:315–318

Meijaard E (2003) Mammals of south-east Asian islands and their Late Pleistocene environments. J Biogeogr 30:1245–1257

Metcalfe I (1996) Gondwanaland dispersion, Asian accretion and evolution of eastern Tethys. Aust J Earth Sci 43:605–623

Metcalfe I (2011) Tectonic framework and Phanerozoic evolution of Sundaland. Gondwana Res 19:3–21

Metcalfe I (2013a) Gondwana dispersion and Asian accretion: tectonic and palaeogeographic evolution of Eastern Tethys. J Asian Earth Sci 66:1–33

Metcalfe I (2013b) Asia: South-East. In: Selley RC, Cocks LRM, Plimer IR (eds) Encyclopedia of Geology. Elsevier, Amsterdam

Micheels A, Bruch AA, Uhl D, Utescher T, Mosbrugger V (2007) A Late Miocene climate model simulation with ECHAM4/ML and its quantitative validation with terrestrial proxy data. Palaeogeogr Palaeoclimatol Palaeoecol 253:251–270

Miller KG, Fairbanks RG (1985) Oligocene to Miocene carbon isotope cycles and abyssal circulation changes. In: Sundquist ET, Broecker WS (eds) The Carbon Cycle and Atmospheric CO: natural variations Archean to present. Tarpon Springs, Washington, pp 469–486

Miller KG, Wright JD, Fairbanks RG (1991) Unlocking the ice house: Oligocene-Miocene oxygen isotopes, eustasy, and margin erosion. J Geophys Res Solid Earth 96:6829–6848

Miller KG, Kominz MA, Browning JV, Wright JD, Mountain GS, Katz ME et al (2005) The Phanerozoic record of global sea-level change. Science 310:1293–1298

Milton K (ed) (1980) The foraging strategies of howler monkeys. Columbia University Press, New York

Molengraaff GAF (1921) Modern deep-sea research in the East Indian Archipelago. Geogr J 57:95–121

Morley RJ (ed) (2000) Origin and evolution of tropical rain forests. Wiley, New York

Morley RJ (2011) Cretaceous and Tertiary climate change and the past distribution of megathermal rainforests. In: Dow K, Dowling TE (eds) Tropical rainforest responses to climatic change. Springer, Berlin, pp 1–34

Myers N, Mittermeier RA, Mittermeier CG, da Fonseca GAB, Kent J (2000) Biodiversity hotspots for conservation priorities. Nature 403:853–858

Nábělek J, Hetényi G, Vergne J, Sapkota S, Kafle B, Jiang M et al (2009) Underplating in the Himalaya-Tibet collision zone revealed by the Hi-CLIMB experiment. Science 325:1371–1374

Naish TR, Wilson GS (2009) Constraints on the amplitude of mid-Pliocene (3.6–2.4 Ma) eustatic sea-level fluctuations from the New Zealand shallow-marine sediment record. Philos Trans R Soc A 367:169–187

Ohlwein C, Wahl ER (2012) Review of probabilistic pollen-climate transfer methods. Quat Sci Rev 31:17e29

Okay AI, Zattin M, Cavazza W (2010) Apatite fission-track data for the Miocene Arabian-Eurasian collision. Geology 38:35–38

Pagani M, Freeman KH, Arthur MA (1999) Late Miocene atmospheric CO_2 concentrations and the expansion of C_4 grasses. Science 285:876–879

Patnaik R, Chauhan P (2009) India at the cross-roads of human evolution. J Biosci 34:729–747

Pelejero C, Kienast M, Wang L, Grimalt JO (1999) The flooding of Sundaland during the last deglaciation: imprints in hemipelagic sediments from the southern South China Sea. Earth Planet Sci Lett 171:661–671

Prell WL, Kutzbach JE (1992) Sensitivity of the Indian Monsoon to forcing parameters and implications for its evolution. Nature 360:647–652

Pollard D, DeConto RM (2009) Modelling West Antarctic ice sheet growth and collapse through the past five million years. Nature 458:329–332

Popov SV, Rögl F, Rozanov AY, Steininger FF, Shcherba IG, Kováè M (2004) Lithological–paleogeographic maps of paratethys. 10 Maps Late Eocene to Pliocene. Cour Forsch-Inst Senckenberg 250:1–46

Reichard UH, Barelli C, Hirai H, Nowak G (2016) The evolution of gibbons and siamang. In: Reichard UH, Hirohisa H, Barelli C (eds) Evolution of gibbons and siamang. Springer, New York, pp 3–41

Rögl F (1998) Palaeogeographic considerations for Mediterranean and Paratethys seaways (Oligocene to Miocene). Ann Naturhistor Mus Wien 99:279–310

Rögl F (1999a) Circum-Mediterranean Miocene paleogeography. In: Rössner G, Heissig K (eds) The Miocene land mammals of Europe. Fritz Pfeil Verlag, Munich, Dr, pp 39–48

Rögl F (1999b) Mediterranean and Paratethys. Facts and hypotheses of an Oligocene to Miocene paleogeography (short overview). Geol Carpath 50:339–349

Rohling EJ, Fenton M, Jorissen FJ, Bertrand P, Ganssen G, Caulet JP (1998) Magnitudes of sea-level lowstands of the past 500,000 years. Nature 394:162–165

Sarnthein M, Pflaumann U, Wang PX, Wong HK (eds) (1994) Preliminary report of the Sonne-95 cruise 'Monitor Monsoon' to the South China Sea. Ber Rep 68 Geol-Paläontol Inst Univ Kiel, Germany

Seluchi ME, Norte FA, Satyamurty P, Chan Chou S (2003) Analysis of three situations of the foehn effect over the Andes (Zonda wind) using the Eta–CPTEC regional model. Weather Forecast 18:481–501

Shevenell AE, Kennett JP, Lea DW (2004) Middle Miocene Southern Ocean cooling and Antarctic cryosphere expansion. Science 305:1766–1770

Shevenell AE, Kennett JP, Lea DW (2008) Middle Miocene ice sheet dynamics, deep-sea temperatures, and carbon cycling: a Southern Ocean perspective. Geochem Geophys 9(2):1–14

Simons WJF, Socquet A, Vigny C, Ambrosius BAC, Haji Abu S, Promthong C et al (2007) A decade of GPS in Southeast Asia: resolving Sundaland motion and boundaries. J Geophys Res 112:B06420

Steiper ME, Young NM, Sukarna TY (2004) Genomic data support the hominoid slowdown and an early Oligocene estimate for the hominoid–cercopithecoid divergence. Proc Natl Acad Sci USA 101:17021–17026

Steppuhn A, Micheels A, Geiger G, Mosbrugger V (2006) Reconstructing the Late Miocene climate and oceanic heat flux using the AGCM ECHAM4 coupled to a mixed-layer ocean model with adjusted flux correction. Palaeogeogr Palaeoclimatol Palaeoecol 238:399–423

Sun J, Liu T (2000) Stratigraphic evidence for the uplift of the Tibetan Plateau between ~ 1.1 and ~ 0.9 myr ago. Quat Res 54:309–320

Sun J, Zhang Z (2008) Palynological evidence for the mid-Miocene climatic optimum recorded in Cenozoic sediments of the Tian Shan Range, northwestern China. Global Planet Change 64:53–68

Sun X, Wang P (2005) How old is the Asian monsoon system?—Palaeobotanical records from China. Palaeogeogr Palaeoclimatol Palaeoecol 222:181–222

Sussman RW, Tab Rasmussen D, Raven PH (2013) Rethinking primate origins again. Am J Primatol 75:95–106

Szalay FS (2007) Ancestral locomotor modes, placental mammals, and the origin of euprimates: lessons from history. In: Ravosa MJ, Dagosto M (eds) Primate Origins: adaptations and evolution. Springer, New York, pp 457–487

Tapponnier P, Zhiqin X, Roger F, Meyer B, Arnaud N, Wittlinger G, Jingsui Y (2001) Oblique stepwise rise and growth of the Tibet Plateau. Science 294:1671–1677

Tang Z, Ding Z, White PD, Dong X, Ji J, Jiang H, Luo P, Wang X (2011) Late Cenozoic central Asian drying inferred from a palynological record from the northern Tian Shan. Earth Planet Sci Lett 302:439–447

Thinh VN, Mootnick AR, Geissmann T, Li M, Ziegler T, Agil M et al (2010a) Mitochondrial evidence for multiple radiations in the evolutionary history of small apes. BMC Evol Biol 10:74

Thinh VN, Mootnick AR, Thanh VN, Nadler T, Roos C (2010b) A new species of crested gibbon, from the central Annamite mountain range. Vietn J Primatol 4:1–12

Thinh VN, Rawson B, Hallam C, Kenyon M, Nadler T, Walter L, Roos C (2010c) Phylogeny and distribution of crested gibbons (genus *Nomascus*) based on mitochondrial cytochrome b gene sequence data. Am J Primatol 72:1047–1054

Thomas CD, Cameron A, Green RE, Bakkenes M, Beaumont LJ, Collingham YC et al (2004) Extinction risk from climate change. Nature 427:145–148

Vincent E, Berger WH (1985) Carbon dioxide and polar cooling in the Miocene: The Monterey hypothesis. In: Sundquist ET, Broecker WS (eds) The Carbon Cycle and Atmospheric CO: natural variations Archean to present. Tarpon Springs, Washington, pp 455–468

Voris HK (2000) Maps of Pleistocene sea levels in Southeast Asia: shorelines, river systems and time durations. J Biogeogr 27:1153–1167

Vrba ES (2007) Role of environmental stimuli in hominid origins. In: Henke W, Tattersall I (eds) Handbook of paleoanthropology, vol 3. Phylogeny of hominidsSpringer, Berlin Heidelberg New York, pp 1441–1481

Wang B, Fan Z (1999) Choice of South Asian summer monsoon indices. Bull Am Meteorol Soc 80:629–638

Wang C, Zhao X, Liu Z, Lippert PC, Graham SA, Coe RS et al (2008) Constraints on the early uplift history of the Tibetan Plateau. Proc Natl Acad Sci 105(13):4987–4992

Wei GJ, Deng WF, Liu Y, Li XH (2007) High-resolution sea surface temperature records derived from foraminiferal Mg/Ca ratios during the last 260 ka in the northern South China Sea. Paleogeogr Paleoclimatol Paleoecol 250:126–138

Westerhold T, Bickert T, Röhl U (2005) Middle to late Miocene oxygen isotope stratigraphy of ODP site 1085 (SE Atlantic): new constrains on Miocene climate variability and sea-level fluctuations. Palaeogeogr Palaeoclimatol Palaeoecol 217:205–222

Whitten AJ, Damanik SJ, Jazanul A, Mazaruddin H (eds) (2000) The ecology of Sumatra. Periplus Editions (HK) Ltd., Singapore

Wilting A, Sollmann R, Meijaard E, Helgen KM, Fickel J (2012) Mentawai's endemic, relictual fauna: is it evidence for Pleistocene extinctions on Sumatra? J Biogeogr 39:1608–1620

Woodruff DS (2010) Biogeography and conservation in Southeast Asia: how 2.7 million years of repeated environmental fluctuations affect today's patterns and the future of the remaining refugial-phase biodiversity. Biodivers Conserv 19:919–941

Woodruff DS, Turner LM (2009) The Indochinese-Sundaic zoogeographic transition: a description and analysis of terrestrial mammal species distributions. J Biogeogr 36:803–821

Wright JD, Miller KG, Fairbanks RG (1992) Early and middle Miocene stable isotopes: implications for deepwater circulation and climate. Paleoceanography 7:357–389

Wu ZY (1988) Hengduan mountain flora and her significance. J Jpn Bot 63:297–311

Yang J, Wang YF, Spicer RA, Mosbrugger V, Li CS, Sun QG (2007) Climatic reconstruction at the Miocene Shanwang Basin, China, using Leaf Margin Analysis, CLAMP, Coexistence Approach, and Overlapping Distribution Analysis. Am J Bot 94:599–608

Zachos J, Pagani M, Sloan L, Thomas E, Billups K (2001) Trends, rhythms, and aberrations in global climate 65 ma to present. Science 292:686–693

Zhisheng A, Kutzbach JE, Prell WL, Porter SC (2001) Evolution of Asian monsoons and phased uplift of the Himalaya-Tibetan Plateau since Late Miocene times. Nature 411:62–66

Zhao S, Salter CL (eds) (1986) Physical geography of China. Wiley, New York

Zhou L, Su YCF, Thomas DC, Saunders RMK (2012) 'Out-of-Africa' dispersal of tropical floras during the Miocene climatic optimum: evidence from *Uvaria* (Annonaceae). J Biogeogr 39:322–335

Part II
Gibbon and Siamang Phylogeny

Chapter 6
Unique Evolution of Heterochromatin and Alpha Satellite DNA in Small Apes

Akihiko Koga and Hirohisa Hirai

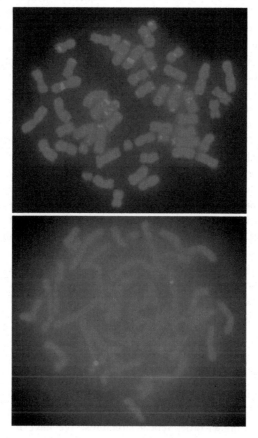

Localization of alpha-satellite DNA (red signals) in *Nomascus siki* (*top*) and Hoolock hoolock (*bottom*)

To the memory of my late friend, Alan R. Mootnick.

A. Koga · H. Hirai (✉)
Primate Research Institute, Kyoto University, Inuyama, Aichi, Japan
e-mail: hirai.hirohisa.7w@kyoto-u.ac.jp

A. Koga
e-mail: koga.akihiko.5n@kyoto-u.ac.jp

© Springer Science+Business Media New York 2016   139
U.H. Reichard et al. (eds.), *Evolution of Gibbons and Siamang*,
Developments in Primatology: Progress and Prospects,
DOI 10.1007/978-1-4939-5614-2_6

Heterochromatin and Alpha Satellite DNA

"The term heterochromatin has proven to be a major biological enigma ever since it was first introduced into the genetic vocabulary by Heitz in 1928" (John 1988). Two classes of heterochromatin, facultative and constitutive, are distinguished in terms of the machinery functions. The former, euchromatin, is facultatively inactivated, and the latter is composed of DNA sequences with distinctive characteristics (John 1988). In this chapter, we focus on the latter, constitutive heterochromatin (C-heterochromatin). C-heterochromatin is found generally at centromeres, telomeres, and the sex chromosome. In small apes, unique positional patterns of C-heterochromatin exist among the four genera (*Symphalangus*, *Nomascus*, *Hylobates*, and *Hoolock*; Fig. 6.1). These genera are grouped into two types (*Symphalangus/Nomascus* and *Hylobates/Hoolock*) according to the overall localization pattern of C-heterochromatin. The former type contains heterochromatin blocks in the centromere and telomere regions. The latter type has relatively small amounts of heterochromatin in the centromere region and often contains interstitial heterochromatin.

The centromere is part of the eukaryotic chromosome and serves as a locomotive by being linked to spindle tubes via the kinetochore for chromosome migration during cell division. The centromere usually contains heterochromatin, of which the DNA component is tandem arrays of repeat units. The nucleotide sequence of the heterochromatin DNA differs among taxa of organisms, and the most abundant sequence in primate centromeres is termed alpha satellite DNA. The nucleotide sequence of the repeat units of alpha satellite DNA varies among different taxa and species within primates and often even among chromosomes within species. Portions of the sequence variation are thought to be linked with centromere function, but a clear picture has not yet been established.

Higher-Order Repeat Structure in Alpha Satellite DNA

Alpha satellite DNA of hominids (humans and great apes) contains sequences that are organized into higher-order repeat structures. We have recently found alpha satellite DNA exhibiting a higher-order structure in a small ape species.

Alpha satellite DNA was first described as a large-scale repetitive sequence in the African green monkey (Maio 1970). Soon after its identification in humans (Musich et al. 1980), structural analyses of alpha satellite DNA nucleotide sequences revealed that the sequences consist of several subsets (Willard 1985; Waye et al. 1987). Some of them were simple repeats of the basic repeat units (monomeric alpha satellite DNA), and others were organized into higher-order repeats (higher-order alpha satellite DNA) in which multiple basic repeats formed a larger repeat unit, and the larger units were repeated in tandem (Fig. 6.2). Subsets of higher-order alpha satellite DNA are specific mostly to chromosomes.

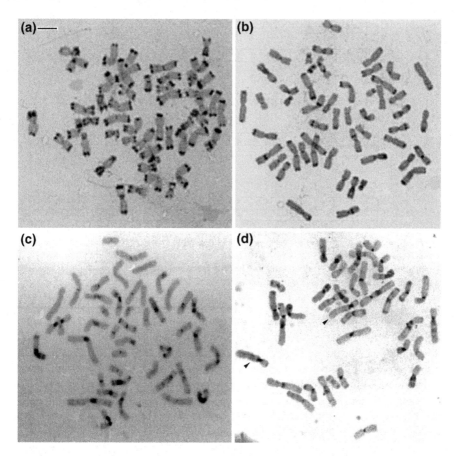

Fig. 6.1 C-heterochromatin pattern difference of four genera of small apes. (**a**) Siamang (*Symphalangus*); (**b**) Concolor gibbon (*Nomascus*); (**c**) Java gibbon (*Hylobates*); (**d**) Hoolock gibbon (*Hoolock*). Dark regions are constitutive heterochromatin (C-heterochromatin) and gray regions are euchromatin. Arrowheads indicate chromosome 4 of the hoolock gibbon. Scale bar: 5 μm

This chromosome specificity is thought to be caused by higher frequencies of sequence shuffling within chromosomes than between chromosomes (Willard 1985; Willard and Waye 1987). There is a clear difference in the chromosomal location between monomeric and higher-order alpha satellite DNAs: the latter occupies the centromeric constriction region and the former is found mostly in pericentric regions (Ikeno et al. 1994). In accordance with this difference, signals that are important for centromere function, such as the CENP-B box (Masumoto et al. 1989), are found more often in higher-order alpha satellite DNA (Ikeno et al. 1994).

In addition to humans, higher-order alpha satellite DNA occurs in great apes, including chimpanzees, gorillas, and orangutans (Baldini et al. 1991; Haaf and Willard 1998). In contrast, there has been only one report of higher-order alpha satellite DNA in small apes, which is described below. For this reason, the most

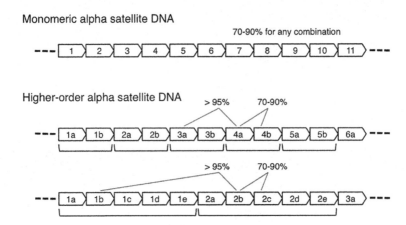

Fig. 6.2 Schematic illustration of two types of alpha satellite DNA. Arrows indicate basic repeat units. Nucleotide similarities between two basic repeats indicated with *lines* are shown as percentages. For higher-order alpha satellite DNA, two examples are shown in which two and five basic repeat units constitute larger repeat units

widely accepted hypothesis regarding the evolutionary origin of higher-order alpha satellite DNA is that it arose in the common ancestor of hominids (family Hominidae). However, we have recently discovered a higher-order repeat structure in the alpha satellite DNA of siamangs (Terada et al. 2013), raising questions about the origin of higher-order alpha satellite DNA; e.g., did higher-order alpha satellite DNA emerge from a common ancestor of hominoids (superfamily Hominoidea)? The higher-order repeat structure found in siamangs contains large repeat units consisting of six or four basic repeat units (Fig. 6.3). However, siamang alpha satellite DNA is contained in the terminal regions of chromosomes in addition to centromeres; therefore, we currently cannot determine from which portion of the chromosome the higher-order repeat sequence originates. For this reason, more analyses are needed to obtain data to address questions regarding the evolutionary origin of higher-order alpha satellite DNA. A direct approach would be to examine the structure of alpha satellite DNA of the species that contain this centromere-specific repetitive DNA.

Difference in Chromosomal Location of Alpha Satellite DNA Among Genera

Alpha satellite DNA is the main component of primate centromeres. In small apes, however, the chromosomal location of alpha satellite DNA varies, e.g., it can also be found in terminal regions and within arms of chromosomes. It is likely that alpha satellite DNA underwent migration to these positions and was subsequently maintained or amplified in these locations.

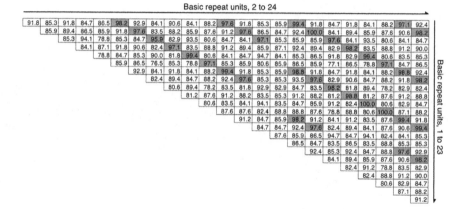

Fig. 6.3 Matrix representation of nucleotide identities between basic repeat units in siamang alpha satellite DNA. The first 24 basic repeat units in the sequence data of the siamang alpha satellite DNA (GenBank accession number AB819921) were placed along the axis and abscissa in the direction indicated by the arrows, and pairwise comparisons among the basic repeat units were made. Each cell contains the nucleotide sequence identity expressed as a percentage. Columns with values >95 % are indicated by shading

Fluorescence in situ hybridization (FISH) analysis of chromosomes using an alpha satellite DNA probe revealed that each genus of small apes has a unique localization pattern (Fig. 6.4). All of the *Symphalangus* chromosomes contain large heterochromatin blocks in the telomere region, and these blocks yielded intense hybridization signals (Fig. 6.4a) (Koga et al. 2012). The majority of *Nomascus* chromosomes also exhibited signals in heterochromatin blocks in both the telomere and the centromere regions (Baicharoen et al. 2012). In addition, strong interstitial signals were observed in four *Nomascus* chromosomes (Fig. 6.4b). The relative signal strength of the telomere regions compared with the centromere regions was considerably more robust in the *Symphalangus* chromosomes than the *Nomascus* chromosomes. In contrast, fluorescent signals were not observed in the telomere region of the other two genera. *Hylobates* showed clear (24 chromosomes) and weak (9 chromosomes) signals only in the centromere regions of a majority of chromosomes (Fig. 6.4c), and *Hoolock* showed strong signals only at the centromere regions of chromosome 4 (Fig. 6.4d) (see also Baicharoen et al. 2012; Koga et al. 2012).

Replacement of Alpha Satellite DNA by a Transposon-Related Sequence

Centromeres usually contain tandem repeat sequences as their main DNA component, and the most abundant sequence in primate centromeres is alpha satellite DNA. At least in hominoids, alpha satellite DNA is found in normal centromeres of

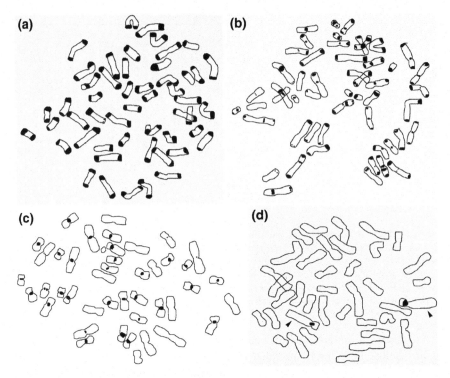

Fig. 6.4 Localization of alpha satellite DNA derived from siamangs on the chromosomes of four genera of small apes. Data are depicted as drawings derived from pictures. (**a**) *Symphalangus*; (**b**) *Nomascus*; (**c**) *Hylobates*; (**d**) *Hoolock*. Black regions are positive signals for alpha satellite DNA. Arrowhead indicates chromosome 4 of the hoolock gibbon

all chromosomes. We have recently found an exceptional situation in the western hoolock gibbon (*Hoolock hoolock*).

As shown in Fig. 6.4, in the FISH analysis of chromosomes using alpha satellite DNA probes, hoolock gibbon chromosomes exhibit strong signals on the centromere region of only one pair of chromosomes (chromosome 4). The C-heterochromatin analysis (Fig. 6.1), however, showed that the centromere regions of all chromosomes contain large amounts of C-heterochromatin. These results indicate that either (i) the nucleotide sequence of alpha satellite DNA has undergone a rapid alteration in chromosomes other than the one pair or that (ii) alpha satellite DNA has been replaced with another sequence in these chromosomes. Cloning and sequencing analyses of C-heterochromatin showed that the latter event had occurred in the lineage preceding the hoolock gibbon and that the new-comer sequence in the majority of the chromosomes had originated from a retrotransposon.

The SVA element is a composite retrotransposon thought to have been formed by the fusion of three solo elements in the common ancestor of hominoids (Ostertag et al. 2003). The solo elements, which are the short interspersed nuclear element (SINE), variable number of tandem repeat (VNTR), and the *Alu* element, comprise the SVA element. The SVA element still retains transposition activity because newly arising insertion mutations that cause human diseases have been reported (Taniguchi-Ikeda et al. 2011). Similar composite transposons consisting of the VNTR region and other components are known (Carbone et al. 2012). The sequence that replaced alpha satellite DNA in the centromere regions of the hoolock gibbon chromosomes was shown to be the VNTR sequence of SVA or similar elements.

To obtain a candidate sequence as a DNA clone, we screened a genomic library of the hoolock gibbon for highly repetitive genomic sequences that were not in the human genome. The method can be described broadly as a comparative genomic hybridization (Fig. 6.5). This method is useful for identifying the differences in the copy number of multicopy sequences between strains, species, or genera. One great advantage of this method is that DNA fragments can be obtained as clones even if their nucleotide sequences are unknown.

Sequencing analyses of the obtained clones revealed long stretches of the VNTR sequence. The genomic library of the hoolock gibbon had been prepared as a collection of genomic DNA fragments approximately 40 kb in length contained in a fosmid vector. The genomic fragments that were ligated into the vector had been prepared by mechanical sheering followed by size selection through agarose gel electrophoresis. In most of the obtained clones, the entire inserts were occupied only by the VNTR sequence. SVA and related transposons are dispersed in the

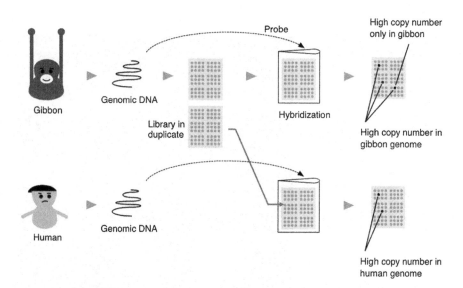

Fig. 6.5 Schematic illustration representing the strategy of the comparative genomic hybridization method. Clones exhibiting strong signals only with the gibbon probe contain repetitive sequences whose copy number is high in the gibbon genome but not in the human genome

genome, and the lengths of their VNTR region vary among copies. The average size of human SVA elements has been estimated to be approximately 0.8 kb (Wang et al. 2005), and the VNTR sequence is part of the SVA element. These results indicate that the VNTR sequence had been elongated to 40 kb or larger in the hoolock gibbon genome.

FISH analyses of the hoolock gibbon chromosomes resulted in strong signals for the VNTR sequence in the centromere regions of 28 of the 38 chromosomes (Fig. 6.6) (Hara et al. 2012). Chromosome 4, which exhibited an intense signal for alpha satellite DNA (Fig. 6.4d), did not show a detectable signal for the VNTR probe. The results of these FISH analyses, in combination with the results of the sequencing analyses, indicate that alpha satellite DNA has been replaced by the VNTR sequence in the centromere regions of the majority of the hoolock gibbon chromosomes. Essentially the same results were obtained by analyzing the chromosomes of eastern hoolock gibbons (*Hoolock leuconedys*) in a study independent of ours (Carbone et al. 2012). The replacement event appears to have occurred in the common ancestor of the two gibbon species.

The mechanism of alpha satellite DNA replacement by the transposon-related sequence is not known. One possible factor to be considered is interspecies hybridization. It has been inferred from vocalization, morphology, and genome analyses that interspecies hybridization occurred frequently in small apes (Myers and Shafer 1979; Marshall and Sugardjito 1986; Hirai et al. 2007), and evolution within the small apes seems to have been profoundly affected by natural hybridization (Arnold and Meyer 2006). Brown and O'Neill (2009) suggested that small apes may have experienced hybridization events within the past 15 million

Fig. 6.6 FISH analysis of chromosomes to establish the chromosomal locations of the VNTR sequence. Strong or weak signals (whitish) were observed at the centromeres of 28 chromosomes, but not observed on 10 chromosomes. The arrowhead indicates chromosome 4 in which a VNTR signal is not detected, but alpha satellite DNA is observed (*see* Fig. 5.4d). Scale bar: 10 μm

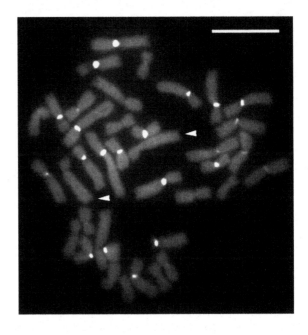

years (Arnold and Meyer 2006), and hybridization-induced chaos of transposable element methylation and stability (O'Neill et al. 1998; Brown et al. 2008) may have induced a process through which such mobile elements cause chromosome rearrangements (Carbone et al. 2009a, b). Interspecies hybridization may also underlie formations of new species.

Application of Alpha Satellite DNA Probes to Genus Identification

The overall localization patterns of alpha satellite DNA (Fig. 6.4) are grouped into two types (*Symphalangus/Nomascus* and *Hylobates/Hoolock*), and the close observation of the patterns facilitates the distinction between respective genera. The manifestation of alpha satellite DNA on the chromosomes was easier and more reliable than that of C-heterochromatin (compare Figs. 6.1 and 6.4). Therefore, we propose that locale of alpha satellite DNA serves as a good identifying marker of each genus, for example, the site of alpha satellite DNA on chromosomes of intergeneric hybrid offspring between *Hylobates* and *Nomascus* primates (Fig. 6.7) that occurred at a zoo in Japan (Hirai et al. 2007). The alpha satellite DNA probes employed in our study can distinguish the parental species chromosome sets of the *Hylobates* and *Nomascus* hybrid offspring by the signal patterns. In fact, the two mixed chromosome sets in the hybrid offspring can be identified easily and classified into two groups: gray chromosomes represent *Hylobates* and white chromosomes with black represent *Nomascus* (Fig. 6.7). Compared with the C-banding method, this technique is superior because it can rapidly identify the parent genus due to genus-specific alpha satellite DNA markers. Although our research did not detect the drastic changes of specific segments, we need to observe regions of repetitive sequence using other techniques, especially interstitial heterochromatin blocks.

As mentioned above, the localization patterns of chromosomal alpha satellite DNA provide good markers to identify the genus of small apes. Thus, FISH analysis with alpha satellite DNA is useful for the first round of examination of hybrid offspring together with the number of chromosomes. In particular, this method is effective for the examination of hybrid offspring when parentage is unknown. Such cases, including the possibility of hybrids, have often occurred in captive circumstances. After the first round of diagnosis using alpha satellite DNA to establish the genus of the parents, species can be identified using other techniques that have already been developed. For instance, G-band analysis can detect specific chromosomal rearrangements, such as complex inversions of *Hylobates* chromosome 8 that were observed in some regional species or races (Stanyon et al. 1987) and the inversions and translocation in chromosomes 1, 7, and 22 of *Nomascus leucogenys* that led to the classification of this species into 3 groups (Couturier and Lernould 1991; Koehler et al. 1995; Carbone et al. 2009b). C-band

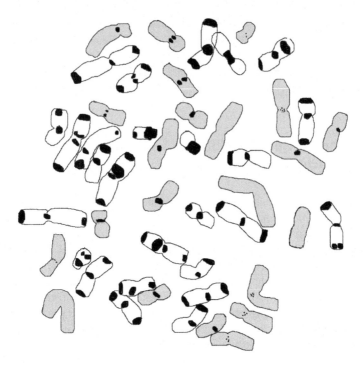

Fig. 6.7 FISH hybridization with alpha satellite DNA on the chromosomes of an intergeneric hybrid offspring between *Hylobates* (*gray*) and *Nomascus* (*white*). Black signals indicate the hybridization of alpha satellite DNA

analysis has also detected chromosome differentiation in *Hylobates* (Hirai et al. 2003). Chromosome painting analysis can detect translocations that have occurred on many chromosomes and thus are good indices to examine the phylogeny and biogeography of small apes (Müller et al. 2003; Hirai et al. 2005). Although these analyses provide credible information, the alpha satellite DNA FISH method described here seems to surpass the other techniques in rapidity and reliability. Therefore, we added a new marker, alpha satellite DNA, to the analytical methodology used for studying small ape chromosome evolution.

Complex differentiation in the chromosome structure is one of the remarkable features of small apes: their chromosome alteration rate is estimated to be 20 times higher than the average rate of alterations in humans, great apes, and other mammals, excluding rodents (Burt et al. 1999; Misceo et al. 2008). To clarify the respective rearrangement events that have occurred in the history of small apes, the availability of appropriate markers for chromosomal segments is needed. Superior markers would be those present in all species but different in the chromosomal location among species. As we have shown in this chapter, Alpha satellite DNA fulfills these requirements.

Acknowledgments For the use of samples in our original investigations, we are grateful to the staff members of the Primate Research Institute, Kyoto University, Japan, the Zoological Park Organization, Thailand, and the WildTeam Cosmos Center, Bangladesh. The original studies were supported by the Japan Society for the Promotion of Science Grants-in-Aid for Scientific Research (24370098 to AK; 22247037 and 24255009 to HH).

References

Arnold ML, Meyer A (2006) Natural hybridization in primates: one evolutionary mechanism. Zoology 109: 261–276

Baicharoen S, Arsaithamkul V, Hirai Y, Hara T, Koga A, Hirai H (2012) In situ hybridization analysis of gibbon suggests that amplification of alpha satellite DNA in the telomere region is confined to two of the four genera. Genome 55:809–812

Baldini A, Miller DA, Shridhar V, Rocchi M, Miller OJ, Ward DC (1991) Comparative mapping of a gorilla-derived alpha satellite DNA clone on great ape and human chromosomes. Chromosoma 101:109–114

Brown JD, Golden D, O'Neill RJ (2008) Methylation perturbation in retroelements within the genome of a Mus interspecific hybrid correlate with double minute chromosome formation. Genomics 91:267–273

Brown JD, O'Neill RJ (2009) The mysteries of chromosome evolution in gibbons: methylation is a prime suspect. PLoS Genet 5:e1000501

Burt DW, Bruley C, Dunn IC, Jones CT, Ramage A, Law AS, Morrice DR, Paton IR, Smitrh J, Windsor D, Sazanov A, Fries R, Waddington D (1999) The dynamics of chromosome evolution in birds and mammals. Nature 402:411–413

Carbone L, Harris RA, Vessere GM, Mootnick AR, Humphray S, Rogers J, Kim SK, Wall JD, Martin D, Juka J, Milosavljievic A, deJong PJ (2009a) Evolutionray breakpoints in the gibbon suggest association between cytosine methylation and karyotype evolution. PLoS Genet 5: e1000538

Carbone L, Mootnick AR, Nadler T, Moisson P, Ryder O, Roos C, de Jong PJ (2009b) A chromosomal inversion unique to the northern white-cheeked gibbon. PLoS ONE 4:e4999

Carbone L, Harris RA, Mootnick AR, Milosavljevic A, Martin DI, Rocchi M, Capozzi O, Archidiacono N, Konkel MK, Walker JA, Batzer MA, de Jong PJ (2012) Centromere remodeling in leuconedys (Hylobatidae) by a new transposable element unique to the gibbons. Genome Biol Evol 4:648–658

Couturier J, Lernould JM (1991) Karyotypic study of four gibbons forms provisionally considered as subspecies of Hylobates (*Nomascus*) concolor (Primates, Hylabtidae). Folia Primatol 56:95–104

Haaf T, Willard HF (1998) Orangutan alpha-satellite monomers are closely related to the human consensus sequence. Mamm Genome 9:440–447

Hara T, Hirai Y, Jahan I, Hirai H, Koga A (2012) Tandem repeat sequences evolutionarily related to SVA-type retrotransposons are expanded in the centromere region of the western gibbon, a small ape. J Hum Genet 57:760–765

Hirai H, Hirai Y, Domae H, Kirihara Y (2007) A most distant intergeneric hybrid offspring (Larcon) of lesser apes, *Nomascus leucogenys* and *Hylobates lar*. Hum Genet 122:477–483

Hirai H, Mootnick AR, Takenaka O, Suryobroto B, Mouri T, Kamanaka Y, Katoh A, Kimura N, Katoh A, Maeda N (2003) Genetic mechanism and property of a whole-arm translocation (WAT) between chromosomes 8 and 9 of agile gibbons (*Hylobates agilis*). Chromosome Res 11:37–50

Hirai H, Wijayanto H, Tanaka H, Mootnick AR, Hayano A, Perwitasari-Farajallah D, Iskandriati D, Sajuthi D (2005) A whole-arm translocation (WAT8/9) separating Sumatran and Bornean agile gibbons, and its evolutionary features. Chromosome Res 13:123–133

Ikeno M, Masumoto H, Okazaki T (1994) Distribution of CENP-B boxes reflected in CREST centromere antigenic sites on long-range alpha-satellite DNA arrays of human chromosome 21. Hum Mol Genet 3:1245–1257

John B (1988) The biology of heterochromatin. In: Verma RS (ed) Heterochromatin: Molecular and structural aspects. Cambridge University Press, Cambridge, pp 1–147

Koehler U, Bigoni F, Weinberg J, Stanyon R (1995) Genomic reorganization in the concolor gibbon (*Hylobates concolor*) revealed by chromosome painting. Genomics 30:287–292

Koga A, Hirai Y, Hara T, Hirai H (2012) Repetitive sequences originating from the centromere constitute large-scale hetgerochromatin in the telomere region in the siamang, a small ape. Heredity 109:180–187

Maio JJ (1970) DNA strand reassociation and polyribonucleotide binding in the African green monkey, *Cercopithecus aethiops*. J Mol Biol 56:579–595

Marshall J, Sugardjito J (1986) Gibbon systematics. In: Ford SM, Swindler DR, Erwin J (eds) Comparative primate biology, vol 1., Systematics, evolution, and anatomyAR Liss, New York, pp 137–185

Masumoto H, Masukata H, Muro Y, Nozaki N, Okazaki T (1989) A human centromere antigen (CENP-B) interacts with a short specific sequence in alphoid DNA, a human centromeric satellite. J Cell Biol 109:1963–1973

Misceo D, Capozzi O, Roberto R, Dell'Oglio MP, Rocchi M, Stanyon R, Archidiacono N (2008) Tacking the complex flow of chromosome rearrangements from the Hominoidea ancestor to extant *Hylobates* and *Nomacus* gibbons by high-resolution synteny mapping. Genome Res 18:1530–1537

Müller S, Hollatz M, Wienberg J (2003) Chromosomal phylogeny and evolution of gibbons (Hylobatidae). Hum Genet 113:493–501

Musich PR, Brown FL, Maio JJ (1980) Highly repetitive component alpha and related alphoid DNAs in man and monkeys. Chromosoma 80:331–348

Myers RH, Shafer DA (1979) Hybrid ape offspring of a mating of gibbon and siamang. Science 205:308–310

O'Neill RI, O'Neill MJ, Graves JA (1998) Undermethylation associated with retroelement activation and chromosome remodeling in an interspecific mammalian hybrid. Nature 393:68–72

Ostertag EM, Goodier JL, Zhang Y, Kazazian HH (2003) SVA elements are nonautonomous retrotransposons that cause disease in humans. Am J Hum Genet 73:1444–1451

Stanyon R, Sineo L, Chiarelli B, Camperio-Ciani A, Haimoff AR, Mootnick EH, Sutarman DRH (1987) Banded karyotypes of the 44-chromosome gibbons. Folia Primatol 48:56–64

Taniguchi-Ikeda M, Kobayashi K, Kanagawa M, Yu CC, Mori K, Oda T, Kuga A, Kurahashi H, Akman HO, DiMauro S, Kaji R, Yokota T, Takeda S, Toda T (2011) Pathogenic exon-trapping by SVA retrotransposon and rescue in Fukuyama muscular dystrophy. Nature 478:127–131

Terada S, Hirai Y, Hirai H, Koga A (2013) Higher-order repeat structure in alpha satellite DNA is an attribute of hominoids rather than hominids. J Hum Genet 58:752–754

Wang H, Xing J, Grover D, Hedges DJ, Han K, Walker JA, Batzer MA (2005) SVA elements: a hominid-specific retroposon family. J Mol Biol 354:994–1007

Waye JS, England SB, Willard HF (1987) Genomic organization of alpha satellite DNA on human chromosome 7: evidence for two distinct alphoid domains on a single chromosome. Mol Cell Biol 7:349–356

Willard HF (1985) Chromosome-specific organization of human alpha satellite DNA. Am J Hum Genet 37:524–532

Willard HF, Waye JS (1987) Chromosome-specific subsets of human alpha satellite DNA: analysis of sequence divergence within and between chromosomal subsets and evidence for an ancestral pentameric repeat. J Mol Evol 25:207–214

Chapter 7
Phylogeny and Classification of Gibbons (Hylobatidae)

Christian Roos

Juvenile white-handed gibbon female maneuvering the canopy, Khao Yai National Park, Thailand.
Photo credit: Ulrich H. Reichard

C. Roos (✉)
Gene Bank of Primates, Primate Genetics Laboratory, German Primate Center,
Leibniz Institute for Primate Research, Kellnerweg 4, 37077 Göttingen, Germany
e-mail: croos@dpz.eu

© Springer Science+Business Media New York 2016
U.H. Reichard et al. (eds.), *Evolution of Gibbons and Siamang*,
Developments in Primatology: Progress and Prospects,
DOI 10.1007/978-1-4939-5614-2_7

151

Gibbons' Place Among Primates

Gibbons or small apes constitute the primate family Hylobatidae and represent the sister lineage to the family Hominidae, which comprises great apes and humans. Together both families are combined in the superfamily Hominoidea (Fig. 7.1; Napier and Napier 1967; Groves 1989, 2001; Fleagle 1999; Geissmann 2003). Gibbons share several synapomorphic characteristics with hominids, for example a broad thorax, dorsally placed scapulae, long clavicles, long forelimbs, a humerus with a spool-shaped trochlea, a reduced lumbar region, a higher number of sacral vertebrae compared to monkeys, loss of the tail, a relatively broad iliac blade (Osilium) and reduced ischial callosities compared to cercopithecine monkeys (e.g., Fleagle 1999; Geissmann 2003). The monophyly of gibbons and their phylogenetic position as the sister lineage of great apes and humans is supported by various comparative studies based on morphology (Sawalischin 1911; Remane 1921; Wislocki 1929, 1932; Schultz 1933; Napier and Napier 1967; Groves 1972; Biegert 1973; Schultz 1973), physiology (Hellekant et al. 1990), cytogenetics (Prouty et al. 1983; Wienberg and Stanyon 1987; Müller et al. 2003; Carbone et al. 2006; Misceo et al. 2008; Stanyon 2013) and molecular data (Dene et al. 1976; Sarich and Cronin 1976; Darga et al. 1984; Sibley and Ahlquist 1984; Felsenstein 1987; Goldman et al. 1987; Sibley and Ahlquist 1987; Goodman et al. 1990, 1998; Bailey et al. 1991; Garza and Woodruff 1992; Hayashi et al. 1995; Hall et al. 1998; Roos and Geissmann 2001; Salem et al. 2003; Takacs et al. 2005; Chatterjee et al. 2009; Fabre et al. 2009; Chan et al. 2010, 2012; Matsudaira and Ishida 2010; Thinh et al. 2010a; Israfil et al. 2011; Kim et al. 2011; Perelman et al. 2011; Meyer et al. 2012; Carbone et al. 2014). According to molecular data, gibbons diverged from great apes and humans in the Early Miocene, 20–16 million years ago (Goodman et al. 1998; Chatterjee et al. 2009; Fabre et al. 2009; Chan et al. 2010, 2012; Matsudaira and Ishida 2010; Thinh et al. 2010a; Israfil et al. 2011; Perelman et al. 2011; Carbone et al. 2014).

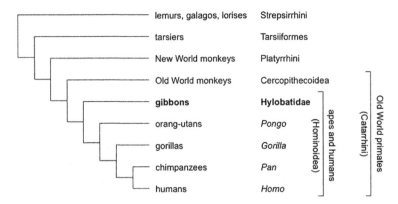

Fig. 7.1 Phylogenetic position of gibbons (Hylobatidae) within the primate order. Figure is redrawn from Rawson et al. (2011)

Classification and Phylogeny Among Major Gibbon Lineages

Broad agreement exists that gibbons should be divided into four major lineages and that these are best recognized as distinct genera (e.g., Groves 2001; Roos and Geissmann 2001; Geissmann 2002a, 2003, 2007; Brandon-Jones et al. 2004; Mootnick and Groves 2005; Mootnick 2006; Thinh et al. 2010a; Chivers 2013; IUCN 2015). However, in early studies, gibbons were divided into just two major lineages or genera, with *Symphalangus* including the siamang, and *Hylobates* all the remaining species (Schultz 1933; Simonetta 1957; Napier and Napier 1967). In the decades following the pioneering work on gibbon taxonomy, more detailed studies revealed that gibbons are better represented as four major lineages, which differ most importantly in their diploid chromosome number (*Hoolock*: $2n = 38$, *Hylobates*: $2n = 44$, *Symphalangus*: $2n = 50$, *Nomascus*: $2n = 52$) (Prouty et al. 1983; Liu et al. 1987; Wienberg and Stanyon 1987; Müller et al. 2003; Stanyon 2013). Such division is further supported by morphological (Groves 1972, 1989, 2001; Creel and Preuschoft 1984), acoustic (Marshall and Sugardjito 1986; Geissmann 1993, 2002b) and molecular differences (Hall et al. 1998; Roos and Geissmann 2001; Takacs et al. 2005; Chatterjee et al. 2009; Fabre et al. 2009; Thinh et al. 2010a; Israfil et al. 2011; Kim et al. 2011; Meyer et al. 2012; Wall et al. 2013; Carbone et al. 2014). Accordingly, these four clades were recognized as subgenera of the genus *Hylobates* (Marshall and Sugardjito 1986; Geissmann 1995; Rowe 1996; Nowak 1999; Groves 2001) or as distinct genera, which is the currently most widely accepted classification (Roos and Geissmann 2001; Geissmann 2002a, 2003, 2007; Brandon-Jones et al. 2004; Mootnick and Groves 2005; Takacs et al. 2005; Carbone et al. 2006, 2014; Mootnick 2006; Roos et al. 2007; Misceo et al. 2008; Chatterjee et al. 2009; Fabre et al. 2009; Chan et al. 2010, 2012, 2013; Matsudaira and Ishida 2010; Thinh et al. 2010a; Israfil et al. 2011; Kim et al. 2011; Meyer et al. 2012; Chivers 2013; Stanyon 2013; Wall et al. 2013; IUCN 2015).

While a common ancestry of the members of each of the four genera is widely accepted now, the phylogenetic relationships among genera are still poorly understood. A wide variety of possible relationships have been proposed (for an overview see Takacs et al. 2005). While some studies suggested that *Symphalangus* was the first genus to diverge from the main stem (Napier and Napier 1967; Groves 1972), others have proposed that either *Nomascus* or *Hoolock* should occupy that position (e.g., Chivers 1977; Haimoff et al. 1982; Creel and Preuschoft 1984; Müller et al. 2003). Unfortunately, the application of molecular techniques in the past two decades, mainly in form of sequencing fragments of the mitochondrial genome, did not provide the expected significant resolution of the branching pattern among gibbon genera (Hall et al. 1998; Roos and Geissmann 2001; Takacs et al. 2005; Roos et al. 2007; Thinh et al. 2010a). However, more promising results have been achieved since the first complete mitochondrial genomes of gibbons have been published (Chan et al. 2010; Matsudaira and Ishida 2010). The recent studies by Chan et al. (2010) and Matsudaira and Ishida (2010), although lacking *Hoolock*, revealed a strong support for a sister group relationship of *Nomascus* to a clade consisting of *Hylobates* and *Symphalangus*. Unfortunately, including *Hoolock* into the phylogeny, the branching pattern among gibbon genera again collapses, so that mitochondrial genome data does not resolve the phylogenetic relationships among gibbon genera (Carbone et al. 2014). Similarly, nuclear sequence data and the presence/absence pattern of retroposon integrations have not been able to resolve the question of how the four gibbon genera are related to each other (Israfil et al. 2011; Kim et al. 2011; Perelman et al. 2011; Chan et al. 2012, 2013; Meyer et al. 2012; Wall et al. 2013). In the end, even complete nuclear genome data revealed no clear branching pattern (Carbone et al. 2014), so that the phylogenetic relationship among the four gibbon genera may remain a long-standing question. As indicated by this "hard" polytomy, the four gibbon genera most likely originated during a short time period. In fact, all molecular data suggest a short radiation period of 2–3 million years sometime in the Late Miocene/Early Pliocene, 9–4 million years ago (Goodman et al. 1998; Chan et al. 2010, 2012; Matsudaira and Ishida 2010; Thinh et al. 2010a; Israfil et al. 2011; Kim et al. 2011; Perelman et al. 2011; Carbone et al. 2014).

Classification and Phylogeny Within Gibbon Genera

The classification of and phylogenetic relationships among members of the four genera are, as the branching pattern among genera, largely disputed. I follow here the most recent classification proposed by Anandam et al. (2013), while I also include the newly described *Hoolock hoolock* subspecies *H. h. mishmiensis* (Choudhury 2013). Accordingly, the gibbon family includes four genera, 19 species and a total of 25 taxa (Table 7.1).

Hoolock gibbons were previously recognized as members of the genus *Bunopithecus*, however, since *Bunopithecus* refers to a fossil, which is not directly related with extant hoolock gibbons, hoolock gibbons were given the new genus name, *Hoolock* (Mootnick and Groves 2005). Traditionally, the genus *Hoolock* contained a single species with two subspecies, *H. h. hoolock* and *H. h. leuconedys*, but due to morphological and genetic differences, both subspecies were elevated to species level and were given the names Western Hoolock (*H. hoolock*) and Eastern Hoolock (*H. leuconedys*: Mootnick and Groves 2005; Geissmann 2007; Thinh et al. 2010a; Anandam et al. 2013). According to mitochondrial sequence data, the two species separated approximately 1.4 million years ago (Thinh et al. 2010a). In 2013, a new hoolock subspecies was described from Arunachal Pradesh and its neighbouring province Assam, India (Choudhury 2013). Choudhury (2013) named this new taxon *H. h. mishmiensis*, but also recognized *H. leuconedys* as a subspecies of *H. hoolock*. Since the range of the new subspecies is close to that of *H. hoolock*, it can be expected that it is indeed a subspecies of *H. hoolock*, although molecular genetic data is required to verify the correct status of *H. h. mishmiensis*.

Table 7.1 Classification of the gibbons (Hylobatidae)

Genus	Species	Subspecies	Common name
Hoolock			Hoolock Gibbons
	H. hoolock		Western Hoolock Gibbon
		H. h. hoolock	Western Hoolock Gibbon
		H. h. mishmiensis	Mishmi Hills Hoolock Gibbon
	H. leuconedys		Eastern Hoolock Gibbon
Hylobates			*Hylobates* Gibbons
	H. pileatus		Pileated Gibbon
	H. lar		Lar Gibbon
		H. l. lar	Malayan Lar Gibbon
		H. l. carpenteri	Carpenter's Lar Gibbon
		H. l. entelloides	Central Lar Gibbon
		H. l. vestitus	Sumatran Lar Gibbon
		H. l. yunnanensis	Yunnan Lar Gibbon
	H. agilis		Agile Gibbon
	H. albibarbis		Bornean White-bearded Gibbon
	H. muelleri		Müller's Gibbon
	H. abbotti		Abbott's grey Gibbon
	H. funereus		East Bornean Grey Gibbon
	H. moloch		Moloch Gibbon
	H. klossii		Kloss's Gibbon
Symphalangus			Siamang
	S. syndactylus		Siamang
Nomascus			Crested Gibbons
	N. hainanus		Hainan Crested Gibbon
	N. nasutus		Eastern Black Crested Gibbon
	N. concolor		Western Black Crested Gibbon
		N. c. concolor	Tonkin Black Crested Gibbon
		N. c. lu	Laotian Black Crested Gibbon
	N. leucogenys		Northern White-cheeked Crested Gibbon
	N. siki		Southern White-cheeked Crested Gibbon
	N. annamensis		Northern Yellow-cheeked Crested Gibbon
	N. gabriellae		Southern Yellow-cheeked Crested Gibbon

The genus *Hylobates*, which encompasses gibbons with 44 chromosomes, is the most specious of all gibbon genera comprising today nine species and a total of 14 taxa (Table 7.1; Thinh et al. 2010a; Anandam et al. 2013). Various taxonomic classifications have been proposed for the genus, with a general trend towards increasing the number of species. This growth in the number of recognized species has mainly been the result of an increasing number of studies on gibbon morphology, ecology, behaviour, acoustics and genetics that all largely expanded our knowledge about them. Recognized species numbers in these classifications vary among two species (Creel and Preuschoft 1984), four species (Napier and Napier 1967), five species (Groves 1984), six species (Chivers 1977; Haimoff et al. 1982; Marshall and Sugardjito 1986; Geissmann 1993, 1995, 2002a; Geissmann et al. 2000), seven species (Groves 2001; Geissmann 2007) and nine species (Thinh et al. 2010a; Anandam et al. 2013). While *Hylobates klossii* has always been recognized as being distinct and therefore separate from other members of the genus, all other representatives of the genus have been grouped in the Lar group and sometimes even lumped into the single species *Hylobates lar* (Creel and Preuschoft 1984). Other authorities gave species status to *Hylobates moloch* and *Hylobates agilis* (Napier and Napier 1967), and also to *Hylobates pileatus* and *Hylobates muelleri* (Chivers 1977; Haimoff et al. 1982; Marshall and Sugardjito 1986; Geissmann 1993, 1995, 2002a; Geissmann et al. 2000). Most recently, also *Hylobates albibarbis* (Groves 2001; Geissmann 2007; Thinh et al. 2010a; Anandam et al. 2013), as well as *Hylobates abbotti* and *Hylobates funereus* (Thinh et al. 2010a; Anandam et al. 2013) have been suggested to represent separate species within the Lar group. While *H. agilis, H. klossii, H. lar, H. moloch, H. muelleri* and *H. pileatus* are now widely recognized as distinct species, the recent classification of *H. albibarbis, H. abbotti* and *H. funereus* as species is preliminary. However, morphological and genetic data show that these three taxa are clearly distinguishable from each other and from their closest relatives *H. agilis* and *H. muelleri* (Groves 2001; Thinh et al. 2010a; Anandam et al. 2013). To further prove this taxonomic splitting, and also to investigate whether *H. agilis* and *H. moloch* are indeed monotypic, additional work is required. Here, *H. agilis* and *H. moloch* are provisionally recognized as monotypic, although genetic data indicate that both may contain two subspecies (*H. a. agilis* and *H. a. unko*: Thinh et al. 2010a; but see Hirai et al. 2009; *H. m. moloch* and *H. m. pongoalsoni*: Andayani et al. 2001). Likewise, the number of *H. lar* subspecies is controversially discussed, particularly whether the now possibly extinct *H. l. yunnanensis* is indeed distinguishable from *H. l. carpenteri*.

The phylogenetic relationships between *Hylobates* gibbons are difficult to resolve. Various attempts to elucidate the branching pattern among *Hylobates* gibbons have been made, but so far no convincing, agreeable pattern has emerged. In early studies it was believed that *H. klossii* was the first species to diverge from the main stem (Chivers 1977; Haimoff et al. 1982; Creel and Preuschoft 1984; Groves 1989), but today it seems more likely that *H. klossii* is best placed among the other species and does not represent the basal lineage (Garza and Woodruff 1992; Takacs et al. 2005; Whittaker et al. 2007; Chatterjee et al. 2009; Fabre et al. 2009; Thinh et al. 2010a; Israfil et al. 2011; Chan et al. 2010, 2012, 2013). In fact, the species that might occupy the basal position among *Hylobates* gibbons may be *H. pileatus* as suggested by recent genetic studies (Chan et al. 2010, 2012, 2013; Israfil et al. 2011). The branching pattern among the remaining species is largely unresolved, although one recent study at least found a strongly supported clade encompassing *H. muelleri, H. abbotti* and *H. funereus* (Thinh et al. 2010a). This, however, is unsurprising since

H. abbotti and *H. funereus* were until recently recognized as subspecies of *H. muelleri* (e.g., Groves 2001; Geissmann 2007). Within the *Hylobates* gibbons, the phylogenetic position of *H. albibarbis* seems particularly interesting, because this species' mitochondrial DNA suggests a closer relationship with the geographically more distant Sumatran *H. agilis* than with the geographically closer Bornean gibbons (Thinh et al. 2010a), while acoustic, morphological and chromosomal data suggest an intermediate position between *H. agilis* and *H. muelleri* (Geissmann 1995; Groves 2001; Hirai et al. 2005, 2009). Therefore, *H. albibarbis* may be interpreted as the result of past hybridization between ancestral Sumatran and Bornean lineages. Indeed, hybridization probably occurred not only in this case, but also between several *Hylobates* gibbon lineages in the past as indicated by genetic data (Kim et al. 2011; Chan et al. 2013). Even today, hybridization between *Hylobates* species occurs in some places where the ranges of two species meet and overlap (i.e., *H. lar* × *H. pileatus*: Brockelman and Gittins 1984; Marshall and Brockelman 1986; Marshall and Sugardjito 1986; Matusaidra et al. 2013; *H. lar* × *H. agilis*: Brockelman and Gittins 1984; *H. albibarbis* × *H. muelleri*: Brockelman and Gittins 1984; Marshall and Sugardjito 1986).

The nearly unresolved branching pattern among *Hylobates* gibbons suggests that extant species most likely emerged as a result of a radiation-like splitting event. Consistent with such interpretation is the fact that genetic studies that estimated divergence times among gibbon species date this radiation into a narrow time window sometime between 4 and 2 million years ago (Chan et al. 2010; Matsudaira and Ishida 2010; Thinh et al. 2010a; Israfil et al. 2011).

The siamang, *S. syndactylus*, is the largest gibbon and the only species of the genus *Symphalangus* (Groves 2001; Geissmann 2002a, 2007; Thinh et al. 2010a; Anandam et al. 2013). I recognize this species as many other authorities as monotypic, but further investigations are needed to address the question whether populations on Sumatra and the Malaysian Peninsular are distinct at a subspecies level.

The genus *Nomascus* encompasses all crested gibbons and currently comprises seven species (Table 7.1; Fig. 7.2; Thinh et al. 2010b; Rawson et al. 2011; Anandam et al. 2013). Although originally crested gibbons were thought to represent the single species *Nomascus concolor* (Simonetta 1957; Napier and Napier 1967; Groves 1972; Chivers 1977; Haimoff et al. 1982; Marshall and Sugardjito 1986), increasing evidence of morphological, genetic, acoustic and behavioural variation collected on different populations during the past decades strongly supports the notion that more than one species should be recognized (Tien 1983; Groves 1984, 1993, 2001; Ma and Wang 1986; Geissmann 1989, 1993, 1995, 2002a, 2007; Groves and Wang 1990; Fooden 1996; Geissmann et al. 2000; Roos 2004; Takacs et al. 2005; Mootnick 2006; Monda et al. 2007; Roos et al. 2007; Thinh et al. 2010a, b, c, 2011; Mootnick and Fan 2011; Rawson et al. 2011). First, all light-cheeked gibbons (todays *Nomascus leucogenys*, *Nomascus siki*, *Nomascus annamensis*, and *Nomascus gabriellae*) have been separated from the black crested gibbons (todays *N. concolor*, *Nomascus nasutus*, and *Nomascus hainanus*) mainly because of anatomical differences, especially in the size of the baculum (Tien 1983; Ma and Wang 1986). Subsequently, a species-level differentiation between

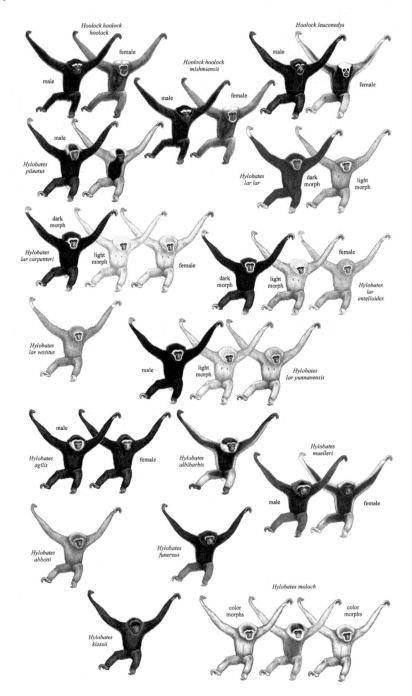

Fig. 7.2 Illustrations of all currently recognized gibbon species (illustrations by Stephen D. Nash, Conservation International and IUCN/SSC Primate Specialist Group)

white-cheeked and yellow-cheeked gibbons was proposed by Groves and Wang (1990) and Groves (1993). Groves and Wang (1990) suggested *N. siki* as subspecies of *N. gabriellae*, but later this taxon was recognized as a subspecies of *Nomascus leucogenys* (Geissmann 1993, 1995; Geissmann et al. 2000; Roos 2004; Roos et al. 2007) or even as a distinct species (Groves 2001; Mootnick 2006; Geissmann 2007; Thinh et al. 2010a, b, c, 2011; Mootnick and Fan 2011; Anandam et al. 2013). Acoustic and genetic data further implied that among light-cheeked crested gibbons, four rather than three taxa should be recognized (Geissmann et al. 2000; Geissmann 2003, 2007; Konrad and Geissmann 2006; Ruppell 2007; Thinh et al. 2010c, 2011) and consequently Thinh et al. (2010b) described this additional taxon as *N. annamensis*. Until the 1990s, gibbons with all-black males were still combined in the single species *N. concolor* (e.g., Groves 1993; Geissmann 1995), but based on prominent differences in vocalizations, a separation of *N. nasutus* (with *N. hainanus* as subspecies) from *N. concolor* was suggested (Geissmann et al. 2000; Geissmann 2002a). In contrast, Groves (2001) elevated *N. hainanus* as its own species, but kept *N. nasutus* as subspecies of *N. concolor*. However, the most recent acoustic and genetic data clearly justify that *N. nasutus* and *N. concolor* should both be recognized as distinct species (Geissmann et al. 2000; Roos 2004; Takacs et al. 2005; Geissmann 2007; Monda et al. 2007; Roos et al. 2007; Thinh et al. 2010a, b, c, 2011) as are *N. hainanus* and *N. nasutus* (Geissmann et al. 2000; Geissmann 2007; Roos et al. 2007; Thinh et al. 2010a, b, c). Within *N. concolor,* four subspecies (*N. c. concolor, N. c. furvogaster, N. c. jingdongensis*, and *N. c. lu*) have previously been recognized based on minor differences in fur length (Ma and Wang 1986; Geissmann 1993, 1995; Groves 2001; Mootnick and Fan 2011). However, because neither morphological (Geissmann 1989; Geissmann et al. 2000) nor acoustic data (Geissmann 2007) support this division, and likewise genetic data suggest a monotypic species *N. concolor*, it seems most appropriate to maintain *N. concolor* as monotypic, although *N. c. lu* might be distinct at a subspecies level (Thinh et al. 2010a, b, c). Whether *N. concolor* is indeed monotypic needs further investigation, but provisionally two subspecies (*N. concolor concolor* and *N. concolor lu*) are recognized (Rawson et al. 2011; Anandam et al. 2013).

Phylogenetic relationships among crested gibbons are, in contrast to unresolved branching patterns among gibbon genera and the *Hylobates* species, well settled, at least if we use mitochondrial sequence data (Roos 2004; Takacs et al. 2005; Roos et al. 2007; Thinh et al. 2010a, b, c). According to mitochondrial cytochrome b data, *N. nasutus* and *N. hainanus* form a sister clade to all other members of the genus, while *N. concolor* is basal to all light-cheeked gibbons. Within *N. concolor, N. c. lu* forms a separate clade, while the other three subspecies group together in a mixed clade. The light-cheeked gibbons further diverged into reciprocally monophyletic white-cheeked and yellow-cheeked gibbon clades, before finally both groups segregated into *N. leucogenys* and *N. siki*, and *N. annamensis* and *N. gabriellae*, respectively. This branching pattern is further substantiated by acoustic data (Geissmann et al. 2000; Thinh et al. 2011) and suggests that the genus originated in the North, perhaps northern Vietnam or southern China, and successively migrated to more southern parts of their current distribution range (Thinh et al. 2010a, c).

According to genetic data, the range expansion might have been a prolonged process that started around 4 million years ago and continued up until approximately 500,000 years ago (Thinh et al. 2010a). Nuclear sequence data of crested gibbons are still limited, but the first analyses suggest that gene flow among crested gibbon species is common and that nuclear diversity in this genus might therefore be much lower than mitochondrial diversity (Kim et al. 2011; Roos unpublished data).

Conclusions

The gibbon family is extraordinarily specious among the apes. Compared to great apes, gibbons are outstanding in the breath of their geographical distribution as much as in their species and subspecies diversity. During the long and complex history of gibbon classification, accompanied with various systematic revisions, the number of species increased steadily and today four genera, 19 species and a total of 25 taxa are recognized. This is mainly due to our greater knowledge of gibbons based on the greater number of studies that focused on small apes, wild and captive, in recent years, and included different aspects of their biology, e.g., behavior, morphology, physiology, ecology, genetics, and vocalizations, to name a few. However, compared to great apes, we still know relatively little about gibbons to fully appreciate their evolutionary history and diversity, and the evolutionary pressures that have shaped them. The latter is of particular interest since gibbons occupy a key position in the primate phylogeny between Old World monkeys and great apes, and thus are crucial to trace and illuminate early hominoid evolution. Likewise, gibbons are in several aspects unique among apes (e.g., brachiation, mainly pair-living, duet singing, karyotype diversity) making them an ideal model organism to study such features. Thus, much more research is necessary, not only to fully understand the evolutionary history and diversity of gibbons, but also to obtain more information on their genomics, general biology and ecology. This knowledge is certainly of interest per se, but in light of the dramatic conservation status of gibbons [all assessed species are classified as threatened (IUCN 2015)], is also of great importance to help save gibbons from extinction.

Acknowledgements I thank Claudia Barelli, Hiro Hirai and Ulrich Reichard for giving me the opportunity to contribute to this important book on small apes.

I am also very grateful as so many times before to Stephen Nash for allowing me to use his wonderful primate illustrations. Illustrations copyright 2013 Stephen D. Nash, Conservation International and IUCN/SSC Primate Specialist Group. Reproduced with permission.

References

Anandam MV, Groves CP, Molur S, Rawson BM, Richardson MC, Roos C et al (2013) Species accounts of Hylobatidae. In: Mittermeier RA, Rylands AB, Wilson DE (eds) Handbook of the mammals of the world, 3, Primates. Lynx Edicions, Barcelona, pp 778–791

Andayani N, Morales JC, Forstner MRJ, Supriatna J, Melnick DJ (2001) Genetic variability in mtDNA of the silvery gibbon: implications for the conservation of a critically endangered species. Conserv Biol 15:770–775

Bailey WJ, Fitch DH, Tagle DA, Czelusniak J, Slightom JL, Goodman M (1991) Molecular evolution of the psi eta-globin gene locus: gibbon phylogeny and the hominoid slowdown. Mol Biol Evol 8:155–184

Biegert J (1973) Dermatoglyphics in gibbons and siamangs. In: Rumbaugh DM (ed) Gibbon and siamang. Karger, Basel, pp 163–184

Brandon-Jones D, Eudey AA, Geissmann T, Groves CP, Melnick DJ, Morales JC et al (2004) Asian primate classification. Int J Primatol 25:97–164

Brockelman WY, Gittins SP (1984) Natural hybridization in the *Hylobates lar* species group: implications for speciation in gibbons. In: Preuschoft H, Chivers DJ, Brockelman WY, Creel N (eds) The lesser apes: evolutionary and behavioural biology. Edinburgh University Press, Edinburgh, pp 498–532

Carbone L, Vessere GM, ten Hallers BF, Zhu B, Osegawa K, Mootnick A et al (2006) A high-resolution map of synteny disruptions in gibbon and human genomes. PLoS Genet 2:e223

Carbone L, Harris RA, Gnerre S, Veeramah KR, Lorente-Galdos B, Huddleston J et al (2014) The gibbon genome provides a novel perspective on the accelerated karyotype evolution of small apes. Nature 513:195–201

Chan YC, Roos C, Inoue-Murayama M, Inoue E, Shih CC, Pei KJ et al (2010) Mitochondrial genome sequences effectively reveal the phylogeny of *Hylobates* gibbons. PLoS ONE 5: e14419

Chan YC, Roos C, Inoue-Murayama M, Inoue E, Shih CC, Vigilant L (2012) A comparative analysis of Y chromosome and mtDNA phylogenies of the *Hylobates* gibbons. BMC Evol Biol 12:150

Chan YC, Roos C, Inoue-Murayama M, Inoue E, Shih CC, Pei KJ et al (2013) Inferring the evolutionary histories of divergences in *Hylobates* and *Nomascus* gibbons through multilocus sequence data. BMC Evol Biol 13:82

Chatterjee HJ, Ho SY, Barnes I, Groves C (2009) Estimating the phylogeny and divergence times of primates using a supermatrix approach. BMC Evol Biol 9:259

Chivers DJ (1977) The lesser apes. In: Bourne GH (ed) Prince Rainier III of Monaco. Primate conservation. Academic Press, New York, pp 539–598

Chivers DJ (2013) Family Hylobatidae (gibbons). In: Mittermeier RA, Rylands AB, Wilson DE (eds) Handbook of the mammals of the world, 3, Primates. Lynx Edicions, Barcelona, pp 754–777

Choudhury A (2013) Description of a new subspecies of hoolock gibbon *Hoolock hoolock* from Northeast India. Newsl J Rhino Found Nat NE India 9:49–59

Creel N, Preuschoft H (1984) Systematics of the lesser apes: a quantitative taxonomic analysis of craniometric and other variables. In: Preuschoft H, Chivers DJ, Brockelman WY, Creel N (eds) The lesser apes: evolutionary and behavioural biology. Edinburgh University Press, Edinburgh, pp 562–613

Darga LL, Baba ML, Weiss ML (1984) Molecular perspectives on the evolution of the lesser apes. In: Preuschoft H, Chivers DJ, Brockelman WY, Creel N (eds) The lesser apes: evolutionary and behavioural biology. Edinburgh University Press, Edinburgh, pp 448–466

Dene HT, Goodman M, Prychodko W (1976) Immunodiffusion evidence on the phylogeny of the primates. In: Goodman M, Tashian RE, Tashian JH (eds) Molecular anthropology. Plenum Press, New York, Genes and proteins in the evolutionary ascent of the primates, pp 171–195

Fabre PH, Rondrigues A, Douzery EJ (2009) Patterns of macroevolution among primates inferred from a supermatrix approach of mitochondrial and nuclear DNA. Mol Phylogenet Evol 53:808–825

Felsenstein J (1987) Estimation of hominoid phylogeny from a DNA hybridization data set. J Mol Evol 26:123–131

Fleagle JG (ed) (1999) Primate adaptation and evolution. Academic Press, New York

Fooden J (1996) Zoogeography of Vietnamese primates. Int J Primatol 17:845–899

Garza JC, Woodruff DS (1992) A phylogenetic study of the gibbons (Hylobatidae) using DNA obtained noninvasively from hair. Mol Phylogenet Evol 1:202–210

Geissmann T (1989) A female black gibbon, *Hylobates concolor* subspecies, from northeastern Vietnam. Int J Primatol 10:455–476

Geissmann T (1993) Evolution of communication in gibbons (Hylobatidae). Dissertation, University of Zurich

Geissmann T (1995) Gibbon systematics and species identification. Int Zoo News 42:467–501

Geissmann T (2002a) Taxonomy and evolution of gibbons. In: Soligo C, Anzenberger G, Martin RD (eds) Anthropology and primatology into the third millennium: the centenary congress of the Zurich Anthropological Institute. Wiley-Liss, New York, pp 28–31

Geissmann T (2002b) Duet-splitting and the evolution of gibbon songs. Biol Rev 77:57–76

Geissmann T (ed) (2003) Vergleichende Primatologie (Comparative primatology). Springer, Heidelberg

Geissmann T (2007) Status reassessment of the gibbons: results of the Asian primate red list workshop 2006. Gibbon J 3:5–15

Geissmann T, Dang NX, Lormée N, Momberg F (2000) Vietnam primate conservation status review 2000. Part 1: gibbons. Fauna and Flora International, Hanoi

Goldman D, Giri PR, O'Brien SJ (1987) A molecular phylogeny of the hominoid primates as indicated by two-dimensional protein electrophoresis. Proc Natl Acad Sci USA 84:3307–3311

Goodman M, Tagle DA, Fitch DHA, Bailey W, Czelusniak J, Koop BF et al (1990) Primate evolution at the DNA level and a classification of hominoids. J Mol Evol 30:260–266

Goodman M, Porter CA, Czelusniak J, Page SL, Schneider H, Shoshani J et al (1998) Toward a phylogenetic classification of primates based on DNA evidence complemented by fossil evidence. Mol Phylogenet Evol 9:585–598

Groves CP (1972) Systematics and phylogeny of gibbons. In: Rumbaugh DM (ed) Gibbon and siamang. Karger, Basel, pp 1–89

Groves CP (1984) A new look at the taxonomy and phylogeny of the gibbons. In: Preuschoft H, Chivers DJ, Brockelman WY, Creel N (eds) The lesser apes: evolutionary and behavioural biology. Edinburgh University Press, Edinburgh, pp 542–561

Groves CP (ed) (1989) A theory of human and primate evolution. Clarendon Press, Oxford

Groves CP (1993) Speciation in living hominoid primates. In: Kimbel WH, Martin LB (eds) Species, species concepts, and primate evolution. Plenum Press, New York, pp 109–121

Groves CP (ed) (2001) Primate taxonomy. Smithsonian Institution Press, Washington

Groves CP, Wang Y (1990) The gibbons of the subgenus *Nomascus* (Primates, Mammalia). Zool Res 11:147–154

Haimoff EH, Chivers DJ, Gittens SP, Whitten AJ (1982) A phylogeny of gibbons (*Hylobates* spp.) based on morphological and behavioural characters. Folia Primatol 39:213–237

Hall LM, Jones DS, Wood BA (1998) Evolution of the gibbon subgenera inferred from cytochrome b DNA sequence data. Mol Phylogenet Evol 10:281–286

Hayashi S, Hayasaka K, Takenaka O, Horai S (1995) Molecular phylogeny of gibbons inferred from mitochondrial DNA sequences: preliminary report. J Mol Evol 41:359–365

Hellekant G, DuBois G, Geissmann T, Glaser D, Van der Weel H (1990) Taste responses of chorda tympani proper nerve in the white-handed gibbon (*Hylobates lar*). In: Døving KB (ed) Proceedings of the 10th international symposium on olfaction and taste, University of Oslo, 16–20 July 1989, pp 115–131

Hirai H, Wijayanto H, Tanaka H, Mootnick AR, Hayano A, Perwitasari-Farajallah D et al (2005) A whole-arm translocation (WAT8/9) separating Sumatran and Bornean agile gibbons, and its evolutionary features. Chromosome Res 13:123–133

Hirai H, Hayanao A, Tanaka H, Mootnick AR, Wijaynto H, Perwitasari-Farajallah D (2009) Genetic differentiation of agile gibbons between Sumatra and Kalimantan in Indonesia. In: Lappan S, Whittacker DJ (eds) The gibbons: new perspectives on small ape socioecology and population biology. Springer, New York, pp 37–49

Israfil H, Zehr SM, Mootnick AR, Ruvolo M, Steiper ME (2011) Unresolved molecular phylogenies of gibbons and siamangs (family: Hylobatidae) based on mitochondrial, Y-linked, and X-linked loci indicate a rapid Miocene radiation or sudden vicariance event. Mol Phylogenet Evol 58:447–455

IUCN (2015) The IUCN red list of threatened species. Version 2015.1. http://www.iucnredlist.org. Accessed 18 June 2015

Kim SK, Carbone L, Becquet C, Mootnick AR, Li DJ, de Jong PJ et al (2011) Patterns of genetic variation within and between gibbon species. Mol Biol Evol 28:2211–2218

Konrad R, Geissmann T (2006) Vocal diversity and taxonomy of *Nomascus* in Cambodia. Int J Primatol 27:713–745

Liu R, Shi L, Chen Y (1987) A study on the chromosomes of white-browed gibbon (*Hylobates hoolock leuconedys*). Acta Theriol Sinica 7:1–7. (Chinese text, English summary)

Ma S, Wang Y (1986) The taxonomy and distribution of the gibbons in southern China and its adjacent region, with description of three new subspecies. Zool Res 7:393–410

Marshall JT, Brockelman WY (1986) Pelage of hybrid gibbons (*Hylobates lar* x *H. pileatus*) observed in Khao Yai National Park, Thailand. Natl Hist Bull Siam Soc 34:145–157

Marshall JT, Sugardjito J (1986) Gibbon systematics. In: Swindler DR, Erwin J (eds) Comparative primate biology. Volume 1: systematics, evolution, and anatomy. Alan R. Liss, New York, pp 137–185

Matsudaira K, Ishida T (2010) Phylogenetic relationships and divergence dates of the whole mitochondrial genome sequences among three gibbon genera. Mol Phylogenet Evol 54:33–37

Matusaidra K, Reichard UH, Malaivijitnond S, Ishida T (2013) Molecular evidence for the introgression between *Hylobates lar* and *H. pileatus* in the wild. Primates 54:33–37

Meyer TJ, McLain AT, Oldenburg JM, Faulk C, Bourgeois MG, Conlin E et al (2012) An *Alu*-based phylogeny of gibbons (Hylobatidae). Mol Biol Evol 29:3441–3450

Misceo D, Capozzi O, Roberto R, Dell'Oglio MP, Rocchi M, Stanyon R et al (2008) Tracking the complex flow of chromosome rearrangements from the Hominoidea ancestor to extant gibbons *Nomascus* and *Hylobates* by high-resolution synteny mapping. Genome Res 18:1530–1537

Monda K, Simmons RE, Kressirer P, Su B, Woodruff DS (2007) Mitochondrial DNA hypervariable Region-1 sequence variation and phylogeny of the concolor gibbons, *Nomascus*. Am J Primatol 69:1–22

Mootnick AR (2006) Gibbon (Hylobatidae) species identification recommended for rescue or breeding centers. Primate Conservation 21:103–138

Mootnick AR, Fan P (2011) A comparative study of crested gibbons (*Nomascus*). Am J Primatol 73:135–154

Mootnick AR, Groves CP (2005) A new generic name for the hoolock gibbon (Hylobatidae). Int J Primatol 26:971–976

Müller S, Hollatz M, Wienberg J (2003) Chromosomal phylogeny and evolution of gibbons (Hylobatidae). Human Genet 113:493–501

Napier JR, Napier PH (1967) A handbook of living primates. Academic Press, London

Nowak RM (1999) Walker's primates of the world. John Hopkins University Press, Baltimore

Perelman P, Johnson WE, Roos C, Seuanez HN, Horvath JE, Moreira MAM et al (2011) A molecular phylogeny of living primates. PLoS Genet 7:e1001342

Prouty LA, Buchanan PD, Pollitzer WS, Mootnick AR (1983) *Bunopithecus*: a genus-level taxon for the hoolock gibbon (*Hylobates hoolock*). Am J Primatol 5:83–87

Rawson BM, Insua-Cao P, Ha NM, Thinh VN, Duc HM, Mahood S, Geissmann T, Roos C (eds) (2011) The conservation status of gibbons in Vietnam. Fauna and Flora International and Conservation International, Hanoi

Remane A (1921) Beiträge zur Morphologie des Anthropoidengebisses. Wiegmann-Archiv für Naturgeschichte 87:1–179

Roos C (2004) Molecular evolution and systematics of Vietnamese primates. In: Nadler T, Streicher U, Long HT (eds) Conservation of primates in Vietnam. Frankfurt Zoological Society, Hanoi, pp 23–28

Roos C, Geissmann T (2001) Molecular phylogeny of the major hylobatid divisions. Mol Phylogenet Evol 19:486–494

Roos C, Thanh VN, Walter L, Nadler T (2007) Molecular systematics of Indochinese primates. Vietn J Primatol 1:41–53

Rowe N (ed) (1996) The püictorial guide to the living primates. Pogonias Press, New York

Ruppell J (2007) The gibbons of Phong Nha-Ke Bang National Park. Gibbon J 3:50–55

Salem AH, Ray DA, Xing J, Callinan PA, Myers JS, Hedges A et al (2003) Alu elements and hominid phylogenetics. Proc Natl Acad Sci USA 100:12787–12791

Sarich VM, Cronin JE (1976) Molecular systematics of the primates. In: Goodman M, Tashian RE, Tashian JH (eds) Molecular anthropology. Plenum Press, New York, Genes and proteins in the evolutionary ascent of the primates, pp 141–170

Sawalischin M (1911) Der Musculus flexor communis brevis digitorum pedis in der Primatenreihe, mit spezieller Berücksichtigung der menschlichen Varietäten. Morphologisches Jahrbuch 42:557–663

Schultz AH (1933) Observations on the growth, classification and evolutionary specialization of gibbons and siamangs. Hum Biol 5:212–255, 385–428

Schultz AH (1973) The skeleton of the Hylobatidae and other observations on their morphology. In: Rumbaugh DM (ed) Gibbon and siamang. Karger, Basel, pp 1–54

Sibley CG, Ahlquist JE (1984) The phylogeny of hominoid primates, as indicated by DNA-DNA hybridization. J Mol Evol 20:2–15

Sibley CG, Ahlquist JE (1987) DNA hybridization evidence of hominoid phylogeny: results from an expanded data set. J Mol Evol 26:99–121

Simonetta A (1957) Catalogo e sinonimia annotata degli ominoidi fossili ed attuali (1758–1955). Atti della Società Toscana di Scienze Naturali, Memorie Serie B 64:53–113

Stanyon R (2013) Cytogenetic studies of small ape (Hylobatidae) chromosomes. Tsitologiia 55:167–171

Takacs Z, Morales JC, Geissmann T, Melnick DJ (2005) A complete species-level phylogeny of the Hylobatidae based on mitochondrial ND3-ND4 gene sequences. Mol Phylogenet Evol 36:456–467

Thinh VN, Mootnick AR, Geissmann T, Li M, Ziegler T, Agil M et al (2010a) Mitochondrial evidence for multiple radiations in the evolutionary history of small apes. BMC Evol Biol 10:74

Thinh VN, Mootnick AR, Thanh VN, Nadler T, Roos C (2010b) A new species of crested gibbon, from the Central Annamite Mountain Range. Vietn J Primatol 4:1–12

Thinh VN, Rawson B, Hallam C, Kenyon M, Nadler T, Walter L et al (2010c) Phylogeny and distribution of crested gibbons (genus Nomascus) based on mitochondrial cytochrome b gene sequence data. Am J Primatol 72:1047–1054

Thinh VN, Hallam C, Roos C, Hammerschmidt K (2011) Concordance between vocal and genetic diversity in crested gibbons. BMC Evol Biol 11:36

Tien DV (1983) On the north Indochinese gibbons (Hylobates concolor) (Primates: Hylobatidae) in north Vietnam. J Hum Evol 12:367–372

Wall JD, Kim SK, Luca F, Carbone L, Mootnick AR, de Jong PJ et al (2013) Incomplete lineage sorting is common in extant gibbon genera. PLoS ONE 8:e53682

Whittaker DJ, Morales JC, Melnick DJ (2007) Resolution of the Hylobates phylogeny: congruence of mitochondrial D-loop sequences with molecular, behaviroal, and morphological data sets. Mol Phylogenet Evol 45:620–628

Wienberg J, Stanyon R (1987) Fluorescent heterochromatin staining in primate chromosomes. Hum Evol 2:445–457

Wislocki GB (1929) On the placentation of primates, with a consideration of the phylogeny of the placenta. Contrib Embryol 20:51–80

Wislocki GB (1932) On the female reproductive tract of the gorilla, with a comparison of that of other primates. Contrib Embryol 23:163–204

Part III
Evolution of Gibbon and Siamang Morphology and Locomotion

Chapter 8
Why Is the Siamang Larger Than Other Hylobatids?

Ulrich H. Reichard and Holger Preuschoft

Adult male white-handed gibbon listening to the duet song of a neighboring gibbon group, Khao Yai National Park, Thailand. Photo credit: Ulrich H. Reichard

U.H. Reichard (✉)
Department of Anthropology and Center for Ecology, Southern Illinois University Carbondale, Carbondale, IL 62901, USA
e-mail: ureich@siu.edu

H. Preuschoft
Former Subdepartment Functional Morphology, Anatomical Institute, Medical Faculty, Ruhr-Universität Bochum, Bochum, Germany
e-mail: Holger.Preuschoft@ruhr-uni-bochum.de

© Springer Science+Business Media New York 2016
U.H. Reichard et al. (eds.), *Evolution of Gibbons and Siamang*,
Developments in Primatology: Progress and Prospects,
DOI 10.1007/978-1-4939-5614-2_8

Introduction

Hyloabtids are not only the most speciose of the hominoids (Harrison 2010; Anandam et al. 2013), but they also are among the species-richest families of all primates. Although hoolocks (~6.9 kg) and crested gibbons (7.3–7.8 kg) are slightly larger than most hylobatids (i.e., gibbons of the lar group share a small, narrow body size range of only 5.3–5.9 kg), the siamang weighing 10.7–11.9 kg (Smith and Jungers 1997)[1] are roughly 1.5–2 times larger than other gibbons. Discrete body size variation of the magnitude seen in hylobatids is rare within primate families and similarly expressed only in the Lorisidae. According to Nekaris and Bearder (2011), the Asian and African species fall into either a 'gracile' or a 'robust' size category, with robust species being anywhere between 2 and 6 times larger than the gracile species. Why hylobatids and lorisids display discrete, punctuated body size variation, while in other primate lineages body size variation, if it does exist, occurs along a continuum across species, is not well understood.

The large body size of siamang appears even more enigmatic if it is assumed that siamangs diverged after the smaller hoolock and/or crested gibbons from the stem hylobatid lineage (Takacs et al. 2005; Chatterjee 2006; Thinh et al. 2010), because it implies that siamangs secondarily increased in size again after stem hylobatids had already undergone substantial body size reduction (Groves 1972, 1984; Tyler 1991; Gebo 2004; Pilbeam and Young 2004; Jablonski and Chaplin 2009).

Because identifying ancestral relationships between species with certainty has proven difficult (Carbone et al. 2014), hylobatid phylogeny is not without controversy. However, past and some recent molecular genetics studies suggest that a basal position of the siamang is unlikely (Chivers 1977; Haimoff et al. 1982; Hayashi et al. 1995; Zhang 1997; Zehr 1999; Roos and Geissmann 2001; Müller et al. 2003; Takacs et al. 2005; Chatterjee 2006; Thinh et al. 2010). This argument is indirectly supported by two observations: (1) extant siamang are restricted to southern latitudes, beginning as far south as the southern peninsula of Malaysia. In fact, current consensus places the origin of hylobatids at northern latitudes, most likely north of the Tropic of Cancer (Chatterjee 2009), where historically only crested gibbons and hoolocks have been identified (Jablonski and Chaplin 2009). However, it is possible that fossil stem siamang have not yet been found at northern latitudes merely because hylobatid fossils are generally rare. (2) Duet songs of siamang are more complex with more derived features than duets of other hylobatids and are therefore, as Geissmann (2002) has argued, more likely to have evolved from a simpler, 'older' hylobatid duet song form than vice versa. The possibility that the siamang secondarily increased in body size, i.e., after stem

---

[1]Body mass data were taken from Smith and Jungers (1997). We recognize that sample sizes for several hylobatid species are small, some even include only a single male and female (e.g., silvery and pileated gibbons). However, because we were interested in broad size differences we felt that the partially inadequate representation of some hylobatid species would be of no consequence to our measurements.

hylobatids had already undergone dwarfing, while all other hylobatids remained small, makes investigating the evolutionary pressures that may have triggered this body size increase an intriguing undertaking. However, despite the size variation seen in hylobatids, both body size classes still fall within the bracket of biomechanically efficient brachiation, which only becomes energetically problematic in species beyond the size of the siamang, such as great apes and humans (Preuschoft et al. 2016).

Even if we assume that the siamang was the basal hylobatid group representing the original body size of the lineage (Groves 1972; Creel and Preuschoft 1984; Garza and Woodruff 1992; Hall et al. 1998; Carbone et al. 2014), which would be more in line with Cope's rule stating that evolving lineages increase rather than decrease in size over time (Hone and Benton 2005; Bonner 2006; Sander et al. 2011), this still does not account for why siamangs are larger than other gibbons. Two scenarios seem plausible: (1) the siamang may represent continuity in body size from a stem hylobatid or even a stem hominoid (Harrison 2016); (2) the siamang may represent the first down-sized step of body size reduction in hylobatids since the split from larger-bodied stem hominoids such as, for example, *Afropithecus* or *Griphopithcus*, which are believed to have exceeded 20 kg (Begun 2013). If the siamang is indeed basal to hylobatid phylogeny and represents a first step in size reduction, the question arises as to why it did not undergo further size reduction like other hylobatids, but became morphologically 'arrested' at a medium body size of 11–12 kg. One possible explanation may be that at its size the siamang had reached a stage at which the cost of further size reduction would no longer be offset by foraging advantages. At the size of a siamang brachiation is energetically efficient (Preuschoft et al. 2016) and its size may have allowed the siamang to economically reach the terminal-branch feeding niche not easily accessible to other, flightless frugivores.

Irrespective of controversies regarding the position of the siamang in the hylobatid phylogeny, the key question of why the siamang is nearly twice the size of small hylobatids remains unresloved. Siamangs are unusual among hylobatids with respect to, not only their discrete body size, but also their sympatric occurrence across their current geographic range. They are sympatric with the small agile gibbons of south Sumatra as well as with white-handed gibbons in north Sumatra and on the Malaysian peninsula (Chatterjee 2009). Considering the ability of siamangs to coexist with other hylobatids, it seems plausible that ecological niche differentiation may have been the underlying reason for both body size variation and sympatry, because if two closely related species occupy the same ecological niche and geographic area, then they will diverge in some dimension of niche use to allow sympatry (Brown and Wilson 1956).

Previously, Raemaekers (1978, 1984) has remarked that Malaysian siamang were slower but physically stronger than the sympatric smaller gibbons, allowing the two species to coexist in the same habitat. The siamang, he assumed, would be able to directly displace the smaller gibbons from contested food trees due to a greater resource holding potential based on their larger size (Morse 1974; Maynard Smith 1982; Abrams 1983), while the more agile, smaller gibbons would indirectly outcompete the larger siamang through faster visitation and re-visitation rates of food sources, longer day ranges, and by maintaining smaller home ranges. Indeed, field studies of the past decades have confirmed that most siamang groups have slightly shorter day ranges and smaller home ranges (MacKinnon and MacKinnon 1980; Bartlett 2011).

Many of the arguments attempting to explain sympatry between the siamang and small gibbons have been built on the gradual difference in foraging and diet, primarily siamangs' greater folivory compared to other gibbons, which has been noted since the earliest siamang field studies (Chivers 1974; MacKinnon 1977; Raemaekers 1977). A larger body size and correspondingly longer intestinal tract may allow for longer, slower digestion, and more efficient nutrient extraction from leaves. Chivers and Raemaekers (1986) found that the colon of the siamang is relatively bulky, suggesting that some bacterial fermentation is likely to occur, but it remains unclear just how much nutritional benefit the siamang can gain from digesting these more fiberous resources.

The idea of ecological niche differentiation is in agreement with observations of less frugivory in the siamang than in both sympatric lar gibbons and long-tailed macaques (MacKinnon and MacKinnon 1980). Alongside the argument of lower fruit consumption in siamang, it has been hypothesized that a more folivorous diet also led to the evolution of a larger size, because folivory is generally related to larger body sizes among closely related, primarily fruit-eating primates (Clutton-Brock and Harvey 1977a; Richards 1985; Harvey et al. 1987). In fact, Clutton-Brock and Harvey (1977b: 8) used hylobatids as an example of the relationship between body size and diet categories (i.e. fruit and leaf foods):

> Among diurnal primates, there is a tendency for terrestrial species to be heavier than arboreal ones and, within the latter two categories, for folivores to be heavier than frugivores. Thus among pairs of related species, the siamang (*Symphalangus syndactylus*) is larger than the more frugivorous gibbons (*Hylobates* spp.) and the gorilla (*Gorilla gorilla*) is larger than the chimpanzee (*Pan troglodytes*).

The notion that proportionally different diets would explain the striking body size difference between the siamang and the relatively smaller gibbons has been uncritically accepted until recently (Elder 2009), despite empirical evidence that the ecological niches of both the siamangs and gibbons overlap to a large degree (Raemaekers 1977, 1979, 1984; Grether et al. 1992; Palombit 1997). Moreover, primate diets should not be represented as static categories (i.e., percent leaves or fruit), since primates generally pursue flexible foraging strategies that include continuous shifting between dietary categories in relation to food availability (Oats 1987). As Gittins (1982) demonstrates, on a single day agile gibbons combine distinct fruit-feeding bouts at large sources with short periods of foraging for insects and other more scattered foods and periods of traveling and resting.

With the development over the past decades of better techniques to directly measure nutrient contents of foods, it is now becoming increasingly clear that dietary categories such as 'fruit' or 'leaf' are too broad to capture the nutritional variation of primate foods (Lambert 2011). Danish et al. (2006), for example, analyzed diets of red colobus and redtail monkeys and found higher sugar concentrations in the leaves eaten by red colobus than in the fruit consumed by redtail monkeys. Thus, categorizing primates as 'frugivore' or 'folivore' neglects overall nutritional and temporal variation in food consumption particularly in primates with overlapping diets (Chapman et al. 2012). These considerations undermine the assumption that subtle differences in the amount of leaves consumed by siamangs and gibbons can be sufficient to explain obvious body size differences. Bartlett's (2011) review of hylobatid feeding behaviors shows that siamang at Ketambe in Sumatra, Indonesia, included as much fruit in their diet as silvery gibbons at Ujong Kulan, Borneo, Indonesia, and more than agile gibbons at Sungai Dal, Malaysia. Conversely, although siamang at Kuala Lompat, Malaysia, included larger amounts of leaves in their diet than most species of the *Hylobates* group, the proportion of leaves in their diet was comparable to those seen in species of crested gibbons. A recent review of wild hylobatid diets, including siamang, likewise found neither a significant difference in the time spent feeding on leaves across the hylobatid family, nor a correspondence between body mass and variation in time spent feeding on leaves (Elder 2009). Elder (2009) concluded, based on a broad comparative sample and detailed analyses, that the diet of siamang does not differ from that of smaller-bodied hylobatids, echoing Palombit's (1997) conclusion that all hylobatids prefer fruit over other foods and spend large proportions of time feeding on figs, regardless of body size. Moreover, using the 'coefficient of gut differentiation', i.e., the ratio of the surface area of the stomach, caecum, and colon to that of the small intestine, the siamang is most similar to other frugivores (Chivers and Hladik 1984). In fact, the most folivorous of the hylobatids are probably crested gibbons, who live at high altitudes in ecologically challenging and highly seasonal mountain habitats where fig trees, as a fall back food, are not available and fruit is generally absent for several months of the year (Haimoff et al. 1986; Bleisch and Chen 1991; Lan 1993). In summary, because of the broad dietary overlap between siamang and sympatric small gibbons, ecological niche separation seems an inadequate explanation for their body size differences.

It has been hypothesized that larger body size may evolve in response to high predation pressure as larger individuals may be better able to escape from or defend themselves against predators (Clutton-Brock and Harvey 1977a, b). It is unclear, however, if this hypothesis can be applied to explain the larger size of siamang as it was developed specifically for terrestrial prey and their predators. Some indirect support for the hypothesis that the larger size of siamang could be related to predation pressure may be drawn from Preuschoft and colleagues' (2016) observations that: due to its larger size and a corresponding longer forelimb length, the siamang is able to produce a greater slapping force than a smaller gibbon, which at least theoretically could translate into an advantage in fending off a potential predator. Not only is the slapping force of the siamang greater, and thus more dangerous compared to a blow from a smaller gibbon, but the heavier siamang body also contains greater kinetic energy ($\frac{1}{2} m \cdot v_2$), which increases its potential fighting power compared to the lighter gibbon in cases of physical contact with a predator. In fact, a small predator will find it more difficult to keep down a larger, compared to a smaller prey animal, when the prey struggles, and the impact of actively 'ramming' a predator will likewise be greater from a larger-bodied animal. The kinetic energy advantage would also apply when two suspended hylobatids come into contact during a competitive situation, because the smaller individual could be displaced more easily by the heavier than vice versa. However, regarding hylobatids, including the siamang, no conclusive evidence of the extent of predation exists with which to evaluate whether predation pressure may have been a driving force for siamangs' larger body size, even if one considers the anecdotal observation of a dead siamang once found in a python's stomach (Schneider 1906). Hylobatid hair has occasionally been detected in the scat of large cats (Rabinowitz 1989) and white-handed gibbons show differential responses to potential predators and non-predators (Uhde and Sommer 2000; Clarke et al. 2006, 2012). These observations suggest that predation may have played a role in hylobatid evolution and may have led to specific behavioral and morphological adaptations. However, no difference in predation risk between the siamang and small gibbons has yet been reported. Hylobatid foraging patterns, the absence of large eagles in southeast Asia, high-speed ricochetal brachiation and exclusive arboreal lifestyle, which aids in detecting predators from a long distance (van Schaik et al. 1983), as well as have led to a consensus among most scholars that hylobatids face low predation risks (Carpenter 1940; Ellefson 1974; Raemaekers and Chivers 1980; Leighton 1987; Grether et al. 1992).

One unexplored explanation of the larger body size of siamang is the metabolic advantage associated with larger size. Raemaekers (1984) first suggested that the siamang may have originally evolved in geographical separation from smaller gibbons. The habitat for which the siamang body size is adaptive may not include only lowland but also montane tropical forests, where a larger size is advantageous due to the relatively smaller surface area of a larger organism. With a relatively smaller surface area per unit body mass, the siamang would lose less heat energy in a cooler climate than a smaller gibbon. Moreover, a parallel shift toward a more leaf-tolerant diet would additionally have allowed the siamang to survive better at higher elevations, where seasonal effects on fruit availability become more pronounced and overall plant species diversity is reduced (Cannon et al. 2007a, b).

While the allometric idea that larger body size entails a more favorable mass-to-surface ratio (i.e., Bergmann's rule; Bergmann 1848; Ashton et al. 2000) is commonly held among zoologists and paleontologists, such a hypothesis has never been tested in primates. It has thus also not been used to explain the larger body size of siamang. However, former work (Preuschoft 2010; Preuschoft et al. 2011) has shown that absolute body size cannot be neglected in this context. A larger body size is associated with relatively lower metabolic costs per unit body weight, because mass is proportional to a third-power function (i.e., volume) while surface area is proportional to a second-power function. Therefore, increases in surface area lags behind increases in mass. Here, we performed simple calculations of body proportions of gibbons and the larger siamang and compared these to orangutan and loris body sizes to test if the larger size of the siamang could be explained by a more favorable mass-to-surface ratio.

Materials and Methods

To test the hypothesis that the body sizes of gibbons and siamang could represent adaptations to different environments with different thermodynamic requirements, we assembled comparative data from orangutans, which are about 3–7 times larger than a 10–12 kg siamang (on average 36 kg in females and 79 kg in males: Smith and Jungers 1997), and Asian lorisids. We added estimates of body mass and body surface of Asian lorisids in order to expand the range of body sizes and to be able to verify the validity of our results for much smaller body sizes than those of orangutan and siamang.

Due to insufficient data of some measures from live specimens or cadavers, we chose to use a simplified body model of a combination of five cylinders: (1) body stem (head, neck, and trunk), (2) two forelimbs, und (3) two hindlimbs. The volume [V] of such a simplified body model can be calculated as

$$
\begin{aligned}
V = & \ (\text{body stem length} * \tfrac{1}{2} \ \text{stem diameter} * \pi) \\
& + 2 \left(\text{forelimb length} * \tfrac{1}{2} \ \text{forelimb diameter}^2 * \pi\right) \\
& + 2 \left(\text{hindlimb length} * \tfrac{1}{2} \ \text{hindlimb diameter}^2 * \pi\right)
\end{aligned}
\tag{8.1}
$$

The surface area [A] of the simplified body can be calculated as

$$
\begin{aligned}
A = & \ (\text{body stem length} * \text{stem diameter} * \pi) \\
& + 2 \,(\text{forelimb length} * \text{forelimb diameter} * \pi) \\
& + 2 \,(\text{hindlimb length} * \text{hindlimb diameter} * \pi)
\end{aligned}
\tag{8.2}
$$

Absolute body measurements were taken from Smith and Jungers (1997; body masses), Napier and Napier (1967; stem lengths [also referred to as sitting height]) and Schultz (1930; trunk and limb lengths; Table 8.1). Schultz (1930) published a remarkable amount of relative measurements, unfortunately without providing absolute sizes, except for the length of the humerus. Therefore, we were compelled to recalculate absolute radius length using the brachial index (radius length = humerus length * brachial index) and, considering humerus and radius lengths, we recalculated the combined length of femur and tibia with the aid of the intermembral index (femur length + tibia length = humerus length + radius length /intermembral index). Data on hand and foot lengths were taken from Schultz (1956). Limb diameters were estimated based on occasional measurements of primate cadavers by H. Preuschoft and data collected by M. Günther, partly at the University of Liverpool, UK, and partly at the Primate Research Institute in Nagoya/Kyoto, Japan. In the case of lorisids, appropriate data were taken from Schulze's homepage (2016, www.loris-conservation.org/database). We assumed that the density of the primate body is sufficiently close to 1, which results in a calculated volume that should be directly proportional to mass (1 cubic cm = 1 g).

Table 8.1 Primate body measurements

	Weight[a]	Stem length[c]	Stem diameter[d]	Arm length[e]	Arm diameter[d]	Leg length[f]	Leg diameter[d]
Slender loris	0.231	22.3	3.0	16.8	0.8	19.7	0.8
Slow loris	0.823	32.7	6.0	20.2	1.3	23.1	1.3
Gibbon	5.690[b]	40.3	10.4	61.4	3.5	50.8	3.8
Siamang	10.700	53.8	13.0	72.6	4.0	53.8	4.8
Orangutan	35.700	76.8	22.0	85.6	8.0	66.2	9.0
Orangutan (M)	78.200	96.5	28.0	96.9	9.0	77.3	10.0

Weights expressed in kilogram and lengths expressed in centimeters
M male
[a]Average female mass after Smith and Jungers (1997)
[b]Excluding Hoolock and crested gibbons
[c]Data from Napier and Napier (1967)
[d]Estimates by H. Preuschoft and M. Günther
[e]Measurement includes the hand, data from Schultz (1939, 1956) and Schulze's database available at www.loris-conservation.org/database
[f]Measurement includes the foot, data from Schultz (1939, 1956) and Schulze's database

Results

Calculated volumes (Table 8.2) were in general agreement with masses reported by Smith and Jungers (1997), validating our base assumption that the density of a primate body is equal to one. As expected, across primate taxa considered here, the increase in mass was steeper than the increase in surface area (Fig. 8.1). Similarly, we found that the small slender loris has a much larger surface area to volume ratio

Table 8.2 Volume and surface calculations of primates

	Body volume [V][a]				Body surface [A][b]				Proportion	
	Stem	2 arms	2 legs	Total	Stem	2 arms	2 legs	Total	V/A	A/V
Slender loris	158	17	20	194	210	84	99	393	0.49	2.03
Slow loris	925	54	61	1040	616	165	189	970	1.07	0.93
Gibbon	3423	1181	1152	5757	1317	1350	1213	3880	1.48	0.67
Siamang	7141	1825	1947	10,913	2197	1825	1623	5644	1.93	0.52
Orangutan	29,194	8605	8423	46,223	5308	4303	3744	13,354	3.46	0.29
Orangutan (M)	59,420	12,329	12,142	83,891	8489	5480	4857	18,825	4.46	0.22

Based on body measures summarized in Table 8.1
M male
[a]Expressed in (cm^3)
[b]Expressed in (cm^2)

Fig. 8.1 Body mass and
surface measurements across
five nonhuman primate
species. *Blue line with
diamonds* represents body
mass (expressed in kg), while
red line with rectangular
represents surface (expressed
in cm²)

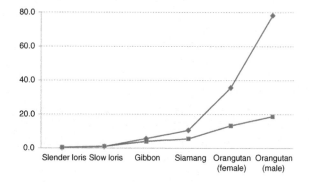

Fig. 8.2 Surface-to-volume
ratio measurements across
five nonhuman primate
species. *Green line with
squares* indicates
surface-to-volume ratio and
*orange line with orange
triangles* volume-to-surface
ratio

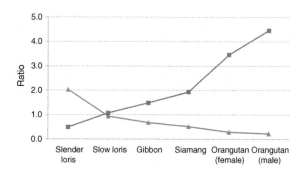

(2.03) compared to the same ratio for the very large male orangutan (0.22), the
species with the smallest relative surface area (Fig. 8.2). If closely related forms
(slender versus slow loris; gibbon versus siamang; female versus male orangutan)
are compared, the larger species invariably have a smaller relative surface area and
a greater relative volume (Fig. 8.2). So far, the general expectation is fulfilled.
However, closer inspection of the differences between the small and the large form
within a lineage revealed that the step from the smaller to the larger
volume-to-surface ratio appeared to be less pronounced among hylobatids than in
either lorisids or from orangutan female to orangutan male.

Discussion

The calculations presented here are based on realistic values of body proportions of
primates in tropical Asia. They represent empirical data suitable for inspecting the
commonly assumed biological explanation for any increases in body size in animal
lineages over time. The general expectation that animals with a greater body mass
would have a smaller relative surface area was confirmed for all primates included
in this analysis, which display a wide range of body masses (i.e., <1–70 kg).

In line with the main result, due to its body mass the siamang has a smaller relative heat dissipating surface area than the smaller gibbons. The more favorable volume-to-surface ratio allows siamangs to maintain physiological homeostasis at a relative lower energetic cost than hylobatids with a larger relative heat dissipating area. Therefore, the result supports the hypothesis that the siamang is better adapted to cooler climates of higher elevation or more northern latitudes, compared to hylobatids of smaller body size. Empirical data of population densities indirectly also support such an argument, because gibbon population densities decline rapidly with increasing altitude (Caldecott 1980; O'Brien et al. 2004). Marshall (2009), for example, has shown the negative impact of elevation on white-bearded gibbons in Borneo, where population density in montane forests (i.e., 750–1100 m.s.l.) was almost ten-times lower than in the next best habitat (i.e., upland granite forest on well drained granite soils at 350–800 m.s.l.). Based on demographic data and modeling, Marshall (2009) concluded that montane habitat was almost certainly sink habitat for the small-bodied, white-bearded gibbon. In contrast, siamang populations seem to survive quite well, although generally at low densities, at higher elevations including the top of mountains (Caldecott 1980; O'Brien et al. 2004). These results are in line with the assumption that the disparity in body size of hylobatids may have been the result of competition between siamang and the sympatric small-bodied gibbons.

However, our calculations do not allow firm conclusions of causal relationships between body size, thermoregulation, and geographic distribution. The energy conservation aspect underlying the assumption that siamang may be better adapted to cooler climates may be invalid because energy cannot be apportioned as a fraction of body mass but is an absolute investment for a particular total body size (McNab 2010). Thus, a larger body, although requiring relatively less energy per kilogram body weight than a smaller one, may still require more absolute energy investment to maintain homeostasis compared to a smaller body. A higher absolute energy requirement of siamang, compared to gibbons, due to its larger size may be compensated by siamangs' ability to digest more leaves, which, compared to fruit, are less challenging to find in tropical forests, allowing siamang to spend more time feeding than foraging for food compared to the smaller gibbons. Consistent with this idea, siamang have shorter average day ranges than smaller hylobatids (Bartlett 2011), which may mean that they can spend more time feeding rather than traveling. MacKinnon and MacKinnon (1980) compared the daily time spent feeding of sympatric siamang and lar gibbons and found that siamang spend 5.75 h feeding while lar gibbons spend only 3.4 h feeding. In agreement with these differences, Raemaekers and Chivers (1980) summarize long-term data of sympatric siamang and lar gibbons that show longer activity time, longer feeding time, and less travel time in siamang compared to lar gibbons. These results may explain how a larger siamang could perhaps compensate for the higher absolute energy requirement when competing with sympatric smaller gibbons, but they do not explain why the siamang is larger in the first place.

Schultz (1931, 1933) measured the density of hylobatid hairs, both in terms of relative density per cm^2 skin as well as total hair counts. Following these observations the siamang[2] has comparatively the lowest relative hair density on the back and chest and the second lowest relative density on the head and total hair count, despite its much larger size, in a sample of approximately 6 hylobatid species. Assuming that denser hair equals better insulation against cold temperatures, it appears that the siamang with its thin pelage would be least adapted to live in colder climates. These observations contrast with the assumption that siamang are better adapted to cooler climates than small gibbons, despite the fact that Pohl (1911) reports, based on museum specimens, that siamang living above 1500 m altitude have thicker and fuller fur than those from lowland animals.

So far, there is no suggestion that siamang have had a successful distribution at northern latitudes, while hoolock and crested gibbons are still found at much northern latitudes today. Interestingly, both hoolock and crested gibbons have a slightly larger anatomy compared to gibbons of the *lar* group (Zihlman et al. 2011). Perhaps, their slightly larger size is related to their more northern latitude distribution compared to the majority of gibbons of the *lar* group. However, both hoolock and crested gibbons are nonetheless considerably smaller than siamangs.

In conclusion, until more stem hylobatid fossils are recovered that help identify the historic biogeographic distribution of siamang and the question as to which genus represents the basal hylobatid body morphology has been resolved, the distinct body size of the siamang remains enigmatic. Although it seems reasonable to speculate that energetic advantages are associated with the larger body size and correspondingly smaller relative surface area in siamang, what kind of adaptive advantage the siamang may have gained is yet entirely unclear. Perhaps, after all no particular adaptive advantage may be associated with the larger body size of siamang, which may simply represent the original hylobatid body size, and future studies may be better advised to question why other gibbon genera are smaller than the siamang.

References

Abrams P (1983) The theory of limiting similarity. Ann Rev Ecol Syst 14:359–376

Anandam MV, Groves CP, Molur S, Rawson BM, Richardson MC, Roos C, Whittaker DJ (2013) Species accounts of Hylobatidae. In: Mittermeier RA, Rylands AB, Wilson DE (eds) Handbook of the mammals of the world, vol 3. Primates. Lynx Edicions, Barcelona, pp 778–791

Ashton KG, Tracy MC, de Queiroz A (2000) Is Bergmann's rule valid for mammals? Am Nat 156:390–415

Bartlett TQ (2011) The hylobatidae: small apes of Asia. In: Campbell CJ, Fuentes A, MacKinnon KC, Bearder SK, Stumpf RM (eds) Primates in perspective. Oxford University Press, Oxford, pp 300–312

---

[2]Six out of the 10 siamang individuals studied by Schultz were juveniles. However, also among juveniles, siamang had the lowest relative and absolute body hair counts.

Begun DR (2013) The Miocene hominoid radiations. In: Begun DR (ed) A companion to paleoanthropology. Blackwell Publishing Ltd, Oxford, pp 397–416

Bergmann C (1848) Ueber die Verhältnisse der Wärmeökonomie der Thiere zu ihrer Grösse. Göttinger Studien. Vandenhoeck und Ruprecht, Göttingen

Bleisch W, Chen N (1991) Ecology and behavior of wild black crested gibbons (Hylobates concolor) in China with a reconsideration of evidence of polygyny. Primates 32:539–548

Bonner JT (ed) (2006) Why size matters: from bacteria to blue whales. Princeton University Press, Princeton

Brown WL, Wilson EO (1956) Character displacement. Syst Zool 5:49–64

Caldecott JO (1980) Habitat quality and populations of two sympatric gibbons (Hylobatidae) on a mountain in Malaya. Folia Primatol 33:291–309

Cannon CH, Curran LM, Marshall AJ, Leighton M (2007a) Long-term reproductive behavior of woody plants across seven Bornean forest types in the Gunung Palung National Park, Indonesia: suprannual synchrony, temporal productivity, and fruiting diversity. Ecol Lett 10:956–969

Cannon CH, Curran LM, Marshall AJ, Leighton M (2007b) Beyond mast-fruiting events: community asynchrony and individual sterility dominate woody plant reproductive behavior across seven Bornean forest types. Curr Sci 93:21–29

Carbone L, Harris RA, Gnerre S, Veeramah KR, Lorente-Galdos B, Huddleston J et al (2014) Gibbon genome and the fast karyotype evolution of small apes. Nature 513:195–201

Carpenter CR (1940) A field study in Siam of the behavior and social relations of the gibbon (Hylobates lar). Comp Psych Monogr 16:1–201

Chapman CA, Rothman JM, Lambert JE (2012) Primate foraging strategies and nutrition: behavioral and evolutionary implications. In: Mitani J, Call J, Kappeler P, Palombit R, Silk J (eds) The evolution of primate societies. University of Chicago Press, Chicago, pp 145–167

Chatterjee HJ (2006) Phylogeny and biogeography of gibbons: a dispersal-vicariance analysis. Int J Primatol 27:699–712

Chatterjee HJ (2009) Evolutionary relationships among the gibbons: a biogeographic perspective. In: Lappan S, Whittaker DJ (eds) The Gibbons, developments in primatology: progress and prospects. Springer, New York, pp 13–36

Chivers DJ (ed) (1974) The Siamang in Malaya: a field study of a primate in tropical rain forest. In: Contributions to primatology, vol 4. Karger, Basel

Chivers DJ (1977) The lesser apes. In: Ranier HP, Bourne G (eds) Primate conservation. Academic Press, New York, pp 539–598

Chivers DJ, Hladik CM (1984) Diet and gut morphology in primates. In: Chivers DJ, Wood BA, Bilsborough A (eds) Food acquisition and processing in primates. Springer, New York, pp 213–230

Chivers DJ, Raemaekers JJ (1986) Natural and synthetic diets of Malayan gibbons. In: Else JG, Lee PC (eds) Primate biology and conservation. Cambridge University Press, Cambridge, pp 39–56

Clarke E, Reichard UH, Zuberbühler KM (2006) The syntax and meaning in wild gibbon songs. PLoS ONE 1:e73

Clarke E, Reichard UH, Zuberbühler KM (2012) The anti-predator behaviour of wild white-handed gibbons (Hylobates lar). Behav Ecol Sociobiol 66:85–96

Clutton-Brock TH, Harvey PH (1977a) Functional aspects of species differences in feeding and ranging behaviour in primates. In: Clutton-Brock TH (ed) Primate ecology: studies of feeding and ranging behaviour in lemurs, monkeys and apes. Academic Press, London, pp 557–579

Clutton-Brock TH, Harvey PH (1977b) Primate ecology and social organization. J Zool 183:1–39

Creel N, Preuschoft H (1984) Systematics of the lesser apes: a quantitative taxonomic analysis of craniometric and other variables. In: Preuschoft H, Chivers DJ, Brockelman WY, Creel N (eds) The lesser apes. Evolutionary and behavioural biology. Edinburgh University Press, Edinpurgh, pp 562–613

Danish L, Chapman CA, Hall MB, Rode KD, O'Discoll Worman C (2006) The role of sugar in diet selection in redtail and red colobus monkeys. In: Hohmann G, Robbins MM, Boesch C (eds) Feeding ecology in apes and other primates: ecological physiological, and behavioural aspects. Cambridge University Press, Cambridge, pp 473–488

Elder AA (2009) Hylobatid diets revisited: the importance of body mass, fruit availability, and interspecific competition. In: Lappan S, Whittaker DJ (eds) The Gibbons, developments in primatology: progress and prospects. Springer, New York, pp 133–159

Ellefson JO (1974) A natural history of white-handed gibbons in the Malayan Penisular. In: Rumbaugh DM (ed) Gibbon and Siamang, vol 3., Natural history, social behavior, reproduction, vocalizationsPrehension. Karger, Basel, pp 1–136

Garza JC, Woodruff DS (1992) A phylogenetic study of the gibbons (*Hylobates*) using DNA obtained noninvasively from hair. Mol Phylogenet Evol 1:202–210

Gebo DL (2004) Paleontology, terrestriality, and the intelligence of great apes. In: Russon AE, Begun DR (eds) The evolution of thought: evolutionary origins of great ape intelligence. Cambridge University Press, Cambridge, pp 320–334

Geissmann T (2002) Taxonomy and evolution of gibbons. Evol Anthropol 1:28–31

Gittins SP (1982) Feeding and ranging in the agile gibbon. Folia Primatol 38:39–71

Grether GF, Palombit RA, Rodman PS (1992) Gibbon foraging decisions and the marginal value model. Int J Primatol 13:1–17

Groves CP (1972) Systematics and phylogeny of gibbons. In: Rumbaugh D (ed) Gibbon and Siamang, vol 1. Karger, New York, pp 2–89

Groves CP (1984) A new look at the taxonomy and phylogeny of the gibbons. In: Preuschoft H, Chivers DJ, Brockelman WY, Creel N (eds) The lesser apes. Evolutionary and behavioural biology. Edinburgh University Press, Edinpurgh, pp 542–561

Haimoff EH, Chivers DJ, Gittins SP, Whitten T (1982) A phylogeny of gibbons (*Hylobates* spp.) based on morphological and behavioural characters. Folia Primatol 39:213–237

Haimoff E, Yang X, He S, Chen N (1986) Census and survey of wild black crested gibbons (*Hylobates concolor concolor*) in Yunnan Province, People's Republic of China. Folia Primatol 46:205–214

Hall LM, Jones D, Wood B (1998) Evolution of the gibbon subgenera inferred from cytochrome b DNA sequence data. Mol Phylogenet Evol 10:281–286

Harrison T (2010) Apes among the tangled branches of human origins. Science 327:532–534

Harrison T (2016) The fossil record and evolutionary history of hylobatids. In: Reichard UH, Hirohisa H, Barelli C (eds) Evolution of gibbons and siamang. Springer, New York, pp 91–110

Harvey PH, Martin RD, Clutton-Brock TH (1987) Life histories in comparative perspective. In: Smuts BB, Cheney DL, Seyfarth RM, Wrangham RW, Struhsaker TT (eds) Primate societies. Chicago University Press, Chicago, pp 181–196

Hayashi S, Hayasaka K, Takenaka O, Horai S (1995) Molecular phylogeny of gibbons inferred from mitochondrial DNA sequences: preliminary report. J Mol Evol 41:359–365

Hone DWE, Benton MJ (2005) The evolution of large size: how does Cope's rule work. Trends Ecol Evol 20:4–6

Jablonski NG, Chaplin G (2009) The fossil record of gibbons. In: Lappan S, Whittaker DJ (eds) The Gibbons, developments in primatology: progress and prospects. Springer, New York, pp 112–130

Lambert JE (2011) Primate nutritional ecology. In: Campbell CJ, Fuentes A, MacKinnon KC, Bearder SK, Stumpf RM (eds) Primates in perspective. Oxford University Press, Oxford, pp 512–522

Lan D (1993) Feeding and vocal behaviors of black gibbons (*Hylobates concolor*) in Yunnan: a preliminary study. Folia Primatol 60:94–105

Leighton DR (1987) Gibbons: territoriality and monogamy. In: Smuts BB, Cheney RM, Seyfarth RW, Wrangham RW, Struhsaker TT (eds) Primate societies. Chicago University Press, Chicago, pp 135–145

MacKinnon JR (1977) A comparative ecology of Asian apes. Primates 18:747–772

MacKinnon JR, MacKinnon KS (1980) Niche differentiation in a primate community. In: Chivers DJ, Raemaeker JJ (eds) Malayan forest primates: ten years' study in tropical rain forest. Plenum Press, New York, pp 167–190

McNab BK (2010) Geographic and temporal correlations of mammalian size reconsidered: a resource rule. Oecologia 164:13–23

Marshall JT (2009) Are montane forests demographic sinks for Bornean white-bearded gibbons *Hylobates albibarbis*? Biotropica 41:257–267

Maynard Smith J (ed) (1982) Evolution and theory of games. Cambridge University Press, Cambridge

Morse DH (1974) Niche breadth as a function of social dominance. Am Nat 108:818–830

Müller S, Hollatz M, Wienberg J (2003) Chromosomal phylogeny and evolution of gibbons (Hylobatidae). Hum Genet 113:493–501

Napier JR, Napier PH (eds) (1967) A handbook of living primates. Academic Press, New York

Nekaris KAI, Bearder SK (2011) The lorisiform primates of Asia and mainland Africa diversity shrouded in darkness. In: Campbell CJ, Fuentes A, MacKinnon KC, Bearder SK, Stumpf RM (eds) Primates in perspective, 2nd edn. Oxford University Press, Oxford, pp 34–54

O'Brien TG, Kinnaird ME, Nurcahyo A, Iqbal M, Rusmanto M (2004) Abundance and distribution of sympatric gibbons in a threatened Sumatran rain forest. Int J Primatol 25:267–284

Oats JF (1987) Food distribution and foraging behavior. In: Smuts BB, Cheney RM, Seyfarth RW, Wrangham RW, Struhsaker TT (eds) Primate societies. Chicago University Press, Chicago, pp 197–209

Palombit RA (1997) Inter- and intra-specific variation in the diets of sympatric siamang (*Hylobates syndactylus*) and lar gibbons (*Hylobates lar*). Folia Primatol 68:321–337

Pilbeam D, Young N (2004) Hominoid evolution: synthesizing disparate data. CR Palevol 3:305–321

Pohl L (1911) Eine Höhenvarietät von Siamanga syndactylus. Desm Zool Anz 38:51–53

Preuschoft H (2010) Selective value of big size and sexual dimorphism in primates. Abstracts, XXIII Congress of the International Primatological Society, Kyoto, Japan

Preuschoft H, Schönwasser K-H, Witzel U (2016) Selective value of characteristic size parameters in hylobatids. A biomechanical approach to small ape size and morphology. In: Reichard UH, Hirohisa H, Barelli C (eds) Evolution of gibbons and siamang. Springer, New York, pp 227–263

Preuschoft H, Hohn B, Stoinski S, Witzel U (2011) Why so huge? Biomechanical reasons for the acquisition of large size in sauropod and theropod dinosaurs. In: Klein N, Remes K, Gee CT, Sander PM (eds) Biology of the Sauropod Dinosaurs understanding the life of giants. Indiana University Press, Bloomington, pp 197–218

Rabinowitz A (1989) The density and behavior of large cats in a dry tropical forest mosaic in Huai Kha Khaeng Wildlife Sanctuary, Thailand. Nat Hist Bull Siam Soc 37:235–251

Raemaekers JJ (1977) Gibbons and trees: comparative ecology of the siamang and lar gibbons. Dissertation, University of Cambridge

Raemaekers JJ (1978) The sharing of food sources between two gibbon species in the wild. Malayan Nat J 31:181–188

Raemaekers JJ (1979) Ecology of sympatric gibbons. Folia Primatol 31:227–245

Raemaekers JJ (1984) Large versus small gibbons: relative roles of bioenergetics and competition in their ecological segregation in sympatry. In: Preuschoft H, Chivers DJ, Brockelman WY, Creel N (eds) The lesser apes: evolutionary and behavioural biology. Edinburgh University Press, Edinburgh, pp 209–218

Raemaekers JJ, Chivers DJ (1980) Socio-ecology of Malayan forest primates. In: Chivers DJ (ed) Malayan forest primates: ten years' study in tropical rain forest. Plenum Press, New York, pp 279–316

Richard AF (ed) (1985) Primates in nature. WH Freeman and Company, New York

Roos C, Geissmann T (2001) Molecular phylogeny of the major hylobatid divisions. Mol Phylogenet Evol 19:486–494

Sander PM, Christian A, Clauss M, Fechner R, Gee CT, Griebeler E-M, Gunga H-C, Hummel J, Mallison H, Perry SF, Preuschoft H, Rauhut OWM, Remes C, Tütken T, Wings O, Witzel U (2011) Biology of the sauropod dinosaurs: the evolution of gigantism. Biol Rev 86:117–155

Schneider G (1906) Ergebnisse zoologischer Forschungsreisen in Sumatra. Zool Jb Abt Syst Geogr Biol Tiere 23:1–172

Schultz AH (1930) The skeleton of the trunk and limbs of higher primates. Hum Biol 2:303–438

Schultz AH (1931) The density of hair in primates. Hum Biol 3:303–321

Schultz AH (1933) Observations on the growth, classification and evolutionary specialization of gibbons and siamangs. Hum Biol 5:212–428

Schultz AH (1939) Notes on diseases and healed fractures in wild apes. Bull Hist Med 7:571–582

Schultz AH (1956) Postembryonic age changes. In: Hofer H, Schultz AH, Starck D (eds) Primatologia. Handbook of primatology, vol I. Karger, Basel, pp 965–1014

Schulze H (2016) Conservation database for lorises (*Loris, Nycticebus*) and pottos (*Arctocebus, Perodicticus*), prosimian primates. http://www.loris-conservation.org/database. Accessed 2 Feb 2016

Smith RJ, Jungers WL (1997) Body mass in comparative primatology. J Hum Evol 32:523–559

Takacs Z, Morales JC, Geissmann T, Melnick DJ (2005) A complete species level phylogeny of the Hylobatidae based on mitochodrial ND3-ND4 gene sequence. Mol Phylogenet Evol 36:456–467

Thinh VN, Mootnick AR, Geissmann T, Li M, Ziegler T, Agil M, Moisson P, Nadler T, Walter L, Roos C (2010) Mitochondrial evidence for multiple radiations in the evolutionary history of small apes. BMC Evol Biol 10:74

Tyler DE (1991) The problems of the Pliopithecidae as a hylobatid ancestor. Hum Evol 6:73–80

Uhde NL, Sommer V (2000) Antipredatory behavior in gibbons (*Hylobates lar*, Khao Yai/Thailand). In: Miller LE (ed) Eat or be eaten: predator sensitive foraging among primates. Cambridge University Press, Cambridge, pp 268–291

van Schaik CP, van Noordwijk MA, Warsono B, Sitriono E (1983) Party size and early detection in Sumatran forest primates. Primates 24:211–221

Zehr SM (1999) A nuclear and mitochondrial phylogeny of the lesser apes (Primates, genus Hylobates). Dissertation, Harvard University

Zhang YP (1997) Mitochondrial DNA sequence evolution and phylogenetic relationships of gibbons. Acta Genet Sinica 24:231–237

Zihlman AL, Mootnick AR, Underwood CE (2011) Anatomical contributions to hylobatids taxonomy and adaptation. Int J Primatol 32:865–877

Chapter 9
Gibbons to Gorillas: Allometric Issues in Hominoid Cranial Evolution

Erin R. Leslie and Brian T. Shea

Adolescent white-handed gibbon female (dark) grooming her mother (dark, with infant) while the groups' adult male (light buff) is nearby, Khao Yai National Park, Thailand. Photo credit: Ulrich H. Reichard

E.R. Leslie (✉)
Department of Anatomy, Midwestern University, Downers Grove, IL 60515, USA
e-mail: eleslie@midwestern.edu

B.T. Shea
Department of Cell and Molecular Biology, Feinberg School of Medicine,
Northwestern University, Chicago, IL 60611, USA

© Springer Science+Business Media New York 2016
U.H. Reichard et al. (eds.), *Evolution of Gibbons and Siamang*,
Developments in Primatology: Progress and Prospects,
DOI 10.1007/978-1-4939-5614-2_9

Introduction

The hominoids present a very challenging case for the morphological systematist. The extant taxa constitute a relatively depauperate superfamily overall, with the hylobatids (small apes) being relatively more diverse than the large "great" apes or hominids (including humans). There is also the fascinating and problematic feature of the major body size disparity between the two primary hominoid subclades, which is captured in a modern, phylogenetic use of the descriptive gradistic tags: "small apes" and "great apes" (Fig. 9.1). We note there has been an unfortunate tendency to ignore the extant small apes in broad studies of hominoid skull form, and we encourage a greater focus on variation both across the entire hominoid radiation, as well as within the hylobatid subclade itself. As there are relatively few strongly corroborated links between Miocene fossil hominoids and extant ape taxa, divergent views have resulted of the biogeography and phylogenetic history of many hominoid lineages (e.g., Begun 1992, 1995, 2002, 2002; Stewart and Disotell 1998; Cote 2004; Folinsbee and Brooks 2007). Understanding hylobatid morphology vis-à-vis large ape morphology is crucial to weighing these scenarios.

In body mass, the hylobatids range from 5.5 kg in the lar gibbons (*Hylobates*) to 11.9 kg in the siamangs (*Symphalangus*), and the hominids range from 39 to 48 kg in *Pan* and 80 to 169 kg in *Gorilla* (Table 9.1). The implications of this extreme range of body size for phylogenetic studies of ape skull form is that any features differentiating the hylobatid and hominid subclades will be either causally (allometrically) or spuriously (coincidentally) correlated with overall size differences. It is noteworthy that many authorities now believe the hylobatid lineage underwent a marked reduction in body size from the primitive hominoid condition subsequent to their divergence 15 to 22 million years ago (Stauffer et al. 2001; Raaum et al. 2005; Chatterjee 2006; Thinh et al. 2010; Israfil et al. 2011), and that this size reduction has been maximized in the lar, hoolock, and concolor gibbons and is associated with their specialized use of ricochetal brachiation (e.g., Groves 1972; Tyler 1991, 1993; Rae 1993, 1999; Pilbeam 1996; Jabonski and Chaplin 2009). This hylobatid dwarfing scenario then suggests that siamangs are likely more similar to the primitive hominoid condition in many morphological features than the even smaller gibbons.

A number of factors combine to make the extant hominoids a difficult group to work with both in terms of recognizing true allometric influences and controlling for these influences in the search for more fundamental nonallometric synapomorphies of the two subclades. In the following Sect. "Scaling in Craniofacial Features", we provide several examples of purported synapomorphies of hominid skull form that illustrate potential size-effects across the hominoids and support the call for further allometric analyses that include the hylobatid taxa. In the Sect. "Hominoid Challenges for Allometric Assessments" we explore issues related to the interpretation of morphological characters to the extreme body size range across the hominoids, and in the Sect. "Big Small Apes and Small Great Apes", we note additional issues related to size variance within the Hominidae and Hylobatidae families. We outline our proposed

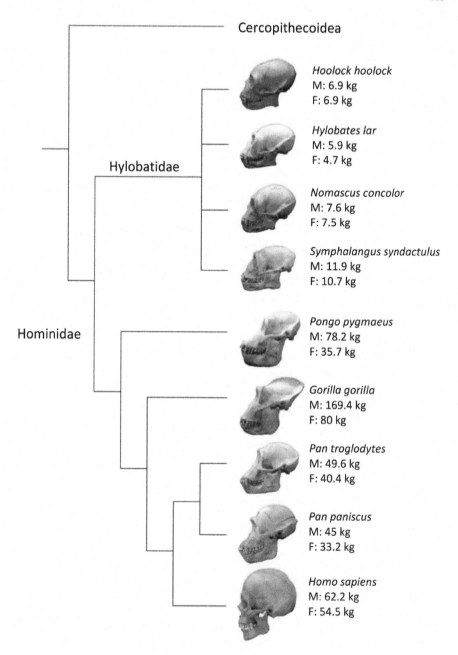

Cercopithecoidea

Hylobatidae

Hoolock hoolock
M: 6.9 kg
F: 6.9 kg

Hylobates lar
M: 5.9 kg
F: 4.7 kg

Nomascus concolor
M: 7.6 kg
F: 7.5 kg

Symphalangus syndactulus
M: 11.9 kg
F: 10.7 kg

Hominidae

Pongo pygmaeus
M: 78.2 kg
F: 35.7 kg

Gorilla gorilla
M: 169.4 kg
F: 80 kg

Pan troglodytes
M: 49.6 kg
F: 40.4 kg

Pan paniscus
M: 45 kg
F: 33.2 kg

Homo sapiens
M: 62.2 kg
F: 54.5 kg

Fig. 9.1 Cladogram depicting the phylogenetic relationships of the hominoid primates. Body masses from Smith and Jungers (1997) are included. From *top* to *bottom*: *Hoolock hoolock* AMNH 112667; *Hylobates lar* NMNH 111970; *Nomascus concolor* NMNH 240492; *Symphalangus syndactylus* NMNH 114495; *Pongo pygmaeus* NMNH 145302; *Gorilla gorilla* NMNH 174715; *Pan troglodytes* NMNH 220327; *Pan paniscus* FMNH 60770; *Homo sapiens* FMNH 42827. *AMNH* American Museum of Natural History; *FMNH* Field Museum of Natural History; *NMNH* Smithsonian National Museum of Natural History

Table 9.1 Published mean hominoid body masses taken from Smith and Jungers (1997)

Taxon	Body mass (kg)		References
	Male	Female	
Hylobates lar group	5.87	5.69	Geissmann (1993), Jungers (1984)
Hoolock hoolock	6.87	6.88	Geissmann (1993)
Nomascus concolor group	7.60	7.47	Geissmann (1993), Ma et al. (1988)
Symphalangus	11.90	10.70	Orgeldinger (1994)
Pongo	78.20	35.70	Jungers (1988, 1997), Markham and Groves (1990)
Gorilla	169.40	80.00	Jungers and Sussman (1984)
Pan pansicus	45.00	33.20	Jungers and Sussman (1984)
Pan troglodytes	49.60	40.40	Jungers and Sussman (1984), Uehara and Nishida (1987), Smith and Jungers (1997)

multistep allometric framework to address the challenges in working with the hominoids in the Sect. titled "HANCOVA (Hominoid ANCOVA)". Finally, we summarize our arguments and urge for routine comparative allometric approaches to studying morphological characters, especially in a paleontological context.

Scaling in Craniofacial Features

The well-preserved skull and various postcranial fragments of *Pierolapithecus catalaunicus*, a genus of Middle Miocene hominoid from the Catalunya Region of Spain, were described by Moyà-Solà et al. (2004). Their phylogenetic assessment of this fossil taxon as a hominid (great ape) was based on the following key cranial features which they claimed to be established synapomorphies of extant, large-bodied hominoids: a short face with the frontal processes of the maxillae, the nasals, and the orbits in the same plane; flat nasal bones which project anteriorly beneath the level of the lower orbital rims; high zygomatic roots; a high nasoalveolar clivus; a deep palate; and, a broad nasal aperture which is widest at the base. Moyà-Solà et al. (2004: p. 1340) noted that "this facial anatomy characterizes extant great apes and must be considered to be shared derived", since it is "absent in the known Early and Middle Miocene forms and in gibbons". However, postcranial features of *Pierolapithecus* signaled a combination of monkey-like hand bones and a possible early form of truncal orthogrady which requires a complex scenario of parallel evolution of the classic modern ape suspensory and climbing features seen in the living small and large apes and possibly in several other extinct taxa (see Larson 1998 for related discussion). The Catalunya deposits have recently yielded two other well-preserved fossil hominoid crania, *Anoiapithecus brevirostris* (Moyà-

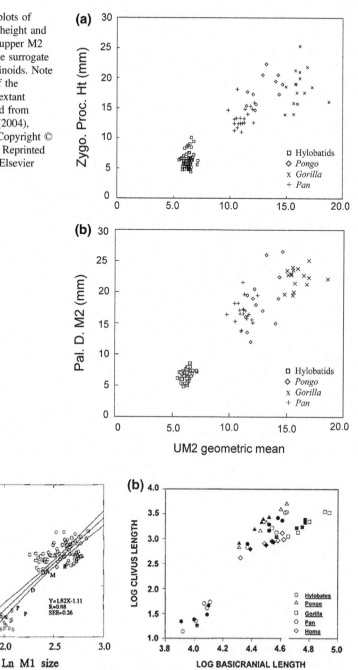

Fig. 9.2 Scatter plots of **a** zygomatic root height and **b** palate depth at upper M2 against a skull size surrogate in the extant hominoids. Note the relationship of the hylobatids to the extant hominids. Adapted from Kunimatsu et al. (2004), Figs. 22 and 23. Copyright © 2004 by Elsevier. Reprinted by permission of Elsevier

Fig. 9.3 Scatter plots of nasoalveolar clivus length against a skull size surrogate in the extant hominoids. **a** Adapted from Begun (1994), Fig. 16b. Copyright © 2005 by John Wiley & Sons, Inc. **b** Reproduced from McCollum and Ward (1997), Fig. 12. Copyright © 1998 by John Wiley & Sons, Inc. Both images reprinted by permission of John Wiley & Sons, Inc

Solà et al. 2009a) and *Dryopithecus fontanii* (Moyà-Solà et al. 2009b), which are also viewed as hominids based on the same suite of purported cranial synapomorphies of the great ape skull (see Shea 2013 for additional discussion).

The clarity of Moyà-Solà et al. (2004, 2009a, 2009b) purported great ape synapomorphies—and their alleged absence in the hylobatids—is less compelling when simple allometric patterns are mapped. We present further examination of three of their purported synapomorphies here. Figure 9.2 illustrates modified scatter plots for **hominoid zygomatic root height** (a) and **palate depth** (b) regressed against a skull size surrogate from Kunimatsu et al. (2004) careful study of *Nacholapithecus*, a Miocene large-bodied hominoid from Kenya. Both plots suggest that zygomatic root height and palate depth are strongly correlated with skull size *across the hominoids*. Likewise, *Hylobates* specimens appear to fall roughly in line with a downward projection of the general trend lines of the extant hominids in each plot. While the hominids are morphologically distinct from the hylobatids in having absolutely higher zygomatic roots and deeper palates, they are in fact not *relatively* so once regressed against a general size surrogate. Another feature previously claimed to clearly separate large apes from the small Asian apes is a **tall, high, or elongated nasoalveolar clivus** (e.g., Harrison 1982; Rae 1993; Begun 1994, 2007). Figure 9.3 illustrates scatterplots of extant hominoid nasoalveolar clivus length regressed against a skull size surrogate from Begun (1994) and McCollum and Ward (1997). The scatter plots suggest the small ape-large ape distinction is not upheld once the *y* variable is roughly "size-corrected". In fact, without explicitly stressing size and allometric issues, Rae (1999, 2004) has discussed the "reversal" of an elongated nasoalveolar clivus (and other apparent changes) in the hylobatid lineage, departing from previous contentions (e.g., Rae 1993) that an elongated nasoalveolar clivus was a novel character in the hominids. Given the preceding assessment of the key features of palate depth, zygomatic height, and nasoalveolar clivus height/length as great ape synapomorphies, what are

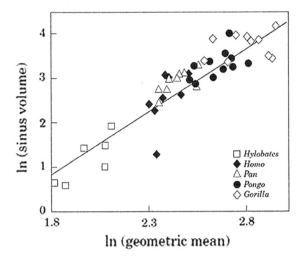

Fig. 9.4 Scatter plot and regression of maxillary sinus volume against a skull size surrogate for a sample of extant hominoids, including humans. The correlation coefficient (r) is 0.89 and the *p*-values was <0.01. The regression slope is not significantly different from isometry at $p < 0.05$. Note the position of hylobatids relative to the large-bodied apes. Adapted from Rae and Koppe (2000), Fig. 5. Copyright © 2000 by Elsevier. Reprinted by permission of Elsevier

we to make of the remaining features on Moyà-Solà et al. (2004, 2009a, 2009b) list or many of the traits otherwise offered by various authorities as clearly differentiating the two subclades?

There are additional examples of "hominid synapomorphies" that clearly warrant an allometric consideration in the context of hylobatid morphology. The **enlarged size (volume) of the maxillary sinus**, once favored as a great ape novelty (e.g., Andrews 1987), has been shown to be strongly correlated with overall skull size in a deliberate study by Rae and Koppe (2000; see Fig. 9.4). We applaud their meticulous approach in this case and stress that it serves as a valid template which paleontologists will hopefully follow before concluding there are fundamental trait shifts characterizing the great ape skull. The presence of a distinct **subarcuate fossa** in hylobatids (and most prosimians and non-hominoid catarrhines) and a faint or absent fossa in the extant hominids and humans, has been seen as a clear separator of the small and great apes (e.g., Moyà-Solà and Köhler 1995). However, the frequency and size of this feature may have a body size-related component as evidenced by the more variable presence of the feature in siamangs compared to the smaller gibbons (Le Gros Clark 1960; Gannon et al. 1988; Spoor and Leakey 1996). Additional support for a size effect on the subarcuate fossa is seen in the absence of this feature in some of the larger cercopithecoid taxa compared to the smaller forms (Spoor and Leakey 1996) and in the giant subfossil lemur, *Megaladapis*, compared to small-bodied prosimians (e.g., Gannon et al. 1988). Further, Spoor and Leakey (1996) have noted the potential for homoplasy in the presence/absence of the subarcuate fossa and urged caution when using this feature for phylogenetic analysis.

Other more general, qualitative features have also been characterized as novel in the great apes in contrast to small apes and these also seem to call for allometric investigation and more careful study of hylobatid morphologies. For example, claims by Bilsborough and Rae (2007) that hylobatids and hominids differ fundamentally in **mandibular robustness** seem to be contradicted by some of Ravosa's (2000) plots of mandibular dimension scaling in the extant and extinct hominoids. These plots suggest that hylobatids typically fall near or below the general strong allometric correlation, which is a pattern also seen in the African apes and a number of Miocene taxa. Additionally, Terhune's (2010) recent dissertation study of **temporomandibular joint morphology** demonstrates that many hylobatid–hominid differences appear to be allometric in origin. Arguments that hylobatids differ from great apes in having a relatively shorter, shallower, or less projecting midface or suborbital height (e.g., Andrews and Martin 1987; Harrison 1987; Benefit and McCrossin 1991; Rae 1993; Begun 2007) also seem likely to be strongly influenced by well-studied allometric patterns of facial growth and proportions (e.g., Corruccini 1981; Shea 1982, 1983a, b, 1984; Demes et al. 1986; Shea and Leslie 2004). As noted above regarding nasoalveolar clivus length, McCollum's (1995), McCollum and Ward (1997) hominoid **nasal floor morphology** studies, which include both hylobatids and hominoid-wide ontogenetic series, also suggest likely allometric influences relevant to hylobatid–hominid comparisons.

Additional qualitative "great ape synapomorphies" are listed in Shea (2013) and include traits such as general cranial robusticity, incisive canal size, relative orbit size (and possibly shape), temporal and nuchal cresting, postorbital constriction, and postglenoid process size. We do not treat these in detail here, or cite specific publications including these features, since our primary goal in this paper is to raise and explicate procedural allometric concerns and not to deal comprehensively with hominoid skull phylogenetics or specific features. This brief overview of the variability of craniofacial features should suffice to justify a call for more frequent inclusion of samples of hylobatids, along with the routine allometric assessment of any proposed novelties hypothesized to define the major hominoid clades.

Hominoid Challenges for Allometric Assessments

We now turn to a key methodological caveat associated with our advocated routine allometric assessment of clade-defining synapomorphies. This might be called the "gibbons-to-gorillas" problem which is endemic to comparative studies of extant hominoid morphology, particularly those based on adult means. The gibbons-to-gorillas problem results from the simple fact that the smallest (gibbons) and largest (gorillas) hominoids occupy positions in morphospace which on the size axis are substantially distanced from the small cluster of intermediate-sized great apes: the orangutans and chimpanzees (and, of course, humans when included). Even though the siamang is nearly twice the size of most gibbon species, it is still dramatically smaller than the next largest living hominoid, the bonobo (Table 9.1), and only about half the estimated body size of the hypothesized ancestral hominoid morphotype (e.g., Pilbeam 2002). It is well known in correlation and regression analyses that such peripheral outliers will exert disproportionate influence (or weight, determination) in generating the correlation coefficient, and also "draw" (highly influence) the regression line toward their x, y coordinate position (e.g., Sokal and Rohlf 2011). This statistical issue has previously been discussed in many places in primate biology, but in particular in a highly relevant craniometric paper by Creel (1986). Creel wrote:

> Regression slopes are determined largely by extreme values. When two major taxa differ greatly in size (as great apes and gibbons do) and there are few groups of intermediate size, the regression slopes must pass through or near the two extreme groups. Unless the two differ only in size, adjustments based on such slopes will remove both size and shape differences (1986: p. 85).

In these cases, the extreme values will tend to lie very close to the regression line itself (with minimal residualized values) and therefore may be interpreted as differing from one another predominantly in overall body size and its allometric correlates, in contrast to more fundamental size-independent shape or feature distinctions. This is an issue that we now raise here with considerable emphasis and caution.

Evidence of the problems associated with extreme regression slope end points is clearly illustrated in McNulty's (2004) geometric morphometric analysis of hominoid crania, which was ultimately based on regression-corrected values (i.e., residualized relative to a regression against overall skull centroid size across the hominoid taxa). McNulty's approach generated an unbiased Mahalanobis D^2 value for *Hylobates* (including siamangs)-*Gorilla* that was "an order of magnitude smaller than any other pair (McNulty 2004: p. 432)" among the hominoid taxa. This inordinate morphological proximity of the hylobatids and the gorillas then became the basis for much of the subsequent discussion in McNulty's paper. Noting that this multivariate shape signal overrides the well-corroborated phylogeny of hominoids, McNulty (p. 432) concluded that "*Hylobates* and *Gorilla* (and *Pan* to a lesser extent) retain a conservative hominoid cranial morphology" and that "the hylobatid skull closely resembles the ancestral hominoid morphotype. Results from this project indicate that *Gorilla* crania may be similarly conservative". This conclusion is perplexing with respect to the numerous and fundamental differences in cranial morphology between the hylobatids and gorillas permeating the orbital and nasal regions, the oral cavity, the vault, the skull base, and in facial hafting (e.g., Schultz 1968; Shea 1988; Leslie 2010; particularly on internal morphology; see also Shea 2013). McNulty overlooked the fact that his regression-based size-correction procedure itself largely generated the surprising similarity between the outlying smallest and largest of the apes, an issue that Creel (1986) clearly admonished some time ago. So, how are we to avoid this problem? It can be done simply by utilizing

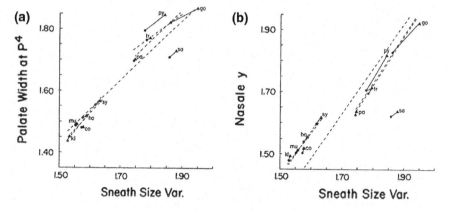

Fig. 9.5 Scatter plots of adult hominoid means for **a** palate width at upper P4, and **b** y distance to nasale point against a skull size surrogate. Key: circles: female; triangles: male. go: *Gorilla gorilla*; sa: *Homo sapiens*; ho: *Hoolock (Hylobates) hoolock*; kl: *Hylobates klossi*; la: *Hylobates lar*; co: *Nomascus (Hylobates) concolor*; pa: *Pan paniscus*; tr: *Pan troglodytes*; py: *Pongo pygmaeus*; sy: *Symphalangus syndactylus*. The solid lines connect female and male means; the short dashed lines are separate regression fits for male and female hominids and hylobatids; and, the *long dashed lines* represent the mean of the female hominid and hylobatid slopes. Adapted from Creel 1986, Figs. 7 and 8. Copyright © 1986 by Oxford University Press. Reprinted by permission of Oxford University Press

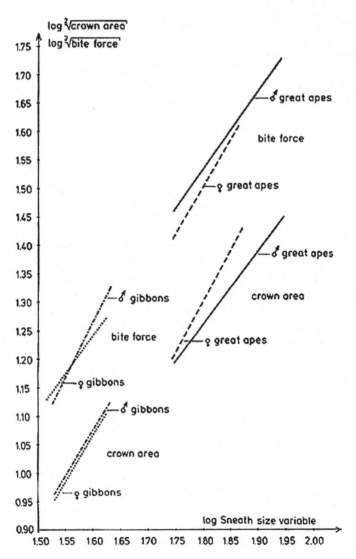

Fig. 9.6 Regression lines of total molar crown area and bite force on a skull size surrogate for the extant hominoids (minus *Homo*). Note that the hylobatids and hominids appear to fall along similar trajectories in both cases. Adapted from Demes et al. 1986, Fig. 2. Copyright © 1986 by Cambridge University Press. Reprinted by permission of Cambridge University Press

the comparative allometric template that we advocate here (see below), which is nicely illustrated by Rae and Koppe (2000) in their analysis of the maxillary sinus that was discussed above (Fig. 9.4).

Big Small Apes and Small Great Apes

The size variance within each of the two major hominoid subclades in fact provides an opportunity to move beyond the preceding first-level assessment of allometric influences. Average body mass differs by a factor of 2.2 among the hylobatids, and by a factor of 2.8 within the hominids (Table 9.1). More compelling cases for causal allometric influences in hominoid skull form can be made if a significant size association is observed independently within the hylobatid and the hominid families, in addition to any broader superfamilial correlation. Statistical equivalence of these within-family size regressions is neither predicted nor required at this level, though any substantial differences could be a basis for further investigation of allometric and nonallometric issues (and see below for further discussion). For example, selected examples of Creel's (1986) craniometric data set are given (Fig. 9.5). These plots reveal strong size correlation of palate width and the relative projection of the midface across all hominoids, but significant associations with size are also apparent within the hylobatids and within the hominids. Similarly, Demes et al. (1986) found that both tooth crown area and calculated bite force increased allometrically with skull size in comparable fashion within each of the respective hominoid families (Fig. 9.6). Further, Ravosa's (2000) investigation of the scaling of mandibular proportions across living and extinct apes considered features including mandibular corpus height and width, dental arch breadth, and symphysis height and width. Ravosa's (2000) plots show within-hylobatid scaling approximating that observed in the extant great apes (usually along what appears to be a common trajectory, though with possible differences for symphysis height), however, additional study of relevant subgroups would be required for confirmation. In the above examples, the inclusion of the larger siamang among the hylobatid taxa samples permits recognition of an association with skull size within and across both families, thus markedly strengthening the likelihood of causal allometric influences.

A qualitative example of potential allometric influences in the hominoid skull is provided by the variance observed in the subarcuate fossa. As noted above, the frequency and size of this feature appears to be related to overall size. Hylobatids as a group (along with other non-hominoid catarrhines) exhibit relatively larger and more prominent fossae than the large apes, but there is also evidence of a significant reduction in fossa frequency and size in the larger siamangs as compared to smaller gibbons (Le Gros Clark 1960; Gannon et al. 1988). This example suggests a potential allometric influence both within and across the hominoid families and, again, emphasizes the value of careful examination of morphological patterning within the hylobatid radiation for hominoid systematics.

HANCOVA (Hominoid ANCOVA)

Some readers will have undoubtedly noted in the preceding discussions that there is a further level of formal allometric execution and statistical precision which would be a desirable addition to our recommendations for (1) plotting size associations across all hominoids, including especially the smallest hylobatids; and (2) noting within-family scaling patterns for hylobatids and hominids separately. Simply put, even when there are strong size associations demonstrable *within* each of the hominoid family clades, and overall across the families, there can still be a novel, derived shift occurring *between* the hominids and the hylobatids. For example, the large apes could be distinguished from the small Asian apes by a marked vertical shift or transposition between comparable family scaling trends, even as this is coincident with whatever shape divergence might result from the larger size values of hominids shifted along a general allometric trend relative to hylobatids. In fact, such a transposition could theoretically also act to minimize or erase shape divergence which would be exaggerated without the shift if the within-family allometries were marked (cf. discussion of "evolution by geometric similarity" in Gould 1971 and Shea 2002). A transposition (or slope divergence) of this nature should still register as a hominid novelty, potentially useful in recognizing fossil taxa as phyletic hominids through their position in morphospace, even if it happened to reduce overall shape divergence between the families.

The testing of slope equivalence in regression trends for the two hominoid families, as well as the existence of vertical transpositions, is in the realm of analysis of covariance (e.g., Sokal and Rohlf 2011). Hence, we use the tag HANCOVA here to signify "Hominoid ANCOVA" and a yet higher level of

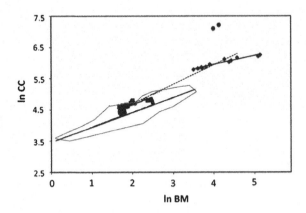

Fig. 9.7 Scatter plot and regression lines of cranial capacity against body mass in catarrhine primates. Key: squares: hylobatids; diamonds: *Pongo, Gorilla,* and *Pan*; circles: *Homo sapiens.* The hominid and hylobatid family-specific regression lines are the short *solid lines* in hominid and hylobatid groupings. The hominoid-wide regression line is the long, *dashed line.* The polygon encloses the cercopithecoid values and their long, solid regression line. Adapted from Alba (2010), Fig. 1b. Copyright © 2010 by David M. Alba. Reprinted by permission of David M. Alba

allometric probing of hominoid cranial morphology. One example of a vertical transposition in morphology between the small and large apes is suggested by Creel's (1986) plot (Fig. 9.5b) demonstrating the y value of the three-dimensional coordinates for his nasale point (the most lateral point on the margin of the nasal aperture), which roughly measures midfacial projection. While the y-values for the nasal point appear to scale allometrically within the hylobatid and hominid families, there also appears to be a shift between the small ape and large ape trends, such that at common skull sizes the hylobatids have a larger nasale y-axis value (more anteriorly projecting nasal aperture) than do the hominids. This may indicate some aspects of midfacial morphology in which hominids share a novel configuration, wherever specific Miocene ape crania may fall. Creel's findings also seem to counter claims that hylobatids have shorter or less prognathic midfaces than do hominids (e.g., Benefit and McCrossin 1991). More detailed analysis is required to also eliminate the possibility that such shifts are allometric in the sense of merely maintaining functional equivalence at varying size (e.g., Gould 1966; see Shea 1981 and 2002 for further discussion).

The allometry of brain versus body size provides an excellent example of an ANCOVA supported shift or transposition between within-family scaling patterns, allowing a more detailed examination of the important claim that hominids may be phylogenetically distinguished from the hylobatids by a significant "upshift" in encephalization (e.g., Begun 1994, 2007; Begun and Kordos 2004). Alba (2010) built on size-correction arguments summarized by Gould (1975) and subsequently applied to hominoids by Williams (2002) and Shea (2005, 2006), to offer a recent and comprehensive allometric reanalysis of brain–body size and encephalization quotients in hominoid evolution. Alba identified an upward grade shift or transposition in encephalization relative to body mass from the hylobatids to the large-bodied apes (Fig. 9.7). He also found an upward shift differentiating the hominoids as a whole from other catarrhine taxa. The shift toward increased encephalization in the large-bodied apes appears to be a derived feature which is further supported by considerable evidence of their increased cognitive abilities relative to the hylobatids (see Alba 2010 for discussion and citations). Thus, careful analysis of brain–body scaling in each of the hylobatids and hominids clarifies the great ape synapomorphy here (though precious few fossil crania will allow adequate determination of these variables in extinct forms). Increased encephalization at the origins of both the hominoids and the hominids is also compatible with reports of extended life history strategies in these groups with the hylobatids being intermediate in relative brain volume and life history parameters between the cercopithecoids and large apes (e.g., Kelley 1997; Shea 2006; Barrickman et al. 2008; Reichard and Barelli 2008). Additional research is required here to resolve this interesting question.

Table 9.2 Proposed stepwise sequence for assessing potential allometric influences on features hypothesized to be hominid novelties

Sample	Data collection should cover the full range of extant hominoid taxa, including especially the hylobatids (for Miocene fossils, include wide range in overall size). Within hylobatids, siamangs and at least one of the small gibbon taxa (genera) should be included. Within hominids, small-to-large African ape skulls should be complemented by orangutans. Assess with and without the highly derived (huge neurocranial, reduced dentofacial) human skulls, which may distort hominid means and disguise scaling trends in correlation/regression analyses.
Scale	Determine whether basic size associations exist at superfamily level. Assess by correlation-regression analyses and exploratory plots for quantitative features (e.g., Fig. 9.3) or by relative frequency or size/robusticity in qualitative traits (such as subarcuate fossa presence/size in several studies). Note: If there is no significant correlation or size association in hominoid-wide plots, then hominid-hylobatid distinctions cannot be allometric in nature.
Family allometries	Determine whether these general size associations also hold within hylobatids (between smaller gibbons and larger siamangs) and/or within hominids (across smaller and larger African apes, with orangutans intermediate in size) (e.g., Figs. 9.6 and 9.7). With qualitative traits, assess frequency or relative size/robusticity within respective families as well as across the entire superfamily sample (as in observed variation in subarcuate fossa frequency/size by Gannon et al. 1988). Code or quantify "qualitative" and robusticity features whenever possible.
HANCOVA	Apply formal analysis of covariance (ANCOVA) to quantitative features across hominoids when both families exhibit significant scaling associations, with taxonomy/phylogeny as the "treatment" (grouping) variable, in order to test for any fundamental differences in the family scaling patterns. Differentiating variables are "allometrically corrected" at the family level, and identify potential hominid novelties (and/or hylobatid synapomorphies) based on out-group comparisons. Note that sampling across selected generalized non-hominoid catarrhines will allow identification of broader allometric patterning and clarify character polarity within hominoids (see the example of relative brain size in this paper). Ontogenetic scaling (cf. Huxley 1932; Gould 1966) will allow further clarification of scaling patterns, especially at the family and genus levels (see Leslie 2010 for examples of angular and bony features within hylobatids and across hominoids).

Conclusions

The "gibbons-to-gorillas" extant hominoid radiation poses particular challenges to the paleontological and the neontological systematist alike, resulting from both the reduced number of living taxa and the disparities in body size and other features between the small specialized hylobatids and the larger hominids. There has been a strong tendency among paleontologists and comparative morphologists to treat the hylobatids as an early specialized offshoot of little relevance to many of the Miocene fossil hominoids. This has encouraged other workers on extinct and extant hominoids to frequently overlook the hylobatids in their comparative morphological studies, and thereby, to let underlying questions of body size disparities and possible allometric influences on those presumed "hominid" features to fall between the cracks. We focused on cranial features and discussed two model studies countering this tendency: Kunimatsu et al. (2004) analysis of gnathic features in fossil hominoids and Rae and Koppe (2000) analysis of maxillary sinus size in extant hominoids. These comparative size-based studies allow us to question presumed, rather than demonstrated, synapomorphies of the great ape skull. We have also examined here a number of other claimed hominid skull novelties from a comparative allometric perspective, demonstrating that careful examination of hylobatid morphology combined with application of basic allometric approaches can contribute significantly to ongoing debates concerning hominoid skull evolution.

We then offered a general framework for allometric investigation of skull features which can be readily applied to the study of fossil and extant ape morphologies, and is summarized in Table 9.2. This stepwise approach begins with simple examination of size associations for either quantitative or qualitative morphological traits, and continues on through examination of allometries within each of the hominoid families, culminating in a formal analysis of covariance and the consideration of inter-taxon criteria of functional equivalence. Fundamental to this allometric approach is the analysis of generous samples of hylobatids, including both the smallest gibbons and the larger-bodied siamangs. In a synthesized overview of hominoid evolution, Pilbeam (1996: p. 160) stressed how "important and understudied" the hylobatids are relative to the other hominoids and favored a marked size reduction as a hallmark of hylobatid evolutionary history (see also Groves 1972; Tyler 1993). Pilbeam (1996) reconstructed the stem hylobatid as between 20 and 40 kg, and listed the absence of a subarcuate fossa, plus several postcranial features and life history parameters, in which the larger siamangs are more similar to the great apes than to the smaller gibbons. Pilbeam (1996: p. 160) added that "it is unclear to me how these could be size related" in the relatively reduced size range between small gibbons and siamangs, but it now seems more likely that at least some of his noted distinctions are indeed allometrically influenced.

Authorities on the diverse array of Miocene hominoid fossils have noted tendencies of "craniophilic" researchers (Ward 1997) to overemphasize their own characters in comparison to those of postcranial locomotor anatomy (e.g., Ward 1997; Pilbeam and Young 2004 for discussion). Pilbeam (2002: p. 310) rather pessimistically suggested that progress in our understanding of hominoid evolution will continue to be impeded by a tendency of individual workers to "continue to pursue their favored characters and ignore or dismiss those of others, so discourse on phylogenetics will continue to resemble ships passing in the night," and this particular quote was later echoed by Brown et al. (2005) for emphasis. But, this unfortunate situation is not without solutions. As quantitatively oriented neontologists, we understand the tendency sometimes exhibited by paleontologists to critically question the direct relevance of detailed comparative morphometrics to their systematic studies of fragmentary fossils. However, we hope that our brief discussion of the size associations of many cranial features that have traditionally been central to hominid phylogenetics will encourage workers to more fully embrace routine comparative allometric approaches in their investigations of extinct hominoid morphologies and taxa. We believe the validity of many of the hominid skull synapomorphies currently argued by various researchers to link their favored Miocene hominoid fossils with the crown hominids or hominines is still open to question and requires additional investigation, particularly of the hylobatids.

References

Alba DM (2010) Cognitive inferences in fossil apes (Primates, Hominoidea): does encephalization reflect intelligence? J Anthropological Sci 88:11–48

Andrews P (1987) Aspects of hominoid phylogeny. In: Patterson C (ed) Molecules and morphology in evolution: conflict or compromise? Cambridge University Press, Cambridge, pp 23–54

Andrews P, Martin L (1987) Cladistic relationships of extant and fossil hominoids. J Hum Evol 16:101–118

Barrickman NL, Bastian ML, Isler K, van Schaik CP (2008) Life history costs and benefits of encephalization: a comparative test using data from long-term studies of primates in the wild. J Hum Evol 54:568–590

Begun DR (1992) Miocene fossil hominids and the chimp-human clade. Science 257:1929–1933

Begun DR (1994) Relations among the great apes and humans: New interpretations based on the fossil great ape *Dryopithecus*. Yrbk Phys Anthropol 37:11–63

Begun DR (1995) Late Miocene European orangutans, gorillas, humans, or none of the above? J Hum Evol 29:169–180

Begun DR (2002) European hominoids. In: Hartwig WC (ed) The primate fossil record. Cambridge University Press, Cambridge, pp 339–368

Begun DR (2007) Fossil Record of Miocene Hominoids. In: Henke W, Tattersall I (eds) Handbook of Paleoanthropology, vol II. Primate evolution and human origins. Springer, Berlin, pp 921–977

Begun DR, Kordos L (2004) Cranial evidence in the evolution of intelligence in fossil apes. In: Russon AE, Begun DR (eds) The evolution of thought: evolutionary origins of great ape intelligence. Cambridge University Press, Cambridge, pp 260–279

Benefit BR, McCrossin ML (1991) Ancestral facial morphology of old world higher primates. Proc Natl Acad Sci USA 88:5267–5271

Bilsborough A, Rae TC (2007) Hominoid diversity and adaptation. In: Henke W, Tattersall I (eds) Handbook of paleoanthropology. Primate evolution and human origins, vol II. Springer, Berlin, pp 1031–1105

Brown B, Kappelman J, Ward SC (2005) Lots of faces from different places: what Craniofacial morphology does(n't) tell us about hominoid phylogenetics. In: Lieberman DE, Smith RJ, Kelley J (eds) Interpreting the past: essays on human, primate, and mammal evolution in honor of David Pilbeam. Brill Academic Publishers Inc, Boston, pp 167–188

Chatterjee HJ (2006) Phylogeny and biogeography of gibbons: a dispersal-vicariance analysis. Int J Primatol 27:699–712

Corruccini RS (1981) Analytical techniques for cartesian coordinate data with reference to the relationship between *Hylobates* and *Symphalangus* (Hylobatidae; Hominoidea). Syst Zool 30:32–40

Cote SM (2004) Origins of the African hominoids and assessment of the palaeobiological evidence. CR Palevol 3:323–340

Creel N (1986) Size and phylogeny in hominoid primates. Syst Zool 35:81–99

Demes B, Creel N, Preuschoft H (1986) Functional significance of allometric trends in the hominoid masticatory apparatus. In: Else JG, Lee PC (eds) Primate evolution. Cambridge University Press, Cambridge, pp 229–237

Folinsbee KE, Brooks DR (2007) Miocene hominoid biogeography: pulses of dispersal and differentiation. J Biogeogr 34:383–397

Gannon PJ, Eden AR, Laitman JT (1988) The subarcuate fossa and cerebellum of extant primates: Comparative study of a skull-brain interface. Am J Phys Anthropol 77:143–164

Geissmann T (1993) Evolution of communication in Gibbons (Hylobatidae). Dissertation, University of Zurich

Gould SJ (1966) Allometry and size in ontogeny and phylogeny. Biol Rev 41:587–640

Gould SJ (1971) Geometric similarity in allometric growth. Am Naturalist 105:113–136

Gould SJ (1975) Allometry in primates, with emphasis on scaling and the evolution of the brain. In: Szalay F (ed) Approaches to primate paleobiology, vol 5. Karger, Basel, pp 244–292

Groves CP (1972) Systematics and phylogeny of the gibbons. In: Rumbaugh DM (ed) Gibbon and Siamang, vol 1. Karger, Basel, pp 1–89

Harrison T (1982) Small-bodied apes from the miocene of East Africa. Dissertation, University of London

Harrison T (1987) The phylogenetic relationships of the early catarrhine primates: a review of the current evidence. J Hum Evol 16:41–80

Huxley JS (ed) (1932) Problems of relative growth. Methuen, London

Israfil H, Zehr SM, Mootnick AR, Ruvolo M, Steiper ME (2011) Unresolved molecular phylogenies of gibbons and siamangs (Family: Hylobatidae) based on mitochondrial, Y-linked, and X-linked loci indicate a rapid Miocene radiation or sudden vicariance event. Mol Phylogenet Evol 58:447–455

Jabonski NG, Chaplin G (2009) The fossil record of Gibbons. In: Lappan S, Whittaker DJ (eds) The Gibbons: new perspectives on small ape socioecology and population biology. Springer, New York, pp 111–131

Jungers WL (1984) Scaling of the hominoid locomotor skeleton with special reference to lesser apes. In: Preuschoft H, Chivers DJ, Brockelman W, Creel N (eds) The lesser apes: evolutionary and behavioural biology. Edinburgh University Press, Edinburgh, pp 146–169

Jungers WL (1988) New estimates of body size in australopithecines. In: Grine FE (ed) Evolutionary history of the "robust" australopithecines. Aldine de Gruyer, New York, pp 115–125

Jungers WL (1997) Orang-utans of the menage scientific expedition to Borneo. Am J Phys Anthropol 24:138–139

Jungers WL, Sussman RL (1984) Body size and skeletal allometry in African apes. In: Sussman RL (ed) The pygmy Chimpanzee. Plenum Press, New York, pp 131–177

Kelley J (1997) Paleobiological and phylogenetic significance of life history in Miocene hominoids. In: Begun DR, Ward CV, Rose MD (eds) Function, phylogeny and fossils: Miocene hominoid origins and adaptations. Plenum Press, New York, pp 173–208

Kunimatsu Y, Ishida H, Nakatsukasa M, Nakano Y, Sawada Y, Nakayama K (2004) Maxillae and associated gnathodental specimens of *Nacholapithecus kerioi*, a large-bodied hominoid from Nachola, northern Kenya. J Hum Evol 46:365–400

Larson SG (1998) Parallel evolution in the hominoid trunk and forelimb. Evol Anthropol 6:87–89

Le Gros Clark WE (ed) (1960) The antecedents of man. Quadrangle Books, Chicago

Leslie ER (2010) Phylogenetic patterning of facial orientation in the hominoids. Dissertation, Northwestern University

Ma S, Wang Y, Poirier FE (1988) Taxonomy, distribution and status of gibbons (*Hylobates*) in Southern China and adjacent areas. Primates 29:277–286

Markham R, Groves CP (1990) Brief communication: weights of wild orang utans. Am J Phys Anthropol 81:1–3

McCollum MA (1995) Palatal thickening and facial form in *Paranthropus*: evaluation of alternative developmental models. Dissertation, Kent State University

McCollum MA, Ward SC (1997) Subnasoalveolar anatomy and hominoid phylogeny: evidence from comparative ontogeny. Am J Phys Anthropol 102:377–405

McNulty KP (2004) A geometric morphometric assessment of hominoid crania: conservative African apes and their liberal implications. Ann Anat 186:429–433

Moya-Sola S, Alba DM, Almecija S, Casanovas-Vilar I, Kohler M, De Esteban-Trivigno S, Robles JM, Galindo J, Fortuny J (2009a) A unique middle Miocene European hominoid and the origins of the great ape and human clade. P Natl Acad Sci USA 106:9601–9606

Moyà-Solà S, Köhler M (1995) New partial cranium of Dryopithecus lartet, 1863 (Hominoidea, Primates) from the upper Miocene of Can Llobateres, Barcelona, Spain. J Hum Evol 29:101–139

Moyà-Solà S, Köhler M, Alba DM, Casanovas-Vilar I, Galindo J, Robles JM, Cabrera L, Garces M, Almecija S, Beamud E (2009b) First partial face and upper dentition of the Middle Miocene hominoid Dryopithecus fontani from Abocador de Can Mata (Valles-Penedes Basin, Catalonia, NE Spain): taxonomic and phylogenetic implications. Am J Phys Anthropol 139:126–145

Moyà-Solà S, Köhler M, Alba DM, Casanovas-Vilar I, Galindo J (2004) Pierolapithecus catalaunicus, a new middle Miocene great ape from Spain. Science 306:1339–1344

Orgeldinger M (1994) Monitoring body weight in captive primates, with special reference to siamangs. International Zoo News 41:17–26

Pilbeam DR (1996) Genetic and morphological records of the hominoidea and hominid origins: a synthesis. Mol Phylogenet Evol 5:155–168

Pilbeam DR (2002) Perspectives on the Miocene hominoidea. In: Hartwig WC (ed) The primate fossil record. Cambridge University Press, Cambridge, pp 303–310

Pilbeam DR, Young NM (2004) Hominoid evolution: synthesizing disparate data. C R Palevol 3:305–321

Raaum RL, Sterner KN, Noviello CM, Stewart CB, Disotell TR (2005) Catarrhine primate divergence dates estimated from complete mitochondrial genomes: concordance with fossil and nuclear DNA evidence. J Hum Evol 48:237–257

Rae TC (1993) Phylogenetic analysis of proconsulid facial morphology. Dissertation, State University of New York at Stony Brook

Rae TC (1999) Mosaic evolution in the origin of the Hominoidea. Folia Primatol 70:125–135

Rae TC (2004) Miocene hominoid craniofacial morphology and the emergence of great apes. Ann Anat 186:417–421

Rae TC, Koppe T (2000) Isometric scaling of maxillary sinus volume in hominoids. J Hum Evol 38:411–423

Ravosa MJ (2000) Size and scaling in the mandible of living and extinct apes. Folia Primatol 71:305–322

Reichard UH, Barelli C (2008) Life history and reproductive strategies of Khao Yai white-handed gibbon females (Hylobates lar). Int J Primatol 29:823–844

Schultz AH (1968) The recent hominoid primates. In: Washburn SL, Jay PC (eds) Perspectives on human evolution, vol 1. Holt, Rinehart and Winston, New York, pp 122—195

Shea BT (1981) Relative growth of the limb and trunk in the African apes. Am J Phys Anthropol 56:179–202

Shea BT (1982) Growth and size allometry in the African Pongidae: cranial and postcranial analyses. Dissertation, Duke University

Shea BT (1983a) Allometry and heterochrony in the African apes. Am J Phys Anthropol 62:275–289

Shea BT (1983b) Size and diet in the evolution of African ape craniodental form. Folia Primatol 40:32–68

Shea BT (1984) An allometric perspective on the morphological and evolutionary relationships between Pygmy (Pan paniscus) and common (Pan troglodytes) Chimpanzees. In: Susman RL (ed) The pygmy Chimpanzee: evolutionary biology and behavior. Plenum Press, New York, pp 89–130

Shea BT (1988) Phylogeny and skull form in the hominoid primates. In: Schwartz J (ed) Orang-utan biology. Oxford University Press, New York, pp 233–246

Shea, BT (2002) Quantitative heterochrony: Are some heterochronic transformations likelier than others? In: Minugh-Purvis N, McNamara KJ (eds) Human evolution through developmental change. Johns Hopkins University Press, Baltimore, pp 79–101

Shea BT (2005) Brain/body allometry: using extant apes to establish appropriate scaling baselines. Am J Phys Anthropol S 126:189

Shea BT (2006) Start small and live slow: encephalization, body size, and life history strategies in primate origins and evolution. In: Ravosa MJ, Dagosto M (eds) Primate origins: adaptations and evolution. Springer, New York, pp 583–623

Shea BT (2013) Cranial evolution in the apes. In: Begun DR (ed) A Companion to paleoanthropology. Wiley-Blackwell, West Sussex, UK, pp 119–135

Shea BT, Leslie ER (2004) Allometric influences on facial form in lesser apes. Am J Phys Anthropol S 123:179

Smith RJ, Jungers WL (1997) Body mass in comparative primatology. J Hum Evol 32:523–559

Sokal RR, Rohlf FJ (eds) (2011) Biometry, 4th edn. WH Freeman & Company, New York

Spoor F, Leakey M (1996) Absence of the subarcuate fossa in cercopithecids. J Human Evol. 31:569–575

Stauffer RL, Walker A, Ryder OA, Lyons-Weiler M, Blair Hedges S (2001) Human and ape molecular clocks and constraints on paleontological hypotheses. J Hered 92:469–474

Stewart CB, Disotell TR (1998) Primate evolution—in and out of Africa. Curr Biol 8:R582–588

Terhune CE (2010) The temporomandibular joint in anthropoid primates: functional, allometric, and phylogenetic influences. Dissertation, Arizona State University

Thinh VN, Mootnick AR, Geissmann T, Li M, Ziegler T, Agil M, Moisson P, Nadler T, Walter L, Roos C (2010) Mitochondrial evidence for multiple radiations in the evolutionary history of small apes. BMC Evol Biol 10:74

Tyler DE (1991) The problems of the pliopithecidae as a hylobatid ancestor. Hum Evol 6:73–80

Tyler DE (1993) The evolutionary history of the gibbon. In: Jablonski NG (ed) Evolving landscapes and evolving biotas of east Asia since the mid-tertiary. Centre for Asian Studies, Hong Kong, pp 228–240

Uehara S, Nishida T (1987) Body weights of wild chimpanzees (*Pan troglodytes schweinfurthii*) of the Mahale Mountains National Park, Tanzania. Am J Phys Anthropol 72:315–321

Ward S (1997) The taxonomy and phylogenetic relationships of *Sivapithecus* Revisited. In: Begun DR, Ward CV, Rose MD (eds) Function, phylogeny, and fossils: miocene hominoid evolution and adaptations. Plenum Press, New York, pp 269–290

Williams MF (2002) Primate encephalization and intelligence. Med Hypotheses 58:284–290

Chapter 10
The Torso-Orthograde Positional Behavior of Wild White-Handed Gibbons (*Hylobates lar*)

Matthew G. Nowak and Ulrich H. Reichard

Adult white-handed gibbon female with infant brachiating in the high canopy, Khao Yai National Park, Thailand. Photo credit: Ulrich H. Reichard

M.G. Nowak (✉)
Department of Anthropology, Southern Illinois University
Carbondale, Carbondale, IL 62901, USA
e-mail: nowak.mg@gmail.com

M.G. Nowak
Sumatran Orangutan Conservation Programme (PanEco Foundation-YEL), Medan, Sumatra 20154, Indonesia

U.H. Reichard
Department of Anthropology and Center for Ecology, Southern Illinois University
Carbondale, Carbondale, IL 62901, USA

© Springer Science+Business Media New York 2016
U.H. Reichard et al. (eds.), *Evolution of Gibbons and Siamang*,
Developments in Primatology: Progress and Prospects,
DOI 10.1007/978-1-4939-5614-2_10

Introduction

All extant hominoids[1] are characterized by a habitual use of torso-orthograde (TO)-positional behaviors when arboreal (Hunt 1991, 2004, 2016; Thorpe and Crompton 2006; Nowak and Reichard 2016), which include (albeit at varying degrees of use) bipedal locomotion/posture, brachiation/forelimb swing, clamber/transfer, forelimb suspension, and vertical climb (Hunt et al. 1996). Compared to large-bodied extant hominids (33–175 kg; Smith and Jungers 1997), however, the TO-locomotor and -postural repertoires of the smaller-bodied hylobatids (5–12 kg; Smith and Jungers 1997) are suggested to be more stereotypical, as their use of brachiation/forelimb swing and forelimb-suspensory posture consistently overshadows the TO-behavioral versatility of this basal extant hominoid family (Hunt 1991, 2004, 2016; Povinelli and Cant 1995; Thorpe and Crompton 2006; but see Nowak and Reichard 2016).

The development of the derived locomotor and postural repertoires present among hylobatids are hypothesized to have resulted from a series of modifications to the ancestral hominoid morphotype, comprising generalized orthograde and semi-orthograde (e.g., bridging and pronograde clambering on angled substrates) positional behavior, which were closely linked to a lineage specific reduction(s) in body size (i.e., phyletic dwarfing) that occurred subsequent to the hylobatid-hominoid divergence (Groves 1972; Andrews and Groves 1976; Tyler 1991; Rose 1997; Gebo 1996, 2004; Pilbeam 1996; Ward 1997, 2015; Larson 1998; Young 2003; Pilbeam and Young 2004; Begun 2007; Crompton et al. 2008; Diogo and Wood 2011; Reichard et al. 2016). The unique reliance on brachiation/forelimb swing and forelimb-suspensory posture in hylobatids is thought to convey a number of ecological and energetic advantages, including, (1) mechanical efficiency when collisional energy loss is minimized; (2) the ability to utilize more direct arboreal pathways when moving between two points; (3) the ability to maximize the number of available travel routes, providing a larger set of pathway options; (4) the ability to circumvent problems associated with balance and branch deformation; (5) the ability to maximize the available feeding area in a food patch; and (6) the ability to more inconspicuously move through the (Grand 1972, 1984; Andrews and Groves 1976; Temerin and Cant 1983; Cant and Temerin 1984, Hollihn 1984; Preuschoft and Demes 1984; Cant 1992; Cannon and Leighton 1994; Bertram 2004; Michilsens et al. 2011, 2012; Chenye et al. 2013). Coupled with relatively small body size, these advantageous characteristics are thought to have allowed hylobatids to diversify and flourish when other larger-bodied hominoids struggled during periodic climate changes that occurred after gibbons and

---

[1]We follow the taxonomy of Harrison (2013), which separates the superfamily Hominoidea into two extant families, the Hylobatidae (gibbons and siamang) and the Hominidae (great apes and humans). Hominids are further separated into two extant subfamilies, including Ponginae (orangutans) and Homininae (gorillas, chimpanzees, bonobos, and humans). In this chapter, use of hominid is restricted to nonhuman members of this taxonomic family.

siamang had diverged from the ancestral hominoid line in the early/middle Miocene, by arboreal canopy reducing their metabolic requirements and enhancing their ability to acquire high quality food resources (Jablonski 1998, 2005; Jablonski et al. 2000; Jablonski and Chaplin 2009; Reichard et al. 2016).

While early field studies of hylobatid positional behavior (Chivers 1972, 1974; Fleagle 1976, 1980; Gittins 1983; Srikosamatara 1984; Cannon and Leighton 1994) were influential in the shift from a primarily qualitative understanding of hylobatid positional behavior to quantitative analyses, early hylobatid studies predate the important methodological breakthrough in categorizing positional behaviors achieved in the 1990s by Hunt et al. (1996), or have neglected those methodological improvements (Sati and Alfred 2002; Fan et al. 2013; Fei et al. 2015). As a result of less rigorous behavioral ethograms,[2] early and recent hylobatid positional behavior studies tend to underestimate the context, diversity, and frequency of gibbon and siamang locomotor and postural behaviors, leaving a major gap in a comprehensive understanding of hylobatid and hominoid behavioral variability. Moreover, the empirical void of the full locomotor and postural repertoire of hylobatids has encouraged speculation about hylobatids having evolved into specialized brachia-tors that are incapable of the (TO)-behavioral flexibility documented among larger-bodied extant hominids and thus has restricted comparative analyses (Hunt 1991, 2004, 2016; Povinelli and Cant 1995; Thorpe and Crompton 2006). As such, despite the fact that hylobatids are the most speciose extant hominoid family, they currently play a minor role in furthering our understanding of the evolution of hominoid torso-orthograde positional behavior.

This study presents new and detailed positional behavior data from a large population of habituated white-handed gibbons (*Hylobates lar*) from Khao Yai National Park, Thailand. Results show how these unique small-bodied apes provide a vital contribution to our understanding of hylobatid and hominoid behavioral evolution. Building off of a wealth of hominoid positional behavior studies and acknowledging the close phylogenetic relationship between hylobatids and great apes, we set out to specifically test the hypothesis that white-handed gibbons are versatile in their locomotor and postural modes, that they share a diverse TO-positional repertoire with extant great apes, and that similar to extant great apes, the hylobatid TO-positional repertoire collectively enhances small apes' navigation of the arboreal canopy and maximizes their ability to acquire food resources. In testing the later part of this hypothesis, we evaluate the utilization of TO-positional behavior modes in different behavioral contexts, and while accessing different-sized substrates, tree canopy areas, and tree canopy heights.

---

[2]Locomotor behaviors included in previous gibbon and siamang positional behavior studies have been limited to the generalized categories of bipedalism, brachiation, climbing, and leaping, whereas postural behaviors were limited to the generalized categories of bipedalism, lie, sit, and suspension.

Methods

Study Site and Subjects

Data were collected from January–July 2007 on wild white-handed gibbons (*Hylobates lar*) at Khao Yai National Park, Thailand. The study site ranges in altitude from 600 to 800 m and consists of seasonally wet evergreen rainforest (Kitamura et al. 2005). Observational study was undertaken on 11 habituated gibbon groups, resulting in 374 h of direct observation on 24 adult individuals (11 females and 13 males; Table 10.1).

Observational Methods

Instantaneous focal animal sampling every 2.5 min was used to sample behavioral data (Altmann 1974). A different animal was chosen each day and followed from night tree to night tree. All behavioral observations were collected by MGN. At every 2.5-minute mark, the following five observational variables were recorded using Leica® 10 × 42 binoculars:

1. *Behavioral context* (i.e., feed, rest, social, or travel);
2. *Positional behavior*, following the primary positional modes of Hunt et al. (1996). Five primary postural modes outlined by Hunt et al. (1996) did not occur during our study: **quadrupedal/tripedal stand**,[3] **tripod, cantilever, tail-suspend**, and **hind limb suspend**. We combined **forelimb-hind limb suspension** with **quadrumanous suspension** (*sensu* Hunt et al. 1996), and referred to each of those two postural modes as **torso-pronograde (TP) suspension**. This mode contrasts with the **torso-orthograde (TO) suspensory** category, which is synonymous with the Hunt et al. (1996) **forelimb-suspend** category. Among the primary locomotor modes outlined by Hunt et al. (1996), **quadrupedal/tripedal walk, quadrupedal/tripedal run, bipedal hop, tail swing, tree sway**, and **ride** were not observed in our study population, and are thus not included in our analyses. **Scoot** was not differentiated in this study, but was instead grouped in our **bipedal walk/run** category. We chose to separate the Hunt et al. (1996) **TO-suspensory locomotion** category into two separate modes, **TO-clamber/transfer** and **TO-brachiate/ forelimb swing**, as many authors have previously utilized these two TO-locomotor modes to differentiate smaller-bodied and larger-bodied hominoids (Hunt 1991, 2004, 2016; Thorpe and Crompton 2006). **Landings** for both **leaps** and **drops** are included in their respective categories (i.e., either **leap** or **drop**);

---

[3]We highlight, in bold text, the positional modes we used that correspond to Hunt et al. (1996) and the minor adjustments to these positional modes utilized in this chapter, so as to differentiate these detailed positional modes from the broad positional modes (i.e., more generalized forms of clambering, climbing, hoisting, and suspension) that are still common in the current literature.

Table 10.1 Focal individual sampling data

Group/individual	Infant[a]	Days followed	# IFSs[b]
Female			
A	N	2	183
B	N	3	123
C	Y	4	263
H	Y	3	138
J	Y	1	58
J_o	N	2	105
M	N/Y[c]	8	490
N	Y	2	203
R	Y	3	148
T	Y	2	133
W	Y	2	193
11	–	32	2037
Male			
A_p	–	2	204
B	–	3	170
B_o	–	1	109
C	–	4	315
E	–	3	246
H	–	3	229
J_p	–	4	193
M	–	4	242
N_p	–	2	96
N_s	–	2	105
R_p	–	6	221
T	–	2	157
W	–	2	187
13	–	38	2474

[a]presence (Y)/absence (N) of infant
[b]IFSs = instantaneous focal samples
[c]Female was observed prior to giving birth, and once she was carrying an infant
$_o$: adult offspring; $_p$: primary male; $_s$: secondary male

3. *Support diameter*, which were estimated visually and classified into three broad size classes, <2 cm (corresponding to small-sized supports), 2–10 cm (corresponding to medium-sized supports), or >10 cm (corresponding to large-sized supports), following Fleagle (1976);
4. *Canopy location*, which was categorized as either tree core (i.e., the inner 2/3 of a tree's canopy, as measured from the tree bole) or tree periphery (i.e., the outer 1/3 of a tree's canopy);

5. *Canopy height*, which was measured to the nearest meter using a Nikon® rangefinder and later classified into three broad height classes, <15, 15–25, or >25 m, following Mittermeier (1978).

We evaluated lar gibbon positional behavior in relation to behavioral context, support diameter, canopy location, and canopy height, as both behavioral and ecological factors are known to impact primate positional behavior, especially among primates that habitually subsist and reproduce in the complex arboreal canopy (Garber 2011). These variables will aid in evaluating the contextual use of lar gibbon TO-locomotion and -posture at Khao Yai, and highlight how their positional repertoire helps to circumvent common problems associated with habitat structure (Cant 1992), including negotiating different-sized substrates in various and structurally diverse zones of the tree canopy during their daily behavioral routine.

Statistical Analyses

Preliminary analyses revealed negligible differences in female and male positional behavior and canopy/substrate use. This result was not surprising since hylobatids are monomorphic and similar male and female bodies can be expected to be compromised by similar physical forces and correspondingly lead to similar positional behavior solutions to a complex arboreal environment. Hylobatids are not only monomorphic with regard to body size, but males and females are also known to often engage in the same behaviors and no data yet exists to suggest that the sexes have a different diet among wild white-handed gibbons. Likewise, when data of females carrying an infant were compared to those not carrying an infant, again no significant differences emerged. Finally, our sample included five males residing in multi-male single-female groups. Following Barelli et al. (2007) males in our sample were identified as primary or secondary male partner of the female in multi-male groups, although preliminary analyses of male positional behavioral failed to show differences related to male social status. Because female reproductive status, male social status, and sex had no detectable impact on positional behaviors, we pooled our individual data points to bolster our sample size and the power of statistical analyses.

In comparisons of behavioral context by postural/locomotor behavior and comparisons of postural/locomotor behavior by support diameter, canopy location, and canopy height (i.e., $r \times c$ tables), we utilized a two-sample randomization procedure (Manly 2007). In particular, the observed absolute frequency difference value between two behavioral contexts for a given positional mode (i.e., the test statistic value) was compared to a distribution of randomly generated absolute frequency difference values, which were recomputed from the original data set. Using 10,000 iterations, our *P*-values represent the proportion of randomly generated absolute values that are larger than the observed absolute statistic value. For each $r \times c$ table, a series of pairwise comparisons was therefore calculated between two columns or two rows, until each column or row had been compared. To correct

Table 10.2 Contextual use of postural behavior in *Hylobates lar* (column %'s)

	Total	Feed[a,b]	Rest[c]	Social
Sit	66.8	54.9	72.2	70.0
Squat	0.5	1.4	0.0	0.6
Cling	<0.1	0.0	0.0	0.1
TO B stand	0.1	0.3	0.1	0.0
TO susp	23.3	40.4	19.8	10.1
TP susp	1.3	2.9	0.7	0.3
Lie	8.0	0.1	7.2	18.6
Bridge	<0.1	0.0	0.0	0.1
n	3537	995	1684	858

[a]Feed versus rest: sit, squat, TO susp, TP susp, and lie all significant at $P < 0.001$
[b]Feed versus social: sit, TO susp, TP susp, and lie all significant at $P < 0.001$
[c]Rest versus social: squat, TO susp, and lie all significant at $P < 0.001$

for multiple pairwise comparisons, we used the Holm sequential Bonferroni procedure (Holm 1979; Hochberg and Tamhane 1987). All reported—values reflect the Holm sequential Bonferroni correction. Statistical significance was identified at $P < 0.05$ for all statistical testing. All randomization procedures were performed in Microsoft Excel 2007 using the PopTools v. 3.1.1 add-in (Hood 2009).

Results

Postural Behavior

Overall, **sit** was the most frequently used postural behavior (66.8 %), followed by **TO-suspension** (23.3 %) and **lie** (8.0 %), which combined to equal >90 % of total postural behavior. Despite the fact that six other primary postural modes were utilized by white-handed gibbons, their frequencies of use were all separately <1.0 % (Table 10.2). During feeding, **sit** (54.9 %), **TO-suspension** (40.4 %), **TP-suspension** (2.9 %), and **squat** (1.4 %) were most frequently utilized. Conversely during rest, **sit** was the most frequently utilized postural behavior (72.2 %), followed by **TO-suspension** (19.8 %), and **lie** (7.2 %). During social contexts, **sit** was the most frequently utilized postural behavior (70.0 %), followed by **lie** (18.6 %), and **TO-suspension** (10.1 %). For both rest and social contexts, all other postural modes were rare.

There were significant differences in the use of postural behavior between behavioral contexts (Table 10.2). Comparing the postures used during feeding with those used during resting revealed that **sit** ($P < 0.001$) and **lie** ($P < 0.001$) were used significantly more frequently during rest, whereas **squat** ($P < 0.001$), **TO-suspension** ($P < 0.001$), and **TP-suspension** ($P < 0.001$) were more frequently used during feeding. **Sit** ($P < 0.001$) and **lie** ($P < 0.001$) were used more

Table 10.3 Postural behavior by support diameter (row %'s)

	Sit/Squat	Cling	B stand	TO susp	TP susp	Lie	Bridge	n
[a,b]=<2 cm	51.2	0.0	0.2	45.6	2.4	0.5	0.1	924
[c]=2–10 cm	72.6	<0.1	0.1	20.3	1.2	5.8	0.0	1627
>10 cm	73.8	0.0	0.1	7.2	0.3	18.6	0.0	986
[a]=<2 × 2–10 cm	<0.001	NS	NS	<0.001	NS	<0.001	NS	
[b]=<2 × >10 cm	<0.001	NS	NS	<0.001	<0.001	<0.001	NS	
[c]=2–10 × >10 cm	NS	NS	NS	<0.001	<0.001	<0.001	NS	

Table 10.4 Postural behavior by canopy location (row %'s)

	Sit/squat	Cling	B stand	TO susp	TP susp	Lie	Bridge	n
[a]=Core	72.6	<0.1	0.1	15.7	0.8	10.8	0.0	2552
Periphery	53.9	0.0	0.2	42.9	2.3	0.6	0.1	984
[a]=Core x periphery	<0.001	NS	NS	<0.001	0.001	<0.001	NS	

frequently during social contexts compared to that of feeding contexts, whereas **TO-suspension** ($P < 0.001$) and **TP-suspension** ($P < 0.001$) were more frequently used during feeding. **Squat** ($P < 0.001$) and **lie** ($P < 0.001$) were used significantly more frequently during social contexts compared to that of rest, whereas **TO-suspension** ($P < 0.001$) was used more frequently during rest.

We also found significant differences in postural behavior on different-sized substrate classes (Table 10.3). **Sit/squat** ($P < 0.001$) and **lie** ($P < 0.001$) were used significantly less frequently on <2 cm sized substrates compared to that of 2–10 cm sized substrates, whereas **TO-suspension** was used significantly more frequently on <2 cm sized substrates ($P < 0.001$). A similar trend was observed when <2 cm sized substrates were compared with >10 cm, except that **TP-suspension** was also used more frequently among substrates <2 cm in diameter ($P < 0.001$). Comparing 2–10 cm with >10 cm sized substrates, **TO-suspension** ($P < 0.001$) and **TP-suspension** ($P < 0.001$) were used more frequently on 2–10 cm sized substrates, whereas **lie** was used more regularly on >10 cm sized substrates ($P < 0.001$).

Similarly, postural behaviors varied significantly among different canopy areas (Table 10.4). **Sit/squat** ($P < 0.001$) and **lie** ($P < 0.001$) were used significantly more frequently in the tree core compared to the periphery, whereas **TO-suspension** ($P < 0.001$) and **TP-suspension** ($P = 0.001$) were used more frequently in the tree periphery.

We also found differences in postural behavior use related to canopy height levels (Table 10.5). **TO-suspension** was used more frequently at heights of <15 m compared to heights of 15–25 m ($P < 0.001$), whereas **lie** was used more frequently at heights of 15–25 m ($P < 0.001$). Similarly, **TO-suspension** was used more

Table 10.5 Postural behavior by canopy height (row %'s)

	Sit/Squat	Cling	B stand	TO susp	TP susp	Lie	Bridge	n
[a,b]=<15 m	62.7	0.2	0.2	32.7	1.9	2.4	0.0	539
[c]=15–25 m	67.5	0.0	0.1	24.4	1.3	6.7	<0.1	2114
>25 m	69.8	0.0	0.1	14.8	0.8	14.5	0.0	884
[a]=<15 × 15–25 m	NS	NS	NS	<0.001	NS	<0.001	NS	
[b]=<15 × >25 m	NS	NS	NS	<0.001	NS	<0.001	NS	
[c]=15–25 × >25 m	NS	NS	NS	<0.001	NS	<0.001	NS	

Table 10.6 Contextual use of locomotor behavior in *Hylobates lar* (column %'s)

	Total	Travel[a,b]	Feed[c]	Social
TO B walk/run	6.9	6.0	10.9	9.6
TO V climb	15.9	12.7	31.1	26.0
TO-clamber and transfer	10.2	10.5	10.9	5.5
TO-brachiate and forelimb swing	47.9	50.4	35.3	42.5
TP susp	0.7	0.5	0.8	2.7
Bridge	2.6	2.7	2.5	1.4
Leap	8.2	9.2	2.5	6.8
Drop	7.6	8.1	5.9	5.5
n	974	782	119	73

[a]Travel versus feed: vertical climb ($P < 0.002$); brachiate/forelimb swing ($P = 0.037$)

[b]Travel versus social: vertical climb ($P = 0.044$)

[c]Feed versus social: no significant pairwise comparisons

frequently at heights of <15 m compared to heights of >25 m ($P < 0.001$), whereas **lie** was used more frequently at heights of >25 m ($P < 0.001$). Lastly, **TO-suspension** was used more frequently at heights of 15–25 m compared to heights of >25 m ($P < 0.001$), whereas **lie** was used more frequently at heights of >25 m ($P < 0.001$).

Locomotor Behavior

For white-handed gibbons at Khao Yai National Park, **TO-brachiate/forelimb swing** (47.9 %) were the most frequently used locomotor modes, followed by **TO-vertical climbing** (15.9 %), **TO-clamber/transfer** (10.2 %), **leaping** (8.2 %), **dropping** (7.6 %), **TO-bipedal walk/run** (6.9 %), **bridging** (2.6 %), and **TP-suspensory locomotion** (0.7 %; Table 10.6). During travel, **TO-brachiate/forelimb swing** (50.4 %) was the most frequently used locomotor mode used by white-handed gibbons, followed by **TO-vertical climbing** (12.7 %),

TO-clamber/transfer (10.5 %), **leaping** (9.2 %), **dropping** (8.1 %), **TO-bipedal walk/run** (6.0 %), **bridging** (2.7 %), and **TP-suspensory locomotion** (0.5 %). Conversely during feeding, **TO-brachiate/forelimb swing** (35.3 %) was the most frequently utilized locomotor mode, followed by **TO-vertical climb** (31.1 %), **TO-clamber/transfer** (10.9 %), **TO-bipedal walk/run** (10.9 %), **dropping** (5.9 %), **leaping** (2.5 %), **bridging** (2.5 %), and **TP-suspensory locomotion** (0.8 %). During social contexts, **TO-brachiate/forelimb swing** (42.5 %) was most frequently utilized, followed by **TO-vertical climb** (26.0 %), **TO-bipedal walk/run** (9.6 %), **leaping** (6.8 %), **dropping** (5.5 %), **TO-clamber/transfer** (5.5 %), **TP-suspensory locomotion** (2.7 %), and **bridging** (1.4 %).

There were significant differences in the use of locomotor behavior between the three behavioral contexts (Table 10.6). When travel and feeding contexts were compared, **TO-vertical climb** was used more frequently during feeding ($P < 0.001$), while **TO-brachiate/forelimb swing** was used more frequently during travel contexts ($P = 0.037$). Comparing travel versus social contexts revealed that **TO-vertical climb** was used more frequently during social contexts ($P = 0.044$). There were no significant pairwise comparisons for locomotor behavior when feeding contexts were compared to that of social contexts.

We also found significant differences in locomotor behavior among different substrate size classes (Table 10.7). Comparing <2 cm sized substrates with 2–10 cm sized substrates showed that **TO-clamber/transfer** ($P < 0.001$) and **bridging** ($P < 0.001$) were used significantly more frequently on small <2 cm sized substrates, while **TO-brachiation/forelimb swing** were more frequently used on 2–10 cm sized substrates ($P < 0.001$). **TO-clamber/transfer** ($P < 0.001$), **bridging** ($P = 0.044$), and **leap/drop** ($P = 0.027$) were all used more frequently on <2 cm sized substrates compared to that of >10 cm sized substrates, whereas **TO-bipedal walk/run** ($P < 0.001$) and **TO-brachiation/forelimb swing** ($P < 0.001$) were used more frequently on >10 cm sized substrates. Among the largest two substrate size classes, only **TO-bipedal walk/run** was used more frequently on >10 cm sized substrates compared to that of 2–10 cm sized substrates ($P = 0.028$).

Locomotor behavior varied significantly among different canopy areas (Table 10.8). **TO-bipedal walk/run** ($P = 0.010$), **TO-vertical climb** ($P = 0.006$), and **TO-brachiation/forelimb swing** ($P < 0.001$) were used more frequently in the tree core area, whereas **TO-clamber/transfer** ($P < 0.001$), **bridge** ($P < 0.001$), and **leap/drop** ($P = 0.003$) were used more frequently in the tree periphery. Lastly, there were no significant differences in locomotor behavior among different canopy height levels (Table 10.9).

Table 10.7 Locomotor behavior by support diameter (row %'s)

	B walk/run	Vertical climb	Clamber/ transfer	Brachiation/ forelimb swing	TP susp	Bridge	Leap/ drop	n
[a,b]=<2 cm	3.9	12.4	19.4	38.7	0.7	5.5	19.4	434
[c]=2–10 cm	6.6	18.2	4.0	55.9	0.5	0.3	14.5	379
>10 cm	15.6	20.0	0.0	53.8	1.2	0.0	9.4	160
[a]=<2 × 2– 10 cm	NS	NS	<0.001	<0.001	NS	<0.001	NS	
[b]=<2 × >10 cm	<0.001	NS	<0.001	<0.001	NS	0.044	0.027	
[c]=2– 10 × >10 cm	0.028	NS	NS	NS	NS	NS	NS	

Table 10.8 Locomotor behavior by canopy location (row %'s)

	B walk/run	Vertical climb	Clamber/ transfer	Brachiation/ forelimb swing	TP Susp	Bridge	Leap/ drop	n
[a]=Core	8.7	19.1	4.5	54.4	0.7	0.0	12.5	551
Periphery	4.5	11.9	17.3	39.4	0.7	5.9	20.2	421
[a]=Core x periphery	0.010	0.006	<0.001	<0.001	NS	<0.001	0.003	

Table 10.9 Locomotor behavior by canopy height (row %'s)

	B walk/run	Vertical climb	Clamber/ transfer	Brachiation/ forelimb swing	TP susp	Bridge	Leap/ drop	n
[a,b]=<15 m	5.6	17.7	14.4	40.9	0.9	3.3	17.2	215
[c]=15–25 m	7.2	15.9	9.3	48.8	0.7	2.6	15.5	611
>25 m	7.5	13.6	6.8	55.1	0.7	1.4	15.0	147
[a]=<15 × 15– 25 m	NS	NS	NS	NS	NS	NS	NS	
[b]=<15 × >25 m	NS	NS	NS	NS	NS	NS	NS	
[c]=15– 25 × >25 m	NS	NS	NS	NS	NS	NS	NS	

Discussion

In support of our hypothesis, we found that white-handed gibbons at Khao Yai utilized a diverse positional repertoire that includes all primary TO-locomotor ($n = 4$ modes) and -postural ($n = 2$ modes) modes defined by Hunt et al. (1996), in addition to five additional locomotor and six additional postural modes. This result contrasts to previous studies of hylobatid positional behavior (Chivers 1972, 1974; Fleagle 1976, 1980; Gittins 1983; Srikosamatara 1984; Cannon and Leighton 1994;

Sati and Alfred 2002; Fan et al. 2013; Fei et al. 2015). The difference between our results and those of previous studies of hylobatids are best explained by methodological differences,[4] because ours is the first study to apply the detailed, systematic positional modes outlined by Hunt et al. (1996). Our results also reveal that even a member of the smallest-sized hylobatid genus (i.e., *Hylobates*) utilizes a positional behavior repertoire as diverse as previously thought to be typical only for large-bodied hominids (Hunt 1991, 2004, 2016; Thorpe and Crompton 2006; Nowak and Reichard 2016). Despite their undoubtedly frequent use of TO-forelimb-suspensory locomotion and posture, the TO-positional repertoire of Khao Yai lar gibbons was collectively utilized while exploiting all areas of the arboreal canopy, suggesting that each of the TO-positional modes utilized by Khao Yai lar gibbons are crucial to their arboreal existence. Furthermore, all other locomotor (*n* = 4 modes) and postural modes (*n* = 6 modes) appear to supplement the core TO-repertoire, and in some cases are also crucial in the context of resource exploitation (e.g., **TP-suspension** during feeding contexts) and navigating the arboreal canopy (e.g., **bridging** and **leaping** during gap crossing contexts).

Contexts of Hylobatid Torso-Orthograde-Postural Behavior

For Khao Yai lar gibbons, **TO-suspension** was significantly important during feeding and was associated with small-sized substrates, the tree periphery, and lower canopy levels. The importance of **TO-suspension** during feeding and its relationship with small-sized peripheral substrates is consistent with the earliest theoretical predictions of the substrate to body size ratio model (Napier 1967), and the need to modify postural behavior in order to avoid balance and branch deformation/fracture-related issues when this ratio is small (Grand 1972, 1984; Cartmill 1974, 1985; Cant 1992; Preuschoft et al. 1995; Dunbar and Badam 2000; Preuschoft 2002). It is also consistent with the idea that hylobatids are particularly well adapted to exploit the terminal branches of tropical forest trees, a resource rich zone where preferred gibbon food items (e.g., flowers, insects, ripe fruits, and young leaves) are typically found.

Similar relationships between **TO-suspension** and feeding contexts, small-sized substrates, and/or the tree periphery have also been documented in previous studies of the Malaysian agile gibbon (*Hylobates agilis*) (Gittins 1983) and siamang (*Symphalangus syndactylus*) (Chivers 1972, 1974; Fleagle 1976, 1980), the Cao-Vit crested gibbon (*Nomascus nasutus*) (Fei et al. 2015), as well as in larger-bodied hominids (Kano and Mulavwa 1983; Cant 1987a, b; Hunt 1992; Doran 1993a; Remis 1995, 1998; Myatt and Thorpe 2011).

---

[4]In the following discussion we note that in some cases it is not possible to make detailed comparisons across all previously studied taxa, as many early studies of hominoid positional behavior utilized divergent less-detailed behavioral ethograms to that presented here and those in more recent hominoid studies.

The current observation that the frequency of **TO-suspension** increased with decreasing canopy level warrants further explanation, as this has so far been largely unexplored. However, this inverse relationship is possibly related to a decrease in the frequency of stable supportive elements at lower canopy levels. In support of this notion, Brockelman et al. (2011) have recently reported for the 30 ha long-term vegetation plot within our study site that the number of small tree stems (i.e., stems between 1 and 10 cm DBH) was roughly 7.4 times higher per hectare (i.e., 3781 stems/ha) than the number of tree stems larger than >10 cm DBH per hectare (i.e., 514 stems/ha). Smaller trees frequently only provide smaller supportive branches and are also of lower absolute height than larger trees. Therefore, with decreasing canopy height the frequency of small trees is disproportionately high, forcing the gibbons to use more **TO-suspension** when they explore the lower levels of the canopy. A similar relationship was recently also found in a study of the postural behavior of Northern muriquis (*Brachyteles hypoxanthus*) in the Brazilian Atlantic Forest, where suspensory behaviors were frequently associated with lower canopy levels (i.e., <15 m), and where a fragile understory flora (i.e., trees <15 cm DBH) was likewise abundant (Iurck et al. 2013).

TO-bipedal stand was not found to vary between behavioral contexts or among different-sized substrates, tree canopy areas, or tree height levels, though this may reflect the relatively few sample points of **TO-bipedal stand** recorded during this study (i.e., 0.1 % of all postural samples). It should be noted, however, that among all contexts of **TO-bipedal stand**, the forelimbs were incorporated above the head for additional support, and in fact the **TO-suspension** submode **TO-forelimb-suspend/stand**[5] (*sensu* Hunt et al. 1996) was used far more frequently (i.e., 3.6 % of all postural samples). Thus, lar gibbons at Khao Yai rarely subject their (partially) extended hind limbs to compressive forces during postural bipedalism, and this is in large part due to their extensive use of their forelimbs during TO-postural behaviors.

A similar relationship between forelimb-suspension and postural bipedalism has been noted for populations of large-bodied apes. In some instances, however, limb specific body mass support (i.e., the specific limbs that are bearing the majority of body mass) differs from the pattern observed in Khao Yai white-handed gibbons (Cant 1987a; Hunt 1992, 1994; Doran 1993a; Remis 1995; Stanford 2002, 2006; Thorpe and Crompton 2006; Tourkakis 2009). For instance, a recent study by Thorpe and Crompton (2006) indicates that Sumatran orangutans (*Pongo abelii*) frequently combine bipedal and forelimb-suspensory postures; however, they observed a greater use of postures emphasizing hind limb weight support, rather than the concurrent forelimb suspensory-component. While these untested differences are likely related to body size differences, context, and/or support availability/use, it is reasonable to argue that postural bipedalism and

---

[5]**TO-forelimb-suspend/stand** is differentiated from **TO-bipedal stand** and one of its submodes **TO-bipedal stand/forelimb-suspend** by a greater emphasis of forelimb suspensory body mass support. This is differentiated in the field by observing the body mass induced deformation of each respective weight bearing substrate (Hunt et al. 1996).

forelimb-suspension in the arboreal setting are not mutually exclusive among both large-bodied and small-bodied extant apes.

Contexts of Hylobatid Torso-Orthograde-Locomotor Behavior

We observed an inverse relationship between **TO-brachiation/forelimb swing** and **TO-vertical climb** across behavioral contexts. This association was most apparent when comparing feeding and traveling contexts. More specifically, **TO-brachiation/forelimb swing** was used more frequently during traveling and less frequently during feeding. Conversely, **TO-vertical climb** was used less frequently during travel locomotion and more frequently during feeding. Fleagle (1976) has documented a similar inverse trend (i.e., more **TO-brachiation/forelimb swing** during travel, and more 'climb'[6] during feeding) among the Malaysian white-handed gibbons (*Hylobates lar*) and siamang. A reduced use of **TO-brachiation/forelimb swing** and increased use of **TO-vertical climb** during feeding may reflect the need to frequently adjust height levels relative to available and/or preferred food resources (Fleagle 1980; Fleagle and Mittermeier 1980; Garber 2011). Conversely, a greater use of **TO-brachiation/forelimb swing** and reduced use of **TO-vertical climb** during traveling contexts is consistent with the hypothesis that primates attempt to straighten bouts of locomotion via known arboreal pathways (Fleagle and Mittermeier 1980; Grand 1984; Cant 1992; Cannon and Leighton 1994, 1996; Garber 2011), a notion supported by a recent study showing that Khao Yai lar gibbons repeatedly use known travel routes to navigate their home ranges (Asensio et al. 2011).

In contrast to **TO-brachiation/forelimb swing** and **TO-vertical climbing**, both **TO-bipedal walk/run** and **TO-clamber/transfer** were used at similar frequencies across all behavioral contexts. The overall contextual use of TO-locomotion documented here is reminiscent of that documented for northeast Bornean orangutans (*Po. pygmaeus morio*), central Bornean orangutans (*Po. p. wurmbii*), and Sumatran orangutans. Although, the previously studied orangutan populations were shown to prefer the locomotor behaviors, **TO-clamber/transfer**, relative to **TO-brachiation/forelimb swing** in contrast to the hylobatid scenario (Cant 1987a; Thorpe and Crompton 2006, 2009; Manduell et al. 2011, 2012). As with all other locomotor modes that supplement the use of **TO-brachiation/forelimb swing** and **TO-vertical climb** in Khao Yai lar gibbons, consistency in the frequency of use of **TO-bipedal walk/run** and **TO-clamber/transfer** in hylobatids (and **TO-bipedal walk/run** and **TO-brachiation/forelimb swing** in the case of orangutans) may

---

[6]We place the words 'climb' or 'climbing' in brackets to recognize that this often utilized category of locomotion (*sensu* Fleagle 1976) is now considered to include functionally different locomotor modes (e.g., **bridging**, **torso-orthograde clamber/transfer**, **torso-orthograde vertical climb**, and **torso-pronograde clambering/scrambling**; Hunt et al. 1996).

signify a smaller yet vital role in their exploitation of the arboreal canopy. This is most apparent in the relationships of Khao Yai lar gibbon locomotor behavior to substrate size, tree crown area, and canopy height discussed below.

While **TO-brachiation/forelimb swing** were used frequently among all substrate classes, they were used significantly less frequently on substrates of <2 cm diameter and significantly more frequently on substrates of >2 cm diameter. Given that minimizing collisional energy loss is suggested to be a major component of efficient **TO-brachiation** (Bertram 2004; Michilsens et al. 2011, 2012), we propose that Khao Yai lar gibbons may utilize larger less pliable supports when possible, in order to minimize energy losses that might originate from branch deformation. A similar use of substrates of >2 cm diameter in association with **TO-brachiation/forelimb swing** has been documented in previous studies of gibbon and siamang positional behavior, suggesting a relatively consistent pattern among hylobatids (Fleagle 1976; Gittins 1983; Sati and Alfred 2002; Fan et al. 2013). Interestingly, substrate use associated with **TO-brachiation/forelimb swing** in previously studied hominids was substantially variable, with northeast Bornean and Sumatran orangutans and western lowland gorillas (*Gorilla gorilla*) commonly using medium- to large-sized substrates (Remis 1995; Thorpe and Crompton 2005, 2009), while eastern chimpanzees (*Pan troglodytes schweinfurthii*), western chimpanzees (*Pa. t. verus*), and bonobos (*Pa. paniscus*) commonly utilize small- to medium-sized substrates (Hunt 1992; Doran 1993a, b), substantiating that among hominoids **TO-brachiation/forelimb swing** can be utilized to access a tremendous range of substrate sizes.

TO-clamber/transfer (and also **bridging**), appeared to be inversely related to **TO- brachiation/forelimb swing**, and were most frequently associated with small-sized substrates (i.e., <2 cm in diameter). These results are consistent with the substrate to body size ratio model, and suggest that even among small-bodied lar gibbons (i.e., $\bar{x} = 5.6$ kg; Smith and Jungers 1997), **TO-clamber/transfer** are beneficial to maintaining balance and spreading body mass among multiple substrates during locomotion (Napier 1967; Grand 1984; Cartmill 1974, 1985; Cartmill and Milton 1977; Cant 1992; Preuschoft et al. 1995; Dunbar and Badam 2000; Preuschoft 2002). A similar relationship between small-sized substrates and the use of **TO-clamber/transfer** has been documented for eastern chimpanzees and Sumatran orangutans (although Sumatran orangutans were also documented to frequently use **TO-clamber/transfer** to access medium-sized substrates), signifying a relatively consistent pattern across a broad hominoid size gradient (Hunt 1992; Thorpe and Crompton 2005, 2009). Our results also highlight the complementary nature of **TO-brachiation/forelimb swing** and **TO-clamber/transfer** in negotiating different-sized substrates, and support Thorpe and Crompton's (2006) notion of a positional behavior 'continuum' that primates, especially the versatile hominoids, implement to help solve problems encountered in the arboreal setting.

TO-bipedal walk/run were most frequently associated with larger-sized substrates (i.e., >10 cm in diameter), as their frequency of use increased with increasing substrate size. A similar association between **TO-bipedal walk/run** and substrates of >10 cm diameter has previously been documented among other hylobatid populations (Fleagle 1976; Gittins 1983; Sati and Alfred 2002; but see Fan et al. 2013). While lar gibbons at Khao Yai were capable of accessing various-sized substrates utilizing **TO-bipedal walk/run**, their greater emphasis of use of larger-sized substrates in association with **TO-bipedal walk/run** may be partially related to the maintenance of balance during these locomotor modes (Napier 1967; Cartmill 1974, 1985), especially since hylobatids frequently employ **TO-bipedal walk/run** with abducted forelimbs that are free from substrate contact (Fleagle 1976; pers. obs.). Furthermore, hylobatids frequently utilize bipedal locomotor bouts in between the end of a bout of **TO-brachiation/forelimb swing** or a **drop** and a subsequent bout of **TO-brachiation/forelimb swing** or **leaping** (pers. obs.), potentially making it more advantageous to utilize a larger-sized substrate, given their larger surface area and relatively more rigid structure compared to a smaller-sized substrate.

The pattern of support use of gibbons and siamang during **TO-bipedal walk/run** appears in direct contrast to that of large-bodied hominids, many of which have been observed to utilize **TO-bipedal walk/run** in conjunction with smaller- to medium-sized substrates (e.g., <10 cm diameter) (Hunt 1992; Doran 1993a, b; Remis 1995; Thorpe et al. 2007; Thorpe and Crompton 2005, 2009; but see Stanford 2006; Tourkakis 2009). Given the size differences between the large-bodied and small-bodied extant apes (Smith and Jungers 1997), the aforementioned use of substrate sizes during **TO-bipedal walk/run** in hylobatids compared to that of hominoids appears counterintuitive (Cant 1987b, 1992). However, as with **TO-bipedal stand,** these differences are likely related to variation in the contextual use of **TO-bipedal walk/run** and/or substrate availability/use among the size-variant hominoid populations. For instance, compared with bouts of **TO-bipedal walk/run** in hylobatids, **TO-bipedal walk/run** in Sumatran orangutans is often associated in a continuum of locomotor bouts with the locomotor mode **TO-clamber/transfer**, which is commonly utilized to traverse small- to medium-sized substrates, and thus **TO-bipedal walk/run** can also aid some apes in accessing fragile terminal branches (Thorpe and Crompton 2005, 2009; Thorpe et al. 2007).

Despite the appearance of a greater use of **TO-vertical climb** on substrates of >2 cm diameter, the use of **TO-vertical climb** did not differ statistically between different support size classes in this study. Among other hylobatid populations, Fleagle (1976) documented that Malaysian siamangs utilized a greater percentage of <10 cm diameter substrates while 'climbing,' whereas Malaysian lar gibbons and eastern black-crested gibbons (*Nomascus nasutus*) from China were found to utilize medium-sized branches (i.e., 2–10 cm) far more frequently (Fleagle 1976; Fan et al. 2013). Conversely, Gittins (1983) observed Malaysian agile gibbons to utilize boughs (i.e., >10 cm) more often in association with 'climbing' locomotion. As previous studies of hylobatid locomotion utilized an expansive 'climb' category, it remains speculative to evaluate what fraction of observed 'climbing' corresponded precisely to **TO-vertical climbing** (*sensu* Hunt et al. 1996), obstructing further comparative interpretations of species differences. For those hominid studies that differentiated **TO-vertical climb**, eastern chimpanzees utilized small- to medium-sized substrates (i.e., <10 cm diameter), whereas northeast Bornean and Sumatran orangutans utilized medium- to large-sized substrates (i.e., >4 cm diameter), which when evaluated with the data available for lar gibbons presented here suggests that substrate use during **TO-vertical climb** is variable across ape populations. Nevertheless, as with other ape species, lar gibbons from Khao Yai were highly capable of **TO-vertical climbing** on all substrate size classes identified in this study.

For lar gibbons at Khao Yai, the tree core was frequently navigated utilizing **TO-bipedal walk/run**, **TO-vertical climb**, and **TO-brachiation/forelimb swing**, whereas the tree periphery was frequently associated with **TO-clamber/transfer**. Eastern chimpanzees displayed a similar pattern of canopy use for both **TO-clamber/transfer** and **TO-vertical climb**, while their use of **TO-bipedal walk/run** and **TO-brachiation/forelimb swing** did not significantly differ between the tree periphery or tree core (Hunt 1992). Differences in the use of locomotor behaviors among different canopy locations are likely associated with variation in substrate size (e.g., small-sized substrates are often associated with the tree periphery, whereas larger-sized substrates are more commonly found within the tree core), some of the observed variation is also likely associated with the requirements of crossing canopy gaps, which are essentially the empty spaces between the woody structures of adjacent tree crowns. For instance, Cannon and Leighton (1994) observed that Bornean white-bearded gibbons (*Hylobates albibarbis*) modified their locomotor behaviors while crossing gaps, commonly using greater frequencies of **bridging** and **leaping**, relative to contexts of non-gap crossing travel. In addition to **TO-clamber/transfer**, we also observed increased frequencies of **bridging** and **leaping** in the tree periphery, which suggests that these locomotor modes are collectively vital to navigating areas of the tree canopy where both small-sized substrates and canopy gaps occur (i.e., the tree periphery).

The use of **TO-brachiation/forelimb swing**, **TO-bipedal/walk run**, and **TO-vertical climbing** in association with the tree core, is consistent with our observations that these locomotor modes are commonly associated with medium-to larger-sized substrates. These results may also suggest that at least for **TO-bipedal walk/run** and **TO-vertical climb**, that they are less frequently utilized in the context of crossing canopy gaps. Consistent with this notion, Cannon and Leighton (1994) observed that Bornean white-bearded gibbons infrequently used **TO-bipedal walk/run** and 'climbing' while crossing canopy gaps. Interestingly, Cannon and Leighton (1994), Thompson (2007), and Cheyne et al. (2013) have shown that **TO-brachiation/forelimb** swing facilitated the crossing of relatively large canopy gaps (i.e., up to 9 m) among Bornean white-bearded gibbons and that gap crossing was most frequently accomplished via **TO-brachiation/forelimb swing**. As such, it is enigmatic that we observed a greater frequency of use of **TO-brachiation/forelimb swing** in association with the tree core as compared to the tree periphery. These conflicting results may indicate that gibbons are capable of avoiding the tree periphery while locomoting, and this may be facilitated by their ability to traverse relatively large distances per swing with their primary locomotor modes **TO-brachiation/forelimb swing**. Alternatively, it is also possible that differences in canopy structure and perhaps particularly in relation to the nature of canopy gaps of the forests of Khao Yai and Borneo may explain the difference in locomotor behavior of gibbons.

Lastly, the locomotor behavior of Khao Yai lar gibbons did not statistically differ between our identified canopy levels, with **TO-brachiation/forelimb swing** being the most frequently used locomotor behaviors at all canopy levels. Despite a lack of statistical significance, **TO-brachiation/forelimb swing** were used less frequently as height level decreased, whereas the opposite was true for **TO-vertical climb** and **TO-clamber/transfer**. As suggested above for **TO-forelimb suspension**, these behavioral trends may be partially related to a lack of adequate-sized supportive elements at lower canopy levels (i.e., below <15 m) at our study site (Brockelman et al. 2011), and the need to spread body mass among multiple smaller substrates while locomoting at these canopy levels. Similarly, Fan et al. (2013) observed that eastern black-crested gibbons increased their use of **bridging** and 'climbing' in the understory (i.e., <5 m in their study), which they suggest was related to the relative lack of boughs (i.e., supports of >10 cm in diameter) at their study site. Thorpe and Crompton (2005) and Manduell et al. (2011) have shown that canopy height impacts orangutan TO-locomotor behavior, but they also note that this is best explained by the relationship between canopy height and substrate use, as discussed above. Overall, the collective use of **TO-brachiation/forelimb swing**, **TO-clamber/transfer**, and **TO-vertical climb** effectively facilitate Khao Yai lar gibbons' access to and navigation of different canopy levels.

Conclusions

The results we have presented reveal that *Hylobates lar* at Khao Yai utilizes a more extensive positional repertoire than was previously known for hylobatids, despite the fact that Khao Yai lar gibbons frequently employ TO-suspensory-locomotion and -posture. Importantly, in contrast to earlier assumptions and similar to larger-bodied extant hominids, Khao Yai lar gibbons utilize all of the TO-positional modes outlined by Hunt et al. (1996), which collectively allow these small Asian apes to effectively negotiate different substrate size classes, tree crown areas, and canopy levels. Furthermore, *Hylobates lar* appears to adjust their positional behaviors in a manner predicted by the body to branch size ratio model (Napier 1967; Cartmill 1974, 1985; Cant 1992; Preuschoft et al. 1995; Dunbar and Badam 2000; Preuschoft 2002), indicating that even relatively small body size plays a significant role in understanding patterns of primate positional behavior.

Our results, when compared with that of previously published studies of hylobatids and hominids, demonstrate that TO-positional behaviors are used variably across contexts, substrate sizes/types, canopy locations, and canopy heights (Chivers 1972, 1974; Fleagle 1976, 1980; Gittins 1983; Kano and Mulavwa 1983; Srikosamatara 1984; Cant 1987a, b; Hunt 1992, 1994; Doran 1993a, b; Cannon and Leighton 1994; Remis 1995, 1998; Sati and Alfred 2002; Stanford 2002, 2006;

Thorpe and Crompton 2005, 2006, 2009; Thompson 2007; Thorpe et al. 2007; Tourkakis 2009; Manduell et al. 2011, 2012; Myatt and Thorpe 2011; Fan et al. 2013; Cheyne et al. 2013; Fei et al. 2015). As with all positional behavior, the variability observed within and among hominoid taxa is related to numerous factors, such as ecology, morphology, physiology, and various socio-behavioral attributes (Garber 2011). While a concise and detailed understanding of all of these factors is beyond the scope of this contribution, we highlight our fundamental finding that a flexible use of TO-locomotion and -posture is shared among hominoids, regardless of also marked differences that exist, and that this uniting behavioral variability has been central to hominoid adaptation and evolution (Thorpe and Crompton 2006; Crompton et al. 2008; Nowak and Reichard 2016). Our results support the notion that suspensory locomotion and posture have been key components to the success of hylobatids in their arboreal environment and that the use of TO-behavioral variability, has played a complementary role in their successful radiation throughout South and Southeast Asia.

Acknowledgments   We kindly acknowledge the National Park Division of the Royal Forestry Department, Bangkok, and the National Research Council of Thailand, Bangkok for their permission to conduct research at Khao Yai National Park, Thailand. We also thank Chaleam Sagnate, Jacqueline Prime, and Surasack Homros for their assistance while in the field. Lastly, many of the ideas discussed in this chapter are also based on field observations of Sumatran *Hylobates agilis* and *Symphalangus syndactylus* that would not have been possible without the assistance of an NSF Doctoral Dissertation Improvement Grant (BCS 1061477 awarded to MGN).

References

Altmann J (1974) Observational study of behavior: sampling methods. Behavior 49:227–267

Andrews P, Groves CE (1976) Gibbon and brachiation. In: Rumbaugh DM (ed) Gibbon and Siamang: a series of volumes on the lesser apes, vol 4., Suspensory behavior, locomotion, and other behaviors of captive gibbons: cognitionKarger, Basel, pp 167–218

Asensio N, Brockelman WY, Malaivijitnond S, Reichard UH (2011) Gibbon travel paths are goal oriented. Anim Cogn 14:395–405

Barelli C, Heistermann M, Boesch C, Reichard UH (2007) Sexual swellings in wild white-handed gibbon females (*Hylobates lar*) indicate the probability of ovulation. Horm Behav 51:221–230

Begun DR (2007) How to identify (as opposed to define) a homoplasy: examples from fossil and living great apes. J Hum Evol 52:559–572

Bertram JEA (2004) New perspectives on brachiation mechanics. Am J Phys Anthropol 125:100–117

Brockelman WY, Nathalang A, Gale GA (2011) The Mo Singto forest dynamics plot, Khao Yai National Park, Thailand. Nat Hist Bull Sim Soc 57:35–55

Cannon CH, Leighton M (1994) Comparative locomotor ecology of gibbons and macaques: selection of canopy elements for crossing gaps. Am J Phys Anthropol 93:505–524

Cannon CH, Leighton M (1996) Comparative locomotor ecology of gibbons and macaques: is brachiation more efficient? Trop Biodiv 3:261–267

Cant JGH (1987a) Positional behavior of female Bornean orangutans (*Pongo pygmaeus*). Am J Primatol 12:71–90

Cant JGH (1987b) Effects of sexual dimorphism in body size on feeding postural behavior of Sumatran orangutans (*Pongo pygmaeus*). Am J Phys Anthropol 74:143–148

Cant JGH (1992) Positional behavior and body size of arboreal primates: a theoretical framework for field studies and an illustration of its application. Am J Phys Anthropol 88:273–283

Cant JGH, Temerin LA (1984) A conceptual approach to foraging in primates. In: Rodman PS, Cant JGH (eds) Adaptations for foraging in nonhuman primates. Columbia University Press, New York, pp 304–342

Cartmill M (1974) Pads and claws in arboreal locomotion. In: Jenkins FA (ed) Primate locomotion. Academic Press, New York, pp 45–83

Cartmill M (1985) Climbing. In: Hildebrand M, Bramble DM, Liem KF, Wake BD (eds) Functional vertebrate morphology. Harvard University Press, Cambridgem, pp 73–88

Cartmill M, Milton K (1977) The lorisiform wrist joint and the evolution of "brachiating" adaptations in the Hominoidea. Am J Phys Anthropol 47:249–272

Cheyne SM, Thompson CJH, Chivers DJ (2013) Travel adaptations of Bornean agile gibbons *Hylobates albibarbis* (Primates: Hylobatidae) in a degraded secondary forest, Indonesia. J Threat Taxa 5:3963–3968

Chivers DJ (1972) The siamang and the gibbon in the Malay peninsula. In: Rumbaugh DM (ed) The gibbon and siamang, vol 1. Karger, Basel, pp 103–135

Chivers DJ (1974) The siamang in Malaya: a field study of a primate in tropical rain forest. Contribution to primatology, vol 4. Karger, Basel

Crompton RH, Vereecke EE, Thorpe SKS (2008) Locomotion and posture from the common hominoid ancestor to fully modern hominins, with special reference to the last common panin/hominin ancestor. J Anat 212:501–543

Diogo R, Wood B (2011) Soft-tissue anatomy of the primates: phylogenetic analyses based on the muscles of the head, neck, pectoral region and upper limb, with notes on the evolution of these muscles. J Anat 219:273–359

Doran DM (1993a) Sex differences in adult chimpanzee positional behavior: the influence of body size on locomotion and posture. Am J Phys Anthropol 91:99–115

Doran DM (1993b) Comparative locomotor behavior of chimpanzees and bonobos: the influence of morphology on locomotion. Am J Phys Anthropol 91:83–98

Dunbar DC, Badam GL (2000) Locomotion and posture during terminal branch feeding. Int J Primatol 21:649–669

Fan P, Scott MB, Hanlan FEI, Changyong MA (2013) Locomotion behavior of cao vit gibbon (*Nomascus nasutus*) living in karst forest in Bangliang Nature Reserve, Guangxi, China. Int Zool 8:356–364

Fei H, Ma C, Bartlett TQ, Dai R, Xiao W, Fan P (2015) Feeding postures of Cao Vit gibbons (*Nomascus nasutus*) living in a low-canopy karst forest. Int J Primatol 36:1036–1054

Fleagle JG (1976) Locomotion and posture of the Malayan siamang and implications for hominid evolution. Folia Primatol 26:245–269

Fleagle JG (1980) Locomotion and posture. In: Chivers DJ (ed) Malayan forest primates: ten years' study in tropical rain forest. Plenum Press, New York, pp 191–207

Fleagle JG, Mittermeier RA (1980) Locomotor behavior, body size, and comparative ecology of seven Surinam monkeys. Am J Phys Anthropol 52:301–314

Garber PA (2011) Primate locomotor behavior and ecology. In: Campbell CJ, Fuentes A, MacKinnon K, Bearder SK, Stumpf R (eds) Primates in perspective, 2nd edn. University of Oxford Press, Oxford, pp 543–560

Gebo DL (1996) Climbing, brachiation, and terrestrial quadrupedalism: historical precursors of hominid bipedalism. Am J Phys Anthropol 101:55–92

Gebo DL (2004) Paleontology, terrestriality, and the intelligence of great apes. In: Russon AE, Begun DR (eds) The evolution of thought: evolutionary origins of great ape intelligence. Cambridge University Press, Cambridge, pp 320–334

Gittins SP (1983) Use of the forest canopy by the agile gibbon. Folia Primatol 40:134–144

Grand TI (1972) A mechanical interpretation of terminal branch feeding. J Mammal 53:198–201

Grand TI (1984) Motion economy within the canopy: four strategies for mobility. In: Rodman PS, Cant JGH (eds) Adaptations for foraging in nonhuman primates. Columbia University Press, New York, pp 54–72

Groves CP (1972) Systematics and phylogeny of gibbons. In: Rumbaugh D (ed) Gibbon and siamang, vol 1. Karger, New York, pp 2–89

Harrison T (2013) Catarrhine origins. In: Begun DR (ed) A companion to paleoanthropology. Blackwell Publishing Ltd, New York, pp 376–396

Hochberg Y, Tamhane AC (1987) Multiple comparison procedures. John Wiley and Sons Inc., New York

Hollihn U (1984) Bimanual suspensory behaviour: morphology, selective advantages and phylogeny. In: Preuschoft H, Chivers DJ, Brockelman WY, Creel N (eds) The lesser apes: evolutionary and behavioral biology. Edinburgh University Press, Edinburgh, pp 85–95

Holm S (1979) A simple sequentially rejective multiple test procedure. Scand J Statistics 6:65–70

Hood, GM (2009) PopTools version 3.1.1. Available online at http://www.cse/csiro.au/poptools. Accessed 29 Dec 2015

Hunt KD (1991) Positional behavior in the Hominoidea. Int J Primatol 12:95–118

Hunt KD (1992) Positional behavior of *Pan troglodytes* in the Mahale Mountains and Gombe Stream National Parks, Tanzania. Am J Phys Anthropol 87:83–105

Hunt KD (1994) The evolution of human bipedality: ecology and functional morphology. J Hum Evol 26:183–202

Hunt KD (2004) The special demands of great ape locomotion and posture. In: Russon AE, Begun DR (eds) The evolution of thought: evolutionary origins of great ape intelligence. Cambridge University Press, Cambridge, pp 172–189

Hunt KD (2016) Why are there apes? Evidence for the co-evolution of ape and monkey ecomorphology. J Anat 228:630–685

Hunt KD, Cant JGH, Gebo DL, Rose MD, Walker SE, Youlatos D (1996) Standardized descriptions of primate locomotor and postural modes. Primates 37:363–387

Iurck MF, Nowak MG, Costa LCM, Mendes SL, Ford SM, Strier KB (2013) Feeding and resting postures of wild northern muriquis (*Brachyteles hypoxanthus*). Am J Primatol 75:74–87

Jablonski NG (1998) The response of catarrhine primates to pleistocene environmental fluctuations in East Asia. Primates 39:29–37

Jablonski NG (2005) Primate homeland: forests and the evolution of primates during the tertiary and quaternary in Asia. Anth Sci 113:117–122

Jablonski NG, Chaplin G (2009) The fossil record of gibbons. In: Lappan S, Whittaker DJ (eds) The gibbons: new perspectives on small ape socioecology and population biology. Springer, New York, pp 111–130

Jablonski NG, Whitfort MJ, Roberts-Smith N, Qinqi X (2000) The influence of life history and diet on the distribution of catarrhine primates during the pleistocene in eastern Asia. J Hum Evol 39:131–157

Kano T, Mulavwa M (1983) Feeding ecology of the pygmy chimpanzees (*Pan paniscus*) of Wamba. In: Susman RL (ed) The pygmy chimpanzee: evolutionary biology and behavior. Plenum Press, New York, pp 275–300

Kitamura S, Suzuki S, Yumoto T, Chuailua P, Plongmai K, Poonswand P, Noma N, Maruhashi T, Suckasam C (2005) A botanical inventory of a tropical seasonal forest in Khao Yai National Park, Thailand: implications for fruit-frugivore interactions. Biodiv Conserv 14:1241–1262

Larson SG (1998) Parallel evolution in the hominoid trunk and forelimb. Evol Anthropol 6:87–99

Manduell KL, Morrogh-Bernard HC, Thorpe SKS (2011) Locomotor behavior of wild orangutans (*Pongo pygmaeus wurmbii*) in disturbed peat swamp forest, Sabangau, Central Kalimantan, Indonesia. Am J Phys Anthropol 145:348–359

Manduell KL, Harrison ME, Thorpe SKS (2012) Forest structure and support availability influence orangutan locomotion. Am J Primatol 74:1128–1142

Manly BFJ (2007) Randomization, bootstrap, and Monte Carlo methods in biology, 3rd edn. Chapman and Hall/CRC, New York

Michilsens F, D'Août K, Aerts P (2011) How pendulum-like are siamangs? Energy exchange during brachiation. Am J Phys Anthropol 145:581–591

Michilsens F, D'Août K, Vereecke EE, Aerts P (2012) One step beyond: different step-to-step transitions exist during continuous contact brachiation in siamangs. Biol Open 1:411–421

Mittermeier RA (1978) Locomotion and posture in *Ateles geoffroyi* and *Ateles paniscus*. Folia Primatol 30:161–193

Myatt JP, Thorpe SKS (2011) Postural strategies employed by orangutans (*Pongo abelii*) during feeding in the terminal branch niche. Am J Phys Anthropol 46:73–82

Napier JR (1967) Evolutionary aspects of primate locomotion. Am J Phys Anthropol 27:333–342

Nowak MG, Reichard UH (2016) Locomotion and posture in ancestral hominoids priorto the split of hylobatids. In: Reichard UH, Hirohisa H, Barelli C (eds) Evolution of gibbons and siamang. Springer, New York, pp 55–89

Pilbeam D (1996) Genetic and morphological records of the Hominoidea and hominid origins: a synthesis. Mol Phylogenet Evol 5:155–168

Pilbeam D, Young N (2004) Hominoid evolution: synthesizing disparate data. C R Palevol 3:305–321

Povinelli D, Cant JGH (1995) Arboreal clambering and the evolution of self-conception. Quart Rev Biol 70:393–421

Preuschoft H (2002) What does "arboreal locomotion" mean exactly and what are the relationships between "climbing", environment and morphology? Z Morph Anthrop 83:171–188

Preuschoft H, Demes H (1984) Biomechanics of brachiation. In: Preuschoft H, Chivers D, Brockelman W, Creel N (eds) The lesser apes: evolutionary and behavioural biology. Edinburgh University Press, Edinburgh, pp 96–118

Preuschoft H, Witte H, Fischer M (1995) Locomotion in nocturnal prosimians. In: Alterman L, Doyle GA, Izard MK (eds) Creatures of the dark: the nocturnal prosimians. Plenum Press, New York, pp 453–472

Reichard UH, Barelli C, Hirai H, Nowak G (2016) The evolution of gibbons and siamang. In: Reichard UH, Hirohisa H, Barelli C (eds) Evolution of gibbons and siamang. Springer, New York, pp 3–41

Remis M (1995) Effects of body size and social context on the arboreal activities of lowland gorillas in the Central African Republic. Am J Phys Anthropol 97:413–433

Remis M (1998) The gorilla paradox: the effects of body size and habitat on the positional behavior of lowland and mountain gorillas. In: Strasser E, Fleagle J, Rosenberger A, McHenry H (eds) Primate locomotion. Plenum Press, New York, pp 95–108

Rose MD (1997) Functional and phylogenetic features of the forelimb in Miocene hominoids. In: Begun DR, Ward CV, Rose MD (eds) Function, phylogeny, and fossils: Miocene hominoid evolution and adaptations. Plenum Press, New York, pp 79–100

Sati JP, Alfred JRB (2002) Locomotion and posture in hoolock gibbon. Ann For 10:298–306

Smith RJ, Jungers WL (1997) Body mass in comparative primatology. J Hum Evol 32:523–559

Srikosamatara S (1984) Ecology of pileated gibbons in South-East Thailand. In: Preuschoft H, Chivers DJ, Brockelman WY, Creel N (eds) The lesser apes. Edinburgh University Press, Edinburgh, pp 242–257

Stanford CB (2002) Brief communication: arboreal bipedalism in Bwindi chimpanzees. Am J Phys Anthropol 119:87–91

Stanford CB (2006) Arboreal bipedalism in wild chimpanzees: implications for the evolution of hominid posture and locomotion. Am J Phys Anthropol 129:225–231

Temerin A, Cant JGH (1983) The evolutionary divergence of old world monkeys and apes. Am Nat 122:335–351

Thompson CJH (2007) Gibbon locomotion in disturbed peat-swamp forest, Sebangau, Central Kalimantan. M.A. Thesis, University of Cambridge

Thorpe SKS, Crompton RH (2005) Locomotor ecology of wild orangutans (*Pongo pygmaeus abelii*) in the Gunung Leuser Ecosystem, Sumatra, Indonesia: a multivariate analysis using log-linear modeling. Am J Phys Anthropol 127:58–78

Thorpe SKS, Crompton RH (2006) Orangutan positional behavior and the nature of arboreal locomotion in Hominoidea. Am J Phys Anthropol 131:384–401

Thorpe SKS, Crompton RH (2009) Orangutan positional behavior. In: Wich SA, Utami Atmoko SS, Mitra Setia T, van Schaik CP (eds) Orangutans: geographic variation in behavioral ecology and conservation. Oxford University Press, Oxford, pp 33–48

Thorpe SKS, Holder RL, Crompton RH (2007) Origin of human bipedalism as an adaptation for locomotion on flexible branches. Science 316:1328–1331

Tourkakis CA (2009) Savanna chimpanzees (*Pan troglodytes verus*) as a referential model for the evolution of habitual bipedalism in hominids. MA Thesis, Iowa State University

Tyler DE (1991) The problems of the Pliopithecidae as a hylobatid ancestor. Hum Evol 6:73–80

Ward CV (1997) Functional anatomy and phyletic implications of the hominoid trunk and hindlimb. In: Begun DR, Ward CV, Rose MD (eds) Function, phylogeny, and fossils: Miocene hominoid evolution and adaptations. Plenum Press, New York, pp 101–130

Ward CV (2015) Postcranial and locomotor adaptations of hominoids. In: Henke W, Tattersall I (eds) Handbook of paleoanthropology, Part 2, 2nd edn. Springer, New York, pp 1363–1386

Young NM (2003) A reassessment of living hominoid postcranial variability: implications for ape evolution. J Hum Evol 45:441–464

Chapter 11
Selective Value of Characteristic Size Parameters in Hylobatids. A Biomechanical Approach to Small Ape Size and Morphology

Holger Preuschoft, K.-H. Schönwasser and Ulrich Witzel

Bipedal locomotion of two adult siamangs at a Zoo in Germany. Photo credit: Holger Preuschoft

H. Preuschoft (✉)
Früher Abteilung Funktionelle Morphologie, Medizinische Fakultät, Anatomisches Institut,
Ruhr-Universität Bochum, Bochum, Germany
e-mail: Holger.Preuschoft@ruhr-uni-bochum.de

K.-H. Schönwasser
Graf Engelbert-Gymnasium, Bochum, Germany

U. Witzel
Arbeitsgruppe Biomechanik, Fakultät für Maschinenbau,
Ruhr-Universität Bochum, Bochum, Germany

© Springer Science+Business Media New York 2016
U.H. Reichard et al. (eds.), *Evolution of Gibbons and Siamang*,
Developments in Primatology: Progress and Prospects,
DOI 10.1007/978-1-4939-5614-2_11

Introduction

The contributions to this volume are related to evolutionary aspects of morphology, taxonomy and cognition of gibbons and siamang (family Hylobatidae). Many of the current conclusions concerning taxonomy disagree with previous results (Creel and Preuschoft 1971, 1976, 1984; Schilling and Preuschoft 1984). The present chapter, however, focuses exclusively on characteristics of hylobatid external morphology and locomotion. Molecular genetic variation between local breeding populations has nothing to do with the physical laws relevant when a medium-sized primate of a specific body mass is transported through an unstable and possibly fragile environment.

The small apes of Asia are with 5–12 kg (Raemaekers 1984; Zihlman et al. 2011) within the size range of many arboreal simians (Smith and Jungers 1997, e.g., Atelinae, Alouattinae; guenons, vervets, patas, some macaques, colobines and some baboon females), and at the upper size limit of extant prosimians (e.g., *Propithecus* and *Indri*). On the other hand, hylobatids are much smaller than African apes and orangutans, in which even adult females may be up to three times the size of an adult siamang, not to speak of size differences between males. Among the apes, hylobatids also seem to possess more specialized morphological traits than other primates (e.g., extremely long forelimbs, long hindlimbs: Schultz 1930, 1933a, b, 1944; unique locomotor mode: Napier and Napier 1967; Carpenter 1940; Morbeck 1979; Preuschoft et al., 1984; Harrison 2016; deep cleft between digital rays I and II, but see Zihlman et al. 2011). The prominent, 'dagger-like' canines of gibbons and siamangs as well as aspects of the less than obvious sexual dimorphism (Leutenegger 1982; Frisch 1973; Creel and Preuschoft 1976, p. 267) and minor differences of proportions between taxonomic subgroups or local populations, as recently reported by Zihlman et al. (2011), are not considered in this chapter. Our focus is on the overall body shape and related locomotor biomechanics of all hylobatids.

Given the particular morphology of hylobatids, the question may be asked whether evolutionary advantages can be identified that have led to their anatomical specializations. Aside from other facts, the laws of physics or mechanical conditions in the locomotor apparatus of an animal offer attractive insights into the causes of body shape variation among primates. These general body shapes usually characterize higher taxonomical units, like subfamilies, or families (Preuschoft 1989). Not primatologists alone, but evolutionary biologists in general are convinced that the acquisition of new traits usually implies a 'selective advantage'. However, what exactly this advantage is remains all too often obscure. Body shape is determined by the locomotor apparatus, thus advantages of changing a body form can be expected to touch on the field of mechanics. Therefore, it seems logical to track assumed advantages of particular body shapes to known and relatively simple rules of physics. If these rules are defined, the trait can be used as an independent variable and its influence on 'fitness', or more precisely on defined parameters found in the individual, can be observed. This is intended in the following chapter. We attempt to quantify the effects of some obvious and general traits present in hylobatids on an individual's chances of survival.

Recently, the selective value of big body size based on biomechanical laws has been investigated in such divergent animals as fossil gigantic sauropods (Preuschoft et al. 2011) and in living primates (Preuschoft 2010). In the latter study, hylobatids were left aside, because they appeared unique, even exceptional, with respect to morphological adaptations and locomotion. To emphasize the peculiarities of gibbon and siamang morphology, we compare them to the orangutan, because the large-bodied Asian apes evolved a very different morphotype even though hylobatids and orangutans live allopatric in roughly the same biotope and have faced the same environmental constraints to which each found adaptive anatomical solutions.

Because the 'small body-size trait' has played an important role in the evolution of hylobatids, it should be kept in mind that 'body size' can be measured exactly as body mass (i.e., measured in 'g' or 'kg'). However, the 'size' of an animal can also be represented in length dimensions (i.e., measured in 'cm' or 'm'), or as an area (i.e., measured in 'mm^2' or 'm^2', as for example when cross sectional areas are measured to estimate muscle or bone strength), or as a volume (i.e., measured in 'mm^3' or 'm^3', which is largely proportional to mass).

Materials and Methods

Three sources of information were used in our analyses: (i) observations of postural behaviour, mostly in zoos, some in labs, and a wealth of TV recordings. Through many years, hundreds of observation hours have been accumulated by the authors. Where available, kinematic studies were also considered (Yamazaki et al. 1979, 1983; Yamazaki 1992, 1985; Yamazaki and Ishida 1984; Chang et al. 2000; Bertram and Chang 2001; Michilsens et al. 2010; Channon et al. 2011). (ii) The authors' own dissections (partly published in Preuschoft 1961, 1963) and measurements (Preuschoft et al. 1998) as well as comparisons with dissection results of other authors (Lorenz 1971; Isler 2002, 2003). (iii) Mechanical conditions and laws of physics, which govern locomotion and posture, were obtained from well-known physics and mechanical engineering textbooks (Lehmann 1974–1977; Dubbel 1981).

For calculations, masses and absolute length measurements are needed (Preuschoft et al. 1998, 2010). While a fine collection of primate body masses is available from Jungers (1985) and Smith and Jungers (1997), absolute body and limb length data are difficult to obtain, which therefore had to be restored from relative data (Schultz 1930, 1933a, b, 1944; Napier and Napier 1967; Jungers 1985; Jouffroy et al. 1993). Data were also taken from Isler (2003), Zihlman et al. (2011) and recently Michilsens (2012).

For calculations presented here, we used the best fitting data available from the literature and our own observations. The mechanical laws and morphological traits related to them are explained in the following. Graphs were calculated for dimensions and include the orangutan for comparison. Parallels in other animals are dealt with in the discussion.

Arboreal Habitat and Substrate Conditions: Why Brachiation?

The peripheral crown canopy (Avis 1962; Napier and Napier 1967; Grand 1972; Ripley 1979; Chivers 1984; Raemaekers 1984) is highly attractive for herbivores and insectivorous animals, because it provides new leaves, buds, blossoms, fruit and insects. These resources can also be found elsewhere in the canopy, but peripheral branches of tropical trees often offer the most nutritious resource. Preuschoft and Demes (1984a, b) have detailed that the terminal branches of a tree can be reached from the outside by birds, bats, and exceptionally tall mammals such as giraffes, but primates must approach the periphery from the interior of a tree. This was recently observed by Fei and colleagues (2015) in white cheeked gibbons and can be generalized. An arboreal animal approaching the periphery of a tree crown from a tree's interior inevitably exerts static forces [F] on a branch or twig (Fig. 11.1), due to gravity [g] and its own body mass [m].

$$F = m * g \qquad (11.1)$$

These forces, multiplied by their lever arms, evoke the bending moments, which stress branches (Fig. 11.1). When the animal moves, however, acting forces also depend on the direction and magnitude of acceleration [a], which the animal itself produces.

$$F = m * a \qquad (11.2)$$

Considerable kinetic forces come into action when an animal starts moving away from the substrate, as well as in slowing down a movement towards a substrate. The problems posed for an animal moving towards the outer foliage are detailed in Michilsens (2012). The notoriously high compliance of branches and twigs slows down the change of speed or direction, and smoothes accelerating and decelerating. Only if these forces or bending moments can be sustained by the substrate, will movement of the animal continue while failure interrupts movement or worse may lead to falling out of the tree.

The rather thin, compliant twigs of the outer foliage bend under the weight of an animal. The increasing weakness of supports as terminal parts of a tree are approached cause several problems. First, only branches or twigs resistant enough to carry an animal's weight can be used as support (Fig. 11.1a, b). Therefore, small animals have better chances to reach the periphery of a tree than large ones. This point is especially important, because 90 % of the locomotor behavior of monkeys is quadrupedal walking on relatively horizontal substrates (Arms et al. 2002; Preuschoft 2002), while they will sit on their ischial callosities when resting (Fig. 11.1). Second, a slender branch may well sustain the weight of an animal provided the animal is capable of balancing, and keeping its centre of mass on top of a rather narrow, round area of support (Fig. 11.2; Cartmill 1974; Preuschoft et al. 1992; Preuschoft 2002). Boughs in the periphery become narrower and the

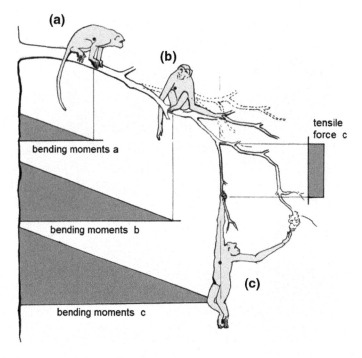

Fig. 11.1 Stress exerted by a primate on a branch becomes visible when branches become thinner. The further a primate moves away from the stem the greater bending moments (*grey*) on the branch become (original branch position indicated by stippling). Position *a* shows a monkey sitting on a branch, *b* shows a hylobatid resting comfortably, and *c* shows a hylobatid suspended from a thin twig. Primates in *a* and *b* have to balance their position to avoid falling while the suspended primate in *c* is in equilibrium. Body weights are assumed equal and bending moments depend solely on the length of lever arms. Bending moments along the branch are limited by flexing terminal twigs downward, which become subject to tensile forces. Stress caused by branch and foliage weight is not shown. The thin branch on which the primate in *c* is suspended does no longer offer a support for sitting or standing. Bending stresses are replaced by tensile stresses (*grey*) against which the material is much more resistant

narrower the substrate, the less safe is the primate's position. Provided a stable angle of deviation from vertical can be maintained, shortening of functional limb length increases safety, because it brings the centre of mass closer to the substrate, while lengthening the limbs will increase the rotation moment and thus increase the risk of losing balance. The decisive factor for equilibrium, however, is not the angle, but the lever arm of body weight against the central axis of the branch. For a pronograde, quadrupedal primate it is costly to maintain balance, because muscle force or energy must be expended. The problem of keeping balance disappears when an animal changes its position to hang under the branch in a below-branch, suspensory posture (Fleagle and Mittermaier 1980; Ishida et al. 1990; Jouffroy and Petter 1990; Jouffroy and Stern 1990). Third, regardless of an animal's position above or below its supports, the branch will bend downward under the primate's

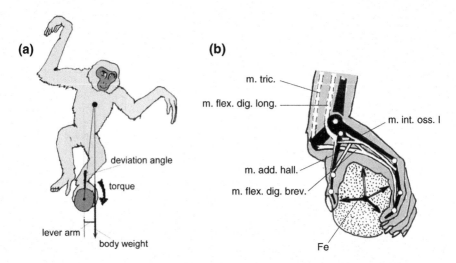

Fig. 11.2 a The hylobatid walking bipedally along a branch is in danger of rotating around the branch. Torque exerted by body weight times lever arm can be balanced by a firm prehensile grip. **b** A prehensile foot can produce high pressure [F_e] between skin and substrate by its intrinsic musculature and digital flexors, and so increase friction without influencing the position of the ankle joint. By using intrinsic muscles [musculus flexor digitorum brevis (m. flex. dig. brev.), musculus adductor hallucis (m. add. hall.), musculus interosseus I (m. int. oss. I)], the grip loses its dependence from the lower leg (=long digital flexor) muscles, which influence the ankle joint as well as the digits. The muscles acting on that joint (musculus triceps (m. tric.), musculus flexor digitorum longus (m. flex. dig. long.)) are shown by broken lines. [F_e] is the reaction force evoked by the flexing toes

weight along a nearly vertical direction, because it does not possess sufficient bending rigidity to maintain a horizontal orientation (Fig. 11.1). Since branches usually offer much higher tensile than bending strength, a twig will continue to carry the weight of the animal, provided the primate is able to secure its handhold. Obviously, downward bending of a support is a greater problem for a larger than for a smaller animal moving on top of the substrate. In contrast, keeping balance is not a problem in a suspended primate, who spends energy alone for securing its grip (Figs. 11.1c and 11.3a).

Prehension

Primates have evolved remarkable (Preuschoft and Chivers 1993) prehensile hands and feet (autopodia) and even prehensile tails in some New World Primates, which can transmit tensile forces, as well as torsional moments rotational moments. The latter ability permits maintaining balance even if the centre of gravity is displaced laterally (Fig. 11.2a; Cartmill 1974, 1985; Preuschoft et al. 1992; Preuschoft 2002). The grip of a primate's autopodium allows pressing the fingers or toes against the

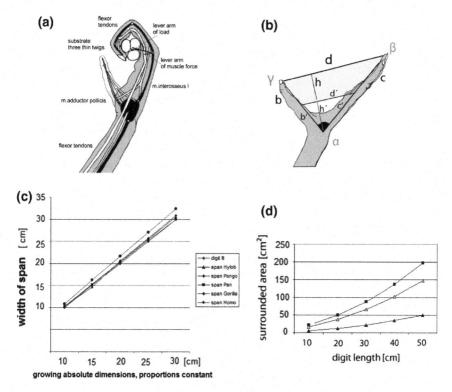

Fig. 11.3 a If digits become very long, they may be curled around a thin substrate so that proximal phalanges and balls act as props for the force exerted by the middle and distal phalanges. Alternatively, the thenar and hypothenar may provide the counterforce. Note the lever ratio between load and force (which is 1:3). **b** The span of the grip depends largely on the lengths of digital rays [*b*] and [*c*]. The *grey area* surrounded by the grip is limited by *d* * *h*/2 minus the part covered by the web between rays I and II. The deep cleft between digital rays I and II in hyloabtids yields a 25 % larger (surrounded by *d′*, *b′* and *c′*) interspace between pollex and lateral fingers than in orangutan, African apes, and humans. **c** The width of span is rather constant among hominoids if absolute dimensions are neglected, and is very large in the human hand, because of human's long thumb. **d** A hylobatid's autopodium can surround a 25 % larger cross sectional area of seized branches or stems, because the depth of the cleft (= height of the triangle) is longer. *Black squares*: areas surrounded by a hylobatid-like hand of increasing length; *open triangles*: areas surrounded by the hand of a human or ape; *black triangles*: not used part of the area between rays I and II if muscle adductor pollicis extends distally, close to the capitula metacarpalium

substrate by activating intrinsic muscles without a need to use long digital flexors, which would otherwise influence the ankle or wrist joint (Fig. 11.2b). As well as the famous hook hand, prehensity allows transmitting tensile forces, which is a necessary precondition for suspending an animal below a branch (Fig. 11.3a). In a suspensory position, if a single twig is too weak, a bundle of twigs can be grabbed (which is impossible, or at least difficult with claws) to distribute the animal's weight.

Prehension has consequences for the morphology of hands and feet. According to Buck and Bär (1993) the friction exerted by a primate hand on various substrates is proportional to the force by which the digits are pressed against the surface when other variables are held constant. The ability to exert high pressure (Fig. 11.2b) allows, in combination with softness of the subdermal connective tissue and the often discussed 'friction skin' of the dermatoglyphics (e.g., Biegert 1961), to produce sufficient frictional resistance to balance the weight force. Preuschoft (2004) has shown that elongation of digits increases contact area and thus reduces the pressure between substrate and palma or planta. To optimize the two variables, i.e., length of digits and small contact area, the autopodium, in particular the digits, need to become narrower (Fig. 11.4a, b). The influence of digital length and breadth on contact area is quantified in Fig. 11.4. If the length of fingers is increased in increments of 0.5 cm (Fig. 11.4c, *black triangles*), the digital area grows linearly (*open diamonds*). If the breadth of the five digits is reduced by 0.5 cm steps (*black circles*), the total contact area of the fingers passes a flat maximum before it becomes smaller than before (*open squares*). The narrow hands and fingers or feet

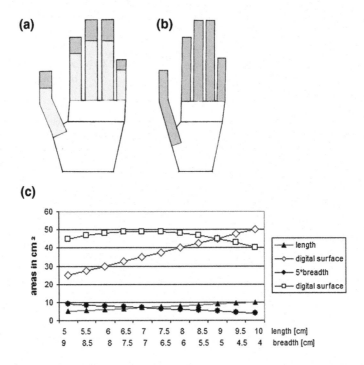

Fig. 11.4 Slender digits exert higher pressure on a substrate than broader ones. If the length of digits increases by ¼, dark finger tips (**a**) then the contact area between hand and substrate becomes larger and the exerted pressure lower. If the increased finger length is maintained, but the width of the hand is reduced by ¼ (**b**), the contact surface of all digits becomes smaller than before. **c** Numerical influence of digital length and breadth on the contact area

and toes respectively, lead to greater contact forces and allow the transmission of greater forces and torques by friction. Digital length, however, also increases the length of the load arm of the reaction force [F_c], which must be counterbalanced by muscular torques requiring extremely strong digital flexors, which inadvertently leads to additional weight.

It seems easy to imagine that the deep cleft between first and all other digital rays in hylobatid hands and feet (Fig. 11.3b) increases the span of the grip, and this has been stated repeatedly (e.g., Napier and Napier 1967). The known morphological facts, however, can be quantified using the following formula, in which the span [d] between rays I and III is given as:

$$d = \sqrt{(b^2 + c^2 - 2bc * \cos \alpha)} \tag{11.3}$$

where [b] is the length of ray I and [c] is the length of ray III. Width of the span follows roughly the *sine* of the angle between rays I and III. For our sample calculation we arbitrarily fixed the angle between rays I and III at 80° (Fig. 11.3d). Further, width depends importantly on the lengths of rays I and III. The longer the digital rays are, the greater the hand span becomes. The influence of ray length on span is shown for an arbitrarily assumed angle of 80° (Fig. 11.3c). The length of ray III varies from 10 to 30 cm like in apes, and thumb length was assumed to resemble hylobatids (ratio I/III = 44.5 %), orangutans (ratio I/III = 35 %), chimpanzees (ratio I/III = 40 %), gorillas (ratio I/III = 47.5 %) and humans (ratio I/III = 62 %: Schultz 1930). Differences between hominoid genera appear negligible and hand span follows the relative length of the lateral fingers as well as that of the thumb. Humans owe their great span to their long thumb. In conclusion, width of span is not the sole decisive selective factor for hand proportions and hand morphology in hylobatids or other hominoids.

In fact, grip efficiency does not depend on the width of span alone. The cross section of objects that can be seized is also determined by the triangular area (Fig. 11.3b) between the lateral digits and their opponent (i.e., digit ray I). It is defined by the span [d] between the tips of digits I and III and the shortest distance from this connection to the deepest point of the cleft, which equals the height [h] of the triangle (Fig. 11.3b). In our example, the surrounded area [A] is defined as:

$$A = d * h/2 \tag{11.4}$$

And the height of the triangle is defined as:

$$h = c * \text{sine } \beta \tag{11.5}$$

While β equals:

$$\beta = \text{arcsine}(b/d * \text{sine } \alpha) \tag{11.6}$$

With [d] being the span, α the angle between rays I and III, and [b] the length of ray I and [c] the length of ray III. The area A surrounded by the lateral rays on one side and the thumb on the other in hylobatids is roughly $A = d * h_d/2$. The longer the digital rays [b, c] the greater area A becomes (Fig. 11.3d). The increase is more than linear, for example, the larger orangutan arrives at a considerable area just by the length of its fingers (Fig. 11.3d). If the adductor muscles of the pollex insert more distally, close to the metacarpal-phalangeal joints of rays I and II (Fig. 11.3a), this area is reduced by approximately the area A', bounded by metacarpal (Mcp) I, Mcp II and the distal margin of the musculus (m.) adductor pollicis, which is proportional to the area calculated above $(A' = d' * h'/2)$. According to Schultz (1930, 1975) the phalanges of ray III in hominoids are constantly 60 % of the length of the whole ray, while in ray I the same ratio is fairly constantly 50 %. In view of this limited variability, the calculation is continued under the simplified assumption that roughly 50 % of rays I and III are covered by the adductor muscle and other soft parts, and the remaining area $[A_{rem}]$ can be calculated as follows:

$$A_{rem} = d * h/2 - d' * h'/2 \qquad (11.7)$$

According to our results (Fig. 11.3), the characteristic separation of the 1st ray in hylobatids allows to enclose a cross sectional area 25 % greater than the energy-saving arrangement of the more distally shifted adductor pollicis in most primates.

Accelerations

Not only the static weight of an animal challenges the strength of the substrate. As soon as an animal moves, positive or negative accelerations come into action. The greatest positive acceleration occurs in taking off for a leap while the greatest negative acceleration (i.e., braking or deceleration) occurs when landing after descending from a higher place or after a leap. The resulting forces do not only challenge the substrate, they also evoke stresses in the animal's body. Not surprisingly, large primates often avoid frequent and high acceleration. This becomes very obvious in the notoriously slow-moving orangutan (Napier and Napier 1967), which manages an arboreal lifestyle in spite of its enormous body mass of an average of 36 kg in females or 79 kg in males (Smith and Jungers 1997). Likewise, the heavier and larger siamang prefers slow climbing and continuous-contact brachiation compared to the lighter, more agile and often ricocheting smaller-bodied gibbons (Fleagle 1984; Raemaekers 1984, Michilsen 2012; Nowak and Reichard 2016). The kinetic energy contained in a moving body (Fig. 11.5a, b) can be formalized as:

$$E_{kin} = 1/2\,m * v^2 \qquad (11.8)$$

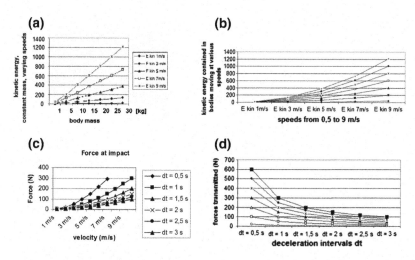

Fig. 11.5 a Kinetic energy plotted against body mass for various locomotor speeds shows a linear dependence on mass. **b** Kinetic energy plotted against speed for varying body mass shows a more-than-linear dependence on speed. **c** The forces exerted at impact by a moved body on the substrate ($F = P/dt$) depend on velocity (*x*-axis) and the deceleration intervals [d*t*]. The shorter [d*t*], the higher the forces. **d** The forces at impact plotted against locomotor speeds for various deceleration intervals [d*t*]. High compliance of thin twigs slows down the deceleration, which reduces the forces exerted at impact

where [*m*] is the body mass and [*v*] is the body's speed. The kinetic energy contained in a body is transmitted to the substrate in the form of a force (Fig. 11.5c). Realistic accelerations [*a*] during locomotion have repeatedly been found to be between 2 and 4 times Earth's acceleration (Preuschoft 1985; Günther 1989), and the force exerted during impact can also be written as:

$$F = m * \Delta v / \Delta t \tag{11.9}$$

where [Δv] is the difference in velocity of the moving animal, and the impact time is [Δt]. The impact force is mass dependent, as well as dependent on the speed of the moving animal and the impact time available for the exchange of forces between the animal and the substrate (Fig. 11.5c).

Under conditions of normal locomotion, the long extremities of hylobatids act as springs, by controlled flexion or extension of the joints, which decelerate the body slowly and so reduce the forces that would otherwise assume too great and potentially harmful values. Additional damping (increase of impact time) may come from substrate compliance. The high flexibility of thin twigs slows down the deceleration process of a moving body and also reduces forces exerted at impact. By contrast, high impact forces may be produced willingly in agonistic interactions between individuals aiming to hit or slap each other.

Suspensory Locomotion by the Forelimbs and Advantages of Limb Length

Brachiating hylobatids follow the physical laws of a pendulum (Fig. 11.6a; Yamazaki and Ishida 1984; Yamazaki 1985, 1990; Preuschoft and Demes 1984a, b, 1985; Demes and Preuschoft 1984; Isler 2003; Michilsen 2012). The centrifugal force Fc acts along the pendulum chord between the gibbon's centre of mass and its handhold. Bertram and Chang (2001) and Bertram (2004) emphasized the effects of various handholds and up- or downward locomotion. The empirical results on reaction forces (Chang et al. 1997; Bertram and Chang 2001, Bertram 2004) are not in conflict with our calculations. However, the physical laws of the pendulum only apply as long as the pendulum's chord does not become too heavy compared to the mass concentrated at the chord's end. Orangutans, for example, need very strong and heavy forelimbs because of their total body mass, and therefore can not profit from the energetically advantageous pendulum movement. The lengths of forelimbs in all large and small apes is in the same size order (Jungers 1985). In contrast, the shortness and low weight of orangutan hindlimbs also reduces both factors make orangutans poor arm-swingers in comparison to hylobatids. The weakness of the hindlimbs may well be related to the unstrate fragility in arboreal environments.

Speed [v] depends on excursion angle [α] and arm length [l], as well as the pendulum period [T] (Fig. 11.6b), following the equation:

$$v = 4y/T, \text{ with } y = l * \text{sine} \, a, \text{ and } T = 2\pi * \sqrt{l/g} \qquad (11.10)$$

where $y = l *$ sine α is the distance covered during each swing, and $T = 2\pi * \sqrt{(l/g)}$.

In ground-living primates and other walking animals the extremity functions alternately as an inverted pendulum (stance phase, during which body weight is supported) and a suspended pendulum (swing phase, during which the free limb is swung forward). The foreswing phase determines the duration, and therefore the frequency of the locomotor cycles. Unlike the ground-walking animals, the weight bearing phase in the suspensory hylobatid follows the laws of the suspended pendulum and determines frequency, while the recovery phase or the foreswing of the free arm seems to be used mainly to shift the animal's centre of gravity away from the handhold, in order to gain speed, with the effect of improving the energy balance.

Like in terrestrial locomotion, at least two gaits can be discriminated in arm-swinging hylobatids (Fig. 11.6a). Either a continuous contact brachiation, by seizing the next handhold before leaving the former (spacing of handholds between 0.8 and 1.2 m: Bertram and Chang 2001), or a gait which includes a phase of aerial, ballistic floating without contact to any support (ricocheting or ricochetal brachiation). The former may be compared to striding and the latter to trotting in a terrestrial mammal. The mechanical essence of ricocheting seems to be that parts of the centre of mass' trajectory represent a flight phase, which yields the greatest gain

(a)

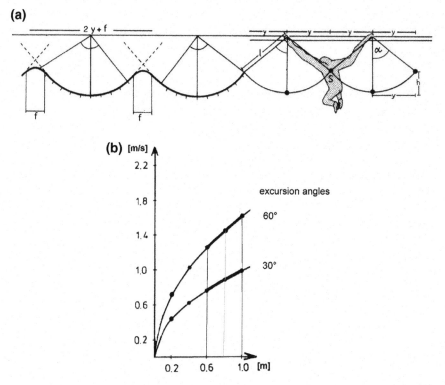

Fig. 11.6 a Hylobatid as a pendulum in two gaits: continuous contact (*right*) and interrupted substrate contacts with phases of free flight (ricocheting, *left*). [*S*]: centre of body mass, [*y*]: distance covered during ¼ of a swinging period (following the *sine* of the excursion angle); [*h*]: height difference between highest and lowest point of trajectory; [*f*]: distance covered in free flight. Along the pendulum chord acts the centrifugal force [*F_c*]. **b** The horizontal speed of a brachiating hylobatid (*vertical axis*) is determined by the length of the pendulum chord (*horizontal axis*), and by the excursion angle. The lengths realized in hylobatid forelimbs (0.55–0.75 m) are between the *vertical bars*. While increasing angles in general make locomotion faster, >60° do not yield much higher velocities than below that value

of horizontal distance (spacing of handholds 1.2–2.25 m: Bertram and Chang 2001), while up and downward movements of the centre of mass are minimized so that the trajectory of the centre of mass becomes flatter and less undulating. The accelerations necessary for free floating, however, cost additional energy. The longer the forelimbs become, the more pronounced this energetically positive effect becomes.

The energy saving and speed increasing effect of great arm length on brachiation suggests that a selection pressure has existed during hylobatid evolution to increase the length of the forelimb to optimize suspensory movements. It should be noted, that genetic evidence related to hylobatid brachiation was recently also discovered (Carbone et al. 2014).

If pendular processes are used in locomotion, only the energy losses occurring during movements must be compensated by muscle activity. In the case of arm-swinging, the most probable and most economic method is whipping down the free forelimb after lift-off (Preuschoft and Demes 1984a, b; Yamazaki 1990). Because of its mass, the downward whipping of the arm shifts the body's centre of mass away from the handhold from 0.7 to 0.8 m and thereby increases the mass moment of inertia during the downswing, while reducing it during the upswing. The model used by Bertram and Chang (2001), in which the body minus supporting forelimb is swinging about the shoulder joint of the support limb, reaching its deepest point during mid-swing, leads to exactly the same result. Very long fore-limbs also favour a rapid escape. If a hylobatid sitting on a high branch perceives danger, it will drop down, catch hold of another branch and move ricocheting away (Fig. 11.7). In this movement, the potential energy contained in the sitting animal equals its body mass [m] times Earth's acceleration [g] times the length of the forelimb [l]

$$E_{pot} = m * g * l \qquad (11.11)$$

which is transformed into kinetic energy:

$$E_{kin} = 1/2\,m * v^2 \qquad (11.8)$$

The transformation takes place, like in arm-swinging, between points B and C (Fig. 11.7). The change of direction from vertical to horizontal requires an input of a centripetal force [F_c]. The small ape loses potential energy and gains horizontal speed up to the value at point C [v_{max}], which can be expressed as follows:

$$1/2\,m * v^2 = m * g * l \qquad (11.12)$$

which is equal to $v_{max}^2 = 2 * g * l$ or $v_{max} = \sqrt{(2 * g * l)}$.

The longer the forelimbs, the higher the horizontal speed becomes (Fig. 11.6b). The centrifugal force [F_c] acting along the forelimb reaches its maximum at point C, where [F_c] is proportional to body mass [m] and the square of speed [v_{max}] divided by the radius [r], which equals forelimb length [l]:

$$F_c = m * v_{max}^2 / l \qquad (11.13)$$

Which can be written as: $F_c = 2 * m * g * l/l = 2 * m * g$ or $F_c = 2 * F_w$.

In addition, the hylobatid is exposed at point C to weight force [F_w] = $m * g$. This force and [F_c] have the same direction, and therefore add up to [F_t] = 3 * [F_w]. The same total force acts on the branch seized by the animal and used as a pivot. The greater the falling height [h] between points A and B, the greater becomes the potential energy which can be transformed into kinetic energy and the greater the initial speed. The maximal velocity [v_{max}] under this condition as shown above (Preuschoft and Demes 1984a, b) is:

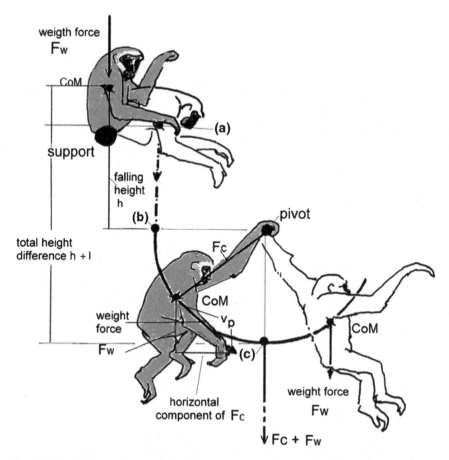

Fig. 11.7 Rapid escape from a resting support. Trajectory of centre of mass [CoM] emphasized by heavy line. Driven by weight force, the animal falls downward (*broken line*) and accumulates speed under the influence of gravity. Gripping hold at a pivot, the hylobatid's forelimb exerts the additional centripetal force [F_c], which redirects the movement. The horizontal component of the force [F_c] becomes greater, so that horizontal speed increases. When the deepest point of the trajectory (*heavy line*) is passed, the weight force begins to slow down progression of the animal

$$v_{\max}^2 = 2 * g * (h+l) \quad \text{or} \quad v_{\max} = \sqrt{2 * g * (h+l)} \qquad (11.14)$$

The centrifugal force [F_c] increases according to the formula (11.13), which means it is possible to rewrite the centrifugal force [F_c] equation as a function of maximal velocity [$v_{\max}$]:

$$F_c = 2 * F_w * (1+h/l) \qquad (11.15)$$

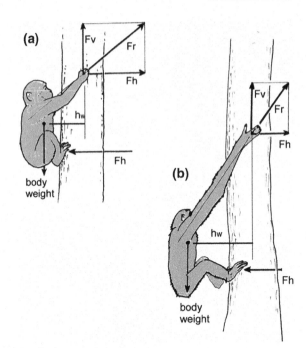

Fig. 11.8 a A short-armed monkey, e.g., *Macaca sylvanus*, and **b** a long-armed ape, e.g., *Hylobates sp.* in forelimb-dominated vertical clinging position. Body weight in both animals is assumed identical, but total force acting on the forelimb is smaller in the long-armed hylobatid. This holds true in spite of the longer lever arm of body weight [h_w] chosen by the small ape. The horizontal force component F_h acting on the hands is balanced by the same force exerted by the feet

As above, the weight force [F_w] must be added. If, for example, [$h = l$], the total force acting against the flexed fingers and applied to the support, is 5 times the body weight (=5 * F_w).

Great length of the forelimbs is also of advantage in vertical clinging and climbing, especially if the climber is forelimb-dominated (Fig. 11.8). In the primates shown in Fig. 11.8, the whole body weight is supported by the forelimbs, and no share of body weight is assumed to be carried on the feet. Weight, which is the vertical component in the drawing, is taken as a constant. The reaction force acting on a hand is proportional to the *sine* of the angle between the long axis of the forelimb and the vertical. The longer the arms, the narrower the angle between forelimb and the vertical becomes, and the smaller the resultant force that must be sustained by the arm, and balanced by (hook-like) flexed fingers. The general shortage of muscular force in larger animals increases the need to keep muscle force on the lowest possible level. Long arms are therefore especially useful for relatively heavy animals.

It is well known that hylobatids also move on top of substrates in an orthograde-bipedal posture, and only exceptionally use pronograde, quadrupedal postures (Orgeldinger 1994; Isler 2002, 2003). However, hylobatid bipedal locomotion is more similar to human running than striding (Okada 1985; Yamazaki et al. 1979, 1983; Yamazaki 1984, 1985; Yamazaki and Ishida 1984). Hip and knee flexion is more pronounced than in walking, and it includes phases of aerial floating, when the body's centre of mass is highest during the swing phases. Elastic properties of musculo-tendinous complexes are probably used. In spite of this, bipedal locomotion of gibbons and siamang is mainly governed by the laws of the pendulum, like that of humans (Preuschoft and Witte 1991; Witte et al. 1991). During the stance phase, the body rotates about the ankle joint like an inverted pendulum. The longer the hindlimbs become, the greater the length of each step. During foreswing, the hindlimb moves like a pendulum suspended at the hip joint and the shorter its chord becomes, the quicker the foreswing. Walking speed is defined, like the speed of the brachiating hylobatid, by step length [l]/pendulum period [T] (see above). Elastic properties of tendons seems to be used (Vereeke and Channon 2013). The inertial forces which the hindlimbs exert on the trunk during the foreswing phase are compensated by the length of the trunk and the long and heavy, either laterally spread or elevated arms (Fig. 11.2). Hindlimb length, however, and the inevitable great weight of especially the lower leg and the foot (Preuschoft et al. 1998) lead to high mass moments of inertia of the hindlimb and therefore to a relatively delayed foreswing, so that stride frequency is relatively low. In addition, gibbons and siamang are good jumpers (Channon et al. 2011). The often overlooked length of their hindlimbs (Schultz 1933a, b) is also of advantage for leaping by providing long acceleration distances (Günther 1989; Preuschoft et al. 1996, 1998).

Forelimb Length in Feeding

One characteristic of primates is their use of the forelimbs to acquire food (Grand 1972). Long forelimbs provide a long reaching distance. Every item within the hemisphere in front of the body can be reached by the hands (Fig. 11.9a). The hemispheres that are reached by both forelimbs overlap ventrally (Fig. 11.10b), and the hemisphere's lateral diameter is a shoulder width larger than the sum of both. The so-defined 'feeding envelope' grows by the 3rd power of forelimb length (Fig. 11.9b). It is completely irrelevant whether rare and tasty, perhaps nutricious fruits are selected or ubiquitous leaf material is harvested. For an animal moving along unstable supports maintaining safe positions is of highest priority and may even be a matter of survival. Any change of feeding position involves a risk of falling and energy expenditure. While the risk of falling is difficult to quantify, energy expenditure for acceleration and deceleration can be calculated. From one safe spot, a long-limbed primate can exploit a much larger volume (Figs. 11.9b and 11.10a, b) than a short-armed primate, and this leads to fewer position changes and by default reduces the potential risk of falling (Fig. 11.10c, d). Staying at one feeding spot for longer saves muscular force (Fig. 11.10d), which reduces energy expenditure. In summary, energy requirements decrease with increasing forelimb length.

(a)

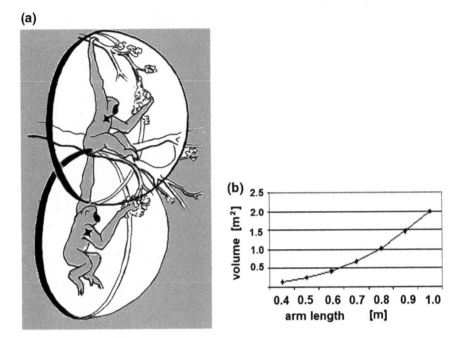

Fig. 11.9 a Reaching distances and feeding envelopes of a hylobatid sitting on a branch (*top*), or hanging suspended below a branch (*bottom*). What can be reached under optical control without moving the trunk is within a hemisphere with the shoulder joint as the centre and arm length as the radius. This figure has been granted and modified from Grand (1972). **b** The feeding volume grows by the third power of forelimb length

Using the Forelimbs in Agonistic Interactions

Carpenter (1940, 1976) has reported that fighting in hylobatids is fast and effective, but rarely involves physical contact. An attacker may approach swiftly, if possible from a higher branch, slap with a hand, kick with a foot, with high impact force due to a lack of damping elements, or deliver a bite and gain distance. Extended wrestling is avoided and biting is usually directed to the limbs (Orgeldinger 1994). Aside from using the impact of the whole body, hand blows are performed in fighting. Essential parameters for the impulse exerted by a slapping hand are the mass moments of inertia of all forelimb segments, the distances between shoulder joint and the centres of mass of all segments and the angular speed of the entire forelimb (Fig. 11.11a, b). The total mass moment of inertia of the hylobatid's forelimbs is $0.4 \ \mathrm{kgm^2}$, and the length of the forelimb alone gives the hand very high peripheral velocities when striking:

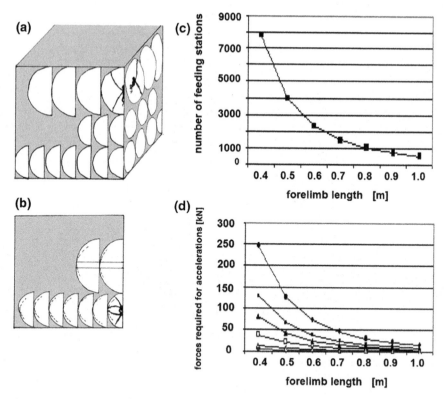

Fig. 11.10 To exploit a given volume of vegetation the hylobatid must change its position ('feeding station'). The cube shown in **a** is the view from above, while in **b** is the ground plan. The ranges of both hands overlap to some degree while at both sides the total feeding range is expanded by the animal's shoulder width. **c** With increasing forelimb length the frequency of necessary changes of position decreases. Any change of feeding stations implies a danger of falling and requires force to accelerate and decelerate the body. **d** This holds true even for very slow accelerations of 0.1 g (*open* and *black triangles*) and more rapid locomotion, e.g., 1.1 g (*black squares* and *open squares*) and 2.1 g (*black diamonds* and *black triangles*). Graphs are drawn for body weights of 15 kg (*rosettes, black squares* and *black triangles*) and 5 kg (*open squares* and *triangles*). The energy-saving effect of arm length in any case becomes smaller with arm lengths greater than 80 cm

$$\text{peripheral speed} = \text{angular velocity} * r \qquad (11.16)$$

The radius [r] is the distance from shoulder joint to the fingers hitting a target, and as [r] increases so does the impact force. A hand with long fingers will therefore produce higher strike forces than a hand with shorter fingers. Neither angular nor peripheral velocities of arm movements have been measured directly in hylobatids, but estimates can be based on the pendulum duration found in arm swinging hylobatids, which vary between 0.8 s (Jungers and Stern 1984) and 0.9 s (Preuschoft and Demes 1984a, b). The full pendulum period implies two foreswings

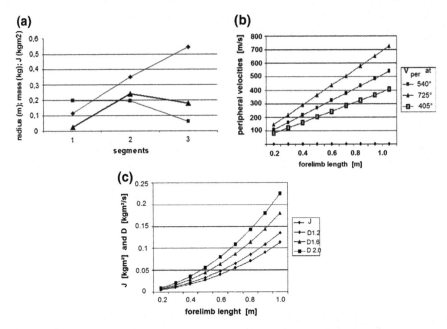

Fig. 11.11 Data relevant for executing a blow. **a** Distances from shoulder to centres of mass of forelimb segments (*diamonds*), masses (*squares*) and mass moments of inertia (*triangles*) of forelimb segments 1 (upper arm), 2 (forearm), and 3 (hand). **b** The 'hardness' or force of a blow is determined by peripheral velocity of the slapping hand, drawn for three angular velocities. **c** Rotational impulse or angular momentum [D] of the hand (*squares, triangles*, and *diamonds*) and mass moment of inertia ([J], *rosettes*) for forelimbs of varying length but constant mass at various angular velocities

of the respective free arm that is a time of 0.4–0.5 s for sweeping through angles of about 280°. From this follows an angular velocity of 560°–700°/s, or 1.6–2 rotations per second. For completeness, a lower speed of 400°/s or 1.2 rotations has also been added (Fig. 11.11b, c). The peripheral speed of a hand movement is directly proportional to the angular momentum or rotational impulse [D]:

$$D = \text{mass moment of inertia} * \text{angular velocity} \qquad (11.17)$$

Angular velocity is measured as $2\pi n$, with [n] being the number of rotations per second. The force [F] exerted at hitting another individual is determined by the deceleration time [Δt], that is the compliance of the body part which is hit—a factor that is very difficult to estimate.

$$F = D/\Delta t \qquad (11.18)$$

Unfortunately, impact time [Δt] is very variable, because it depends on the softness or rigidity of the part to which the blow is directed, and on the target's movement. Therefore, data in this chapter were restricted to factors which determine the blow, not the force transmitted. The moved mass is considerable, in spite of the slender appearance of hylobatid limbs. The outstanding arm length that is characteristic of gibbons and siamang contributes importantly to the force that is produced, and therefore a blow from a hylobatid's hand should by no means be underestimated.

Limitations Set to Body Mass

Advantages related to forelimb length raise questions about limits of forelimb elongation and body size in hylobatids. As reported above, the arboreal habitat sets extrinsic limitations to body mass (Figs. 11.1 and 11.5), because the superstratum has limited strength. If the branches cannot sustain the forces exerted by the animal such segments of the biotope are inaccessible, although orangutans, which are the heaviest arboreal mammals, demonstrate that tropical rainforests are capable of supporting animals of extraordinary mass. Another limiting factor on arm length in a suspensory primate is the space requirement associated with arm-swinging (Preuschoft and Demes 1985). To perform energy efficient brachiation, gibbons and siamangs need a clear space between current handhold and the next lower branch that equals arm length + trunk length + lower limb length. If this open 'window' cannot be found in the three-dimensional mesh of branches, swinging movements become disturbed or impossible. Similarly, Michilsens (2012) noted that the density of supports preferred by a slow-moving siamang does not facilitate arm-swinging, and we add that the much heavier orangutan also primarily climbs and clambers through dense vegetation where arm-swinging would be uneconomical.

In addition to the increased danger of fracturing supports when body weight increases, 'intrinsic' factors of the locomotor apparatus also limit body size. The simplest way to suspend a body is by forming a hook hand (Fig. 11.12). Weight force is much smaller than muscle force, because the latter has a lever arm about one third of the former. The ratio between body weight and the force that must be provided by the finger flexors is 1:3. At this condition, the weight of the forearm flexor muscles is low enough to allow the entire forelimb to function like a weightless chord of a mathematical pendulum. The respective graph in Fig. 11.12b indicates a root function for the available force and a linear function for the necessary force (detailed above under 'suspensory locomotion'). Any increase of total body mass requires a parallel increase of the muscle force (and muscle weight!) available in the forelimb in relation to body weight, resulting in deviation from the most favourable proportion. Given the proportions of hylobatids (Preuschoft et al. 1998), the available muscle force is smaller than the muscular force required at the widest possible excursion angle for supporting body masses of 30 kg or more. The distance between the graphs is maximal at small weights of up to 6 kg (Fig. 11.12b), which equals the weight of the small-bodied Asian apes, and becomes smaller at body masses of up to 12 kg, the weight of siamangs. The available force $[F_m]$ increases less than the necessary force $[F_r]$, both intersect at

30 kg. Evidently, the smaller-bodied hylobatids are in a size class where arm lengths of 60–70 cm are large enough to allow reasonably fast absolute speed, while the $[F_m/F_r]$ ratio is favourable and even the heavier siamang with a slightly longer forelimb length of 70–80 cm maintains a favourable $[F_m/F_r]$ ratio. The ratio between substrate reaction (F_c and F_g) and muscle force (F_m) becomes smaller with increasing body mass. This may be the main reason for the limited use made of arm-swinging in the large apes, especially in adults. Sub-adult orangutans and young chimpanzees sometimes show arm-swinging locomotion over limited distances (as can be seen in German TV, like WDR 'Quarks & Co' on August 4, 2015). Like other functions of the same type (Preuschoft 2010) the relationship between body mass and $[F_m/F_r]$ ratio asymptotically approaches a point of minimal return (Fig. 11.12c).

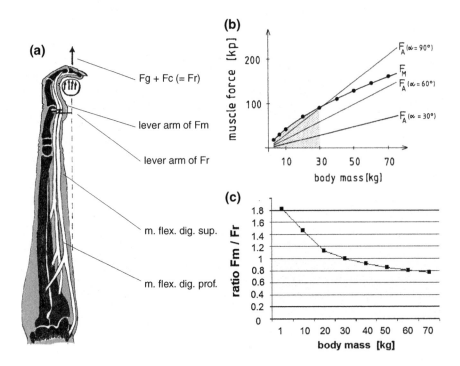

Fig. 11.12 Limitation of body mass by muscle force. **a** A hook hand with load arms of body weight $[F_g]$ and central force $[F_c]$ at metacarpo-phalangeal joint, and force arms of flexor tendons (m. flex. dig. sup., m. flex. dig. prof.). The carpal joint is controlled by the musculi flexors and extensores carpi (not shown). **b** The centrifugal forces plus body weight, which act against the fingers, increase linearly with excursion angle (=speed). The muscle force available in the flexor muscles $[F_m]$ lags behind the force necessary for balancing the external forces $[F_g]$ and $[F_c]$ at body masses exceeding 30 kg. Forces are given in kp (1 kp = 10 N) to make comparisons to body mass easier. **c** The ratio between available muscle force $[F_m]$ of the digital flexors and substrate reaction force $[F_r]$ changes with body mass, as long as size increase is isometric. Note that values <30 kg are >1, while those above are smaller. Gibbons and siamangs (10 kg and less) are able to produce relatively more force than larger apes

Limitations to Elongation of the Forelimbs

Since static torque caused by gravity is the product of length and mass, arm length is limited by the torque which the laterally extended arm exerts by its weight. Increased torques require stronger muscles and stronger bones, both being equivalents to increased weight (Fig. 11.13a, b). An increase of limb length requires an increase of muscle strength and muscle weight, both entailing a rapidly growing increase of limb mass, by factors of up to 4, depending on the joint under consideration (Preuschoft and Demes 1984b, 1985). The resistance of the limb against being moved is described by the mass moment of inertia [J] defined as:

$$J = \text{segment mass} * \text{lever arm length}^2 \tag{11.19}$$

Resistance, which must be overcome by muscle force, grows by one exponent more than in static postures. This has two consequences: first, arm and shoulder muscles under kinetic conditions need to be substantially stronger and heavier as forelimb length increases (Preuschoft and Demes 1985). Second, movements become slower as a consequence of increased forelimb length (Fig. 11.13b), as long as muscular torques remain constant. To maintain speed of movements as the forelimb increases, the shortage of muscular torques [M] must be compensated by an increase of muscle force and strength of skeletal elements leading again to increased body weight. An elongation of the lever arms of the muscles cannot compensate muscular torque, because a long force arm is equivalent to slow excursions of the load arm. Body weight itself is limited by the muscle force available for gripping.

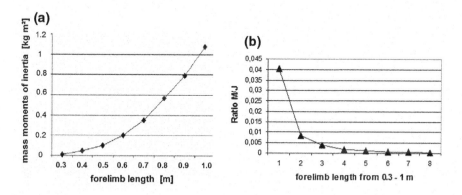

Fig. 11.13 a Mass moment of inertia [J] grows with absolute length of an extremity. This leads to slower movements, because constant muscle torques lag behind the growing moment of inertia. **b** The deficiency of muscular torques in relation to mass moment of inertia [M/J] can be compensated by an increase in muscle force, but this implies an increase of weight and requires additional bone strength, which also increases weight. Compensation by lengthening lever arms of the muscles would imply, however, a retardation of movements

The relationship of forelimb length to body mass is dictated by conditions illustrated under static conditions (Fig. 11.14). A hylobatid sitting on a branch extends his left arm towards a distant target. The extended forelimb exerts a torque at the shoulder $[F_a * h_a]$ which must be balanced by muscle contractions. The forelimb shifts the animal's centre of mass towards the target and rotates the animal about a pivot, which is the foot (or ischial callosity) closest to the target. The animal can maintain its balance on the branch only if the torque exerted by the extended limb is balanced by the torque of the remaining body weight after subtracting the extended arm length $[F_r]$ multiplied by its lever arm $[h_r]$:

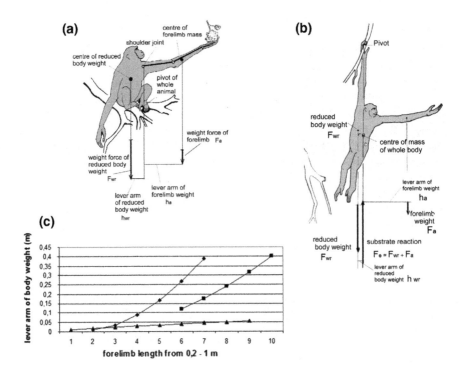

Fig. 11.14 a Equilibrium in a hylobatid sitting on a branch and reaching for food with the left hand. The torque exerted by the extended forelimb $[F_a] * [h_a]$ shifts the animal's centre of mass towards the left foot. If this torque is not counterbalanced, the hylobatid will lose balance and rotate about its left foot. Body weight, reduced by forelimb weight $[F_{wr}]$ needs a lever arm $[h_{wr}]$ to keep the animal in its place. **b** Static equilibrium in a hylobatid suspended on a slender, flexible branch. Like in sitting, the extended left arm leads to a shift of the centre of mass to the hylobatid's left side causing the animal to swing to its right side until the lever arm of reduced body weight is long enough to establish equilibrium at the handhold as pivot. **c** The lever arms of body weight, reduced by the stretched out forelimb, plotted against forelimb length. *Triangles* show conditions in a primate, the forelimbs of which are in a constant relationship to body weight as in many old-world monkeys. *Diamonds* show the lever arm when body weight is kept constant at 5 kg (like in small-sized hylobatids), while forelimb length grows. From a forelimb length of 0.6 m or more a body weight of 10 kg is assumed (roughly as in siamangs, *black squares*)

$$F_a * h_a = h_r * F_r \qquad (11.20)$$

This can be written as

$$h_r = F_a * h_a / F_r \qquad (11.21)$$

If the forelimb becomes longer, either the centre of gravity of the body minus arm weight (=reduced body weight $[F_r]$) must be shifted farther away from the extended arm or the pivot must be shifted towards the target. The lever arm $[h_r]$ must increase in order to keep the animal in balance, if arm weight or arm length increase (Fig. 11.14c). Any other shift of the pivot towards the arm (e.g., carrying loads like fruit in a stretched out hand) leads to the same result as elongation of the forelimb. Shifting the (reduced) body's centre of gravity is possible only within narrow limits, unless a contra-lateral extremity is extended into the opposite direction. Counterbalancing by extension of another extremity costs considerable energy and so does stretching out the foot which forms the pivot or anchoring the animal with the free, right hand. Forelimb extension will be energetically economical only when weight and length of the forelimb remain at an appropriate relation to body weight and its lever arm.

We also investigated the conditions in the suspended posture so typical for hylobatids (Fig. 11.14b; Fei et al. 2015). The body's centre of mass assumes a position perpendicularly below the handhold. If a long, heavy forelimb is extended laterally, the centre of mass of the whole animal is shifted towards the extended arm and the trunk will swing away from the extended arm. The lever arm of reduced body weight $[h_r]$ is the same as in the sitting animal. The pivot of this movement may be a flexing of the superstratum or the handhold. The animal has a choice among four options. First, it may rotate about its handhold or the carpal joint of the supporting limb. In this case, the angle between the vertical and the trunk axis must be maintained by muscular activity. Second, it may rotate about the shoulder joint of the carrying forelimb so that its centre of mass is perpendicularly below the support. In this case, the support arm must be adducted by muscular activity, so that the trunk axis, and with it the centre of gravity, moves to the other side. The muscles of the shoulder girdle possess long lever arms so that only small forces are required. Third, if the hylobatid rotates about the shoulder of the carrying arm it may shift its centre of gravity to, e.g., the left side by flexing one or both hip joints. This moves the hindlimb(s) away from the vertical, shifts the centre of mass towards the right, and so increases the torque which balances the extended arm. The torque of hindlimbs grows with the sine of the flexing angle in the hip joint. The disadvantage of this choice is again the necessity to exert muscle force for flexing the hip joints against gravity. Fourth, balance can often be established by bracing a hindlimb against an additional support (e.g., a nearby branch), but this option is also only possible by exerting muscular force. Overall, forces acting during movements do not principally diverge from the static conditions discussed above, they are just greater.

Since all the above listed options for keeping balance depend on the stiffness of the superstratum and the grip, and also require additional input of muscle force, the energy balance can best be restored by a shift of the centre of reduced body weight to the right side of the animals. If body weight is assumed to be constant, a suitable measure for comparisons is the lever arm of body weight, reduced by the stretched out forelimb $[h_r]$ (Fig. 11.14c). If the increase of arm length is paralleled by an increase of body mass, as in most monkeys, the lever arm length $[h_r]$ grows slowly. If, however, body weight is kept constant, for example at the level of a small-bodied hylobatid, and the arm alone is elongated, the length of the lever arm $[h_r]$ grows very rapidly, which makes balancing without additional energy input impossible. If the body weight of a siamang is assumed at ~ 10 kg, the greater body weight of a siamang needs a much shorter lever arm than in gibbons. The lever arm $[h_r]$ necessary for balance grows less, but still at a rather steep, uneconomical rate.

During movement, passive resistance of the forelimb against being moved requires greater, but not principally different torques of the trunk to keep the body in balance. To conclude, our investigation indicates clearly that arm length and arm mass are limited by body mass.

Discussion

Identifying advantages and disadvantages of a small versus large body size or of selected body size dimensions allows judging the selective value of body size characteristics. Proximate advantages of selected size parameters positively influence survival. Therefore, such proximate, mechanical advantages are undoubtedly proportional to the selective pressures which drive the evolution of the respective traits. One attraction of the biomechanical approach presented here is the possibility to quantify advantages connected to the expression of such traits as long arms or a particular body weight. Therefore, biomechanics provides a rare opportunity to quantify selective pressures associated with such traits.

The relevance of maintaining a safe position in a three-dimensional, unstable, and potentially unreliable environment (Michilsen 2012) becomes evident, if the frequency of lesions due to falling is taken into consideration. In large primate skeletal collections a substantial proportion of individuals show traces of healed fractures of long bones, specifically one third in hylobatids (Schultz 1956b), one fourth in pongids (Lovell 1987) and macaques (Buikstra 1975), and four fifths in baboons (Bramblett 1967). Like other mammals of the same size order (e.g., sloths, tamandua), gibbons and siamang are able to exploit the outer foliage, and are outstanding experts in suspensory behaviour and arm-swinging locomotion. Their body proportions allow them to master various tasks with minimal energy input and high efficiency. Hylobatids engage in bipedal walking, bimanual brachiation, as well as slow, amoeba-like climbing and clambering, which are also performed by orangutans despite of the two apes' divergent morphologies. The usually high tensile strength of branches and twigs alleviates access to the peripheral, terminal areas of a tree crown for relatively large animals. Hylobatids are exceptional among extant apes with their ability to reach the outermost branches of tree crowns not accessible to the other suspensory Asian ape, the orangutan, at least partially due to very different biomechanical conditions brought about by a much larger body mass.

In many lineages of organisms a tendency of increasing body size over evolutionary times has been observed, which has been identified as Cope's rule (Sander et al. 2010). Several physiological (Sander et al. 2010; Clauss 2011) and biomechanical arguments (sauropod dinosaurs: Preuschoft et al. 2011; primates: Preuschoft 2010) suggest advantages for an animal group to become larger over time. Important advantages related to an increase of the length of limbs, discussed in this chapter, have been a longer reach to collect food, faster locomotion without spending additional energy, greater fighting power, and less vulnerability due to a greater body volume. These postulated benefits are compatible with the lengthening of forelimbs in hylobatids. On the other hand, bigger size equals a big body mass, which comes with the cost of energy expenditure for locomotion increasing disproportionately fast and, for arboreal primates, increasing body masses are strongly associated with increasing risks of losing balance and falling as well as substrate failure. Adverse effects of large body size are particularly significant during acceleration or deceleration, which may explain why the larger siamang shows more slow-climbing compared to small-bodied gibbons (Fleagle 1974, 1976; Raemaekers 1984; Michilsens 2012). The same relationship has to be applied to the still much heavier orangutan, which uses by and large the same biotope as hylobatids, but for whom the costs associated with large body mass are exponentially greater than for hylobatids.

Although hylobatids weigh much less than African apes and orangutans, their forelimbs are nearly of the same length (Jungers 1985). The combination of low weight and long arms has clearly been an advantage in hylobatid evolution, because less food is required for maintenance of the long-arms-low-weight body and the access to the periphery of the canopy is mechanically easier with efficient brachiation as well as safer, because suspensory locomotion annihilates the above-branch balance problem. On the other hand, their small body mass makes gibbons and siamang theoretically more vulnerable to predation—or prone to loosing disputes over food resources against larger primates. While predation risk can be compensated by agility, the latter would compel hylobatids to give way as soon as the heavier and stronger orangutan appears (P. Gittins and D. Chivers pers. comm.).

Interestingly, small Asian apes fit exactly the range of body masses (5–12 kg) and forelimb lengths energetically optimal for arm-swinging locomotion (see Fig. 11.12), while apes beyond 30 kg body mass are less efficient brachiators. Hollihn's (1984) observation that large apes (i.e., ≥ 30 kg) avoid brachiating over distances of more than 8 or 10 m is consistent with our observation of the biomechanical advantages of small-to-medium body weights. Large apes' inability to brachiate efficiently over long distances is related to a shortness of muscle force in digital flexors (Preuschoft and Demes 1985). Jungers (1985) notes that forelimb-length/body weight ratio reaches values of 26 for the siamang and 28 for lar-gibbons. The measures in hylobatids are high compared to ratios in *Pan* (i.e., ratio 17). Unfortunately, no value is provided for *Pongo*, but if the orangutan body mass data from Jungers (1985) and forearm length data from Schultz (1930) and

Martin (1928) are used, oranguntans equally reach a suboptimal ratio not more than 17–20. This is still considerably lower compared to hylobatids and supports the argument that gibbons and siamang body proportions are ideal for brachiation and from a biomechanical point of view, the hylobatid morphotype seems nearly perfectly adapted to the terminal branch fruit feeding, arboreal tropical forest niche.

If natural selection optimized locomotion by arm-swinging in hylobatids, long length of their arms provided additional selective advantages in the efficiency of foraging, in competitive situations and predator avoidance (See Reichard and Preuschoft 2016).

Neither suspensory posture nor the possession of prehensile autopodia is exclusive to gibbons and siamang. In principle, suspension can also be achieved by claws or a rigid 'hook hand', but neither claws nor a rigid hook hand can effectively be used when moving along thin twigs. It is precisely the ability of flexing digits tightly around slender objects of various diameters and texture, and the softness of palms and soles that provide a safe hold on vertical substrates, such as thin twigs, liana or dangling ropes even when these are wet as it is often the case in tropical rainforests. According to Kümmell (2009) grasping autopodia are found in ancient synapsids since the Permian. Therefore, it is not surprising that various forms of prehension are observed in living animals as diverse as chameleons, iguanids, varanids, birds, marsupials (*Phascolarctos*), sloths, rodents, bears and in all primates with the exception of humans who have no prehensile feet. Suspension is common in geckos, birds (e.g., woodpeckers, tits, tree creepers, parrots), rodents, marsupials, sloths, lorisids (Jouffroy and Petter 1990; Jouffroy and Stern 1990; Ishida et al. 1990) Atelines (Stern et al. 1977, 1980; Jungers and Stern 1981; Larson and Stern 1989; Larson 1993; Martin 2003), several leaf monkeys (e.g., Morbeck 1979; Rose 1979) and large apes, the so-called 'brachiators'. The link between the strength and compliance of substrates, large body weight, and suspensory locomotion is supported by the notion that some of the largest New World primates (subfamily *Atelinae*), at least some of the colobines (*Colobus*), including the largest of the group (*Nasalis*), as well as the large apes all belong to a category loosely identifiable as 'suspensory climbers'.

The well-known deep cleft between digits I and II in hylobatid hands and feet enlarges the span of the grip and increases the freedom of thumb or hallux movements. If, however, the angle between rays I and III is maintained and the length of the lateral digits and thumb or hallux constant, the width of the span increases linearly with the lengths of rays I and III elongation of the lateral digits may compensate for shortness of the thumb or hallux, like in the orangutan. This is further illustrated by the negligible differences of the ray I/III—ratio between African ape genera and, vice versa, the length of the thumb or hallux may compensate for a shortness of the lateral digits, as seen in *Homo*. It seems, however, that width of span is not the decisive factor for hand morphology and proportions. More important is the cross sectional area of branches that can be enclosed by the grip. Shear length of digits brings an advantage, like in orangutans. The area surrounded by the lateral digits and the first ray, however, is about 25 % greater, if the well-known cleft between rays I and II of hylobatids is present and the adductor muscle of the thumb-or hallux inserts not close to the metapodio-phalangeal joint but near the basis of the second ray as in hylobatids. This allows the medium-sized hylobatids to grasp thicker branches or stems than monkeys of the same size can grasp.

Suspension and suspensory locomotion can be performed by two, four or even five extremities, if also the tail is used. The characteristics of suspended locomotion have been thoroughly investigated (Jouffroy and Petter 1990; Jouffroy and Stern 1990; Ishida et al. 1990). Clearly, the more extremities are used, the safer is the support for the animal, and using the tail additionally frees forelimbs to gather food. On the other hand, using the tail may interfere with the rhythm of locomotion of the four limbs (Martin 2003), in contrast to tailless organism like apes or, for example, sloths (Nyakatura and Andrada 2013), which may make suspended locomotion energetically more expensive.

It seems plausible, that suspension by the forelimbs is connected with a particularly strong development of forelimbs at the expense of hindlimbs. However, if the head of a quadrupedal climber is to be kept on top of the body, then shifting a major share of body weight towards the hindlimbs becomes an option, as seen in prosimians (Preuschoft et al. 1998). If a quadrupedal posture is the starting point, a shift of the bigger part of body weight to the hindlimbs is inevitable (Preuschoft 2002). In fact, Kimura (2002, 2010) has shown that the hindlimbs of more arboreal than terrestrial monkeys are stronger than their forelimbs while more ground-living monkeys have slightly stronger forelimbs. In hylobatids, the masses of the forelimbs oscillate around the masses of the hindlimbs (Zihlman et al. 2011). In this regard, gibbons and siamangs deviate from monkeys as well as from the short-legged and long-armed orangutan and the African apes (Preuschoft et al. 1998).

While most considerations about brachiation are concentrated on horizontal and 'steady state' locomotion, Bertram and Chang (2001) and Bertram (2004) have focussed on the difference between continuous contact brachiation (which is inverse to bipedal striding, although biomechanically very similar) and ricocheting. They emphasize divergences to trotting in the use made of elastic elements, but unfortunately their single experimental animal (a female *H. lar*) was with nearly 8 kg unusually heavy, and it thus remains unclear how representative their results are.

According to Isler (2003) the siamang forelimb contributes 7.2 % to total body weight while the average forelimb weight of small gibbons was 11 % of total body weight. These data contrast with Zihlman et al. (2011), who measured the relative weight of siamang forelimbs to be 10.6 % of total body weight which exceeded that of smaller gibbon species at only 9.3 % of total body weight. However, considerable variation was found between left and right forelimb of the same individual (Isler 2003), which might be influenced by their living in captivity.

Hylobatids, like the large-bodied apes, have hands and forearms as well as feet and lower legs that are more than twice as heavy as the autopodia of most other mammals, and they are clearly heavier than the distal segments of other simians (Preuschoft et al. 1998). Especially the share of forearm and hand of total body weight is nearly as great in hylobatids as it is in *Pongo*, and slightly greater than in

Pan, despite the much smaller body weight of hylobatids (Preuschoft et al. 1998). The greater share of forelimb weight in the heavier orangutan compared to hylobatids (Isler 2003; Zihlman et al. 2011) is simply a mechanical necessity given their different body weights. By contrast, the shortness of the hindlimbs in large apes reduces absolute body weight, which perhaps compensates for the already overall heavy forelimbs and provides additional safety in cases of falling out of the tree, combined with hitting the ground, or a lower branch (Preuschoft 1989). Shortness of hindlimbs is combined with low hindlimb weight, accounting in orangutans for only 7.5 % of total body weight while in hylobatids the hindlimbs contribute 11.3 % to total body weight (Preuschoft et al. 1998).

Using the laws of the pendulum for keeping energetic demands of locomotion low is not unique to hylobatids and by no means unusual among mammals. Bipedal primates (Mochon and McMahon 1980, 1981; Yamazaki 1984, 1985; Yamazaki and Ishida 1984; Yamazaki et al. 1979, 1983) including humans (Witte et al. 1991; Preuschoft and Witte 1993; Preuschoft et al. 1992) as well as quadrupeds (Witte et al. 1995a, b; Preuschoft et al. 1994) make use of the energetically beneficial motion physics of the pendulum. There is, however, a characteristic difference in how this applies to ground-dwelling quadrupeds and bipeds on one side and arboreal brachiators on the other. In ground-dwelling animals the extremity functions alternately as inverted pendulum (stance phase, during which body weight is supported) and suspended pendulum (swing phase, during which the free limb is swung forward: Fig. 11.5). The swing phase determines the duration, and therefore the frequency of the locomotor cycles. In the suspensory hylobatid (Fig. 11.6), however, the weight bearing phase follows the laws of the suspended pendulum and determines the frequency of the locomotor cycle, while the recovery phase or the foreswing of the free arm seems to be used mainly to shift the animal's centre of gravity away from the handhold to gain speed, with the effect of improving the animal's overall energy balance. The biomechanical analyses presented here are in agreement with electromyographical (EMG) data obtained by Jungers and Stern (1980, 1981, 1984).

Yamazaki (1990) simulated movements of the body segments during arm-swinging on the computer and found a convincing reason for the existing proportions of forelimbs in relation to total body weight in hylobatids. First, he found that shorter forelimbs and longer hindlimbs do not lead to progression and that natural proportions of hylobatids lead to optimal speed and periodicity. Second, longer forelimbs combined with shorter hindlimbs would not allow the arm-swinger to reach the height of the next higher handhold. If segment mass is varied instead of segment length, Yamazaki (1990) also found that forelimb masses below 5 % of body mass make the foreswing too slow to reach the next handhold while a too heavy forelimb, i.e., 13.1 % of body mass, would result in higher velocity but also lead to an excessive, unnecessary height gain above the height of the next handhold and thus a waste of kinetic energy.

In an emergency escape from a predator, a long-armed brachiator can accumulate more potential energy while falling and can therefore reach higher horizontal speed for a rapid escape than a shorter-armed brachiator could. While monkeys of 5–10 kg usually have forelimbs of about 40–50 cm length, gibbons reach arm lengths of 60–70 cm, and a siamang's arm stretches 70–80 cm. A longer forearm length increases the height difference in the beginning of an escape manoeuver (Fig. 11.7) and leads to greater peripheral speed. It implies, however, also greater centripetal forces and therefore requires additional strength in the digital flexors for securing handholds.

A rather convenient measure for the limit of forelimb lengthening is to assess lever arm length necessary for the body to maintain equilibrium. This is equivalent for the shift of the body's centre of mass that becomes necessary when a forelimb is stretched out laterally or ventrally. While the extension of the contralateral forelimb for balancing and fixing the body costs energy, a moderate shift of the centre of mass is energetically cheap. If the ratio between forelimb length and reduced body mass is kept constant, lever arms grow linearly regardless of the primate's posture (sitting, standing or suspended). If, however, body mass is kept constant while forelimb length is assumed to grow, the increase of the necessary lever arm is greater than linear which sets clear boundaries to forelimb lengthening. Under conditions of kinetics, the growth of the necessary lever arms is still more pronounced.

The impressive forelimb length of gibbons and siamang should not distract attention from the considerable length of their hindlimbs (Schultz 1933b, 1944; Tuttle 1972). Hylobatids obviously have not developed short and weak hindlimbs, probably because they make use of them during bipedal walking, running (Carpenter 1940; Ishida et al. 1984; Jungers 1984), and leaping (Channon et al. 2011). The tendency of hylobatids to walk bipedally has since long been related to the extreme length of their forelimbs. Seen superficially, it seems to represent convergent evolution to human bipedalism. Since speed depends on stride length times stride frequency, the elongation of stride length may compensate for the length of the foreswing phases caused by the high weight of legs and feet or forearms and hands, respectively. A similar, though not identical bipedal tendency has been recognized in orangutans (Crompton et al. 2003). Orangutans use bipedal locomotion supported and 'camouflaged' by their forelimbs grasping holds above their heads to secure equilibrium (Crompton et al. 2003). This characteristic locomotion entails a knee morphology characterized by the medial and lateral condyles being nearly equal in size, which contrast to the situation in humans versus African apes, as often discussed in the context of human bipedal evolution (Preuschoft 1971).

Aggression among neighbouring individuals of gibbons have been reported since Carpenter's (1940, 1976) seminal observations of wild white-handed gibbons to occur frequently, although often without physical contact. In cases of contact aggression, slapping and kicking seem common and bites are often directed to the extremities (Orgeldinger 1994). The longer the forelimb is, the greater becomes the limb's peripheral velocity and the rotational impulse of the hand, which determine the force of a blow. Hand speeds of chimpanzees during nut-cracking found by Günther and Boesch (1993) allow a comparison to a hylobatid slapping an opponent, using mass proportions of chimpanzees (Preuschoft et al. 1998), body weights from Jungers (1985) and dimensions from Schultz (1956a). The estimated angular speeds of 540–725° used for calculations are in the same size order as observed jaw movements (Preuschoft and Witzel 2005). The values seem to be high compared to those observed by Günther and Boesch (1993), but they may still be less than the angular velocity used in slapping. The important aspect for this study is not the exact determination of the force exerted on an opponent, but identification of the factors directly proportional to force and limb length.

In conclusion, biomechanics helps to identifying the advantages of specific proportions such as relatively long forearms in combination with an overall moderate body size for exploiting the outer foliage of the terminal-branch niche that was critical in gibbon and siamang evolution.

Acknowledgements We feel deeply indebted to Dr. U. Reichard for the invitation to write a contribution to this book and for his very valuable comments on several drafts. He has compelled us to describe conditions precisely and to express our conclusions clearly. We also acknowledge the comments of an unknown reviewer who contributed much to the readability of this text.

References

Arms A, Voges D, Preuschoft H, Fischer M (2002) Arboreal locomotion in small New-World monkeys. In: Okada M, Preuschoft H (eds) Arboreal locomotor adaptation in primates and its relevance to human evolution. Zeitschrift für Morphologie und Anthropologie, Schweizerbart, Stuttgart, pp 243–263

Avis V (1962) Brachiation: the crucial issue for man's ancestry. Southwestern J Anthrop 18:119–148

Bertram JEA (2004) New perspectives on brachiation mechanics. Yearb Phys Anthropol 47:100–117

Bertram JEA and Chang YH (2001) Mechanical energy oscillations of two brachiation gaits: Measurement and simulation. Am J Phys Anthropol 115:319–326

Biegert J (1961) Volarhaut der Hände und Füsse. In: Hofer H, Schultz AH, Starck D (eds) Primatologia II, part 1. Karger, Basel and New York, 3/1–3/326

Bramblett CA (1967) Pathology of the Darajani baboon. Amer J Phys Anthrop 26:331–340

Buck C, Bär H (1993) Investigations on the biomechanical significance of dermatoglyphic ridges. In: Preuschoft H, Chivers DJ (eds) Hands of primates. Springer, New York, Wien, pp 285–306

Buikstra JA (1975) Healed fractures in *Macaca mulatta*: age, sex and symmetry. Folia Primatol 23:140–148

Carbone L, Harris RA, Gnerre S, Veeramah KR, Lorente-Galdos B, Huddleston J, Meyer TJ, Herrero J, Roos C, Aken B, Anaclerio F, Archidiacono N, Baker C, Barrell D, Batzer MA, Bea K, Blancher A, Bohrson CL, Brameier M, Campbell MS, Capozzi O, Casola C, Chiatante C, Cree A, Damert A, de Jong PJ, Dumas L, Fernandez-Callejo M, Flicek P, Fuchs NV, Gut I, Gut M, Hahn MW, Hernandez-Rodriguez J, Hillier LW, Hubley R, Ianc B, Izsva'k Z, Jablonski NJ, Johnstone LM, Karimpour-Fard A, Konkel MK, Kostka D, Lazar NH, Lee SL, Lewis LR, Liu Y, Locke DP, Mallick S, Mendez FL, Muffato M, Nazareth LV, Nevonen KA, O'Bleness M, Ochis C, Odom DT, Pollard KS, Quilez J, Reich D, Rocchi M, Schumann GG, Searle S, Sikela JM, Skollar G, Smit A, Sonmez K, ten Hallers B, Terhune E, Thomas GWC, Ullmer B, Ventura M, Walker JA, Wall JD, Walter L, Ward MC, Wheelan SJ, Whelan CW, White S, Wilhelm LJ, Woerner AE, Yandell M, Zhu B, Hammer MF, Marques-Bonet T, Eichler EE, Fulton L, Fronick C, Muzny DM, Warren WC, Worley KC, Rogers J, Wilson RK, Gibbs RA (2014) Gibbon genome and the fast karyotype evolution of small apes. Nature 513:195–201

Cartmill M (1974) Pads and claws in arboreal locomotion. In: Jenkins P (ed) Primate locomotion. Academic Press, New York, pp 45–83

Cartmill M (1985) Climbing. In: Hildebrand M, Bramble DM, Liem KF, Wake DB (eds) Functional vertebrate morphology. Harvard University Press, Cambridge, MA, pp 73–88

Carpenter CR (1940) A field study in Siam of the behavior and social relations of the gibbon (*Hylobates lar*). Comp Psychol Monogr 16:38–206

Carpenter CR (1976) Suspensory behaviour of gibbons *Hylobates lar*. Gibbon and siamang 4:167–218

Chang YH, Bertram JE, Ruina A (1997) A dynamic force and moment analysis system for brachiation. J Exp Biol 200:3013–3020

Chang YH, Bertram JEA, Lee DV (2000) External forces and torques generated by the brachiating white-handed gibbon (H. lar). Am J Phys Anthropol 113:201–216

Channon AJ, Günther MM, Crompton RH, d'Aout K, Preuschoft H, Vereeke E (2011) The effect of substrate compliance on the mechanics of gibbon leaps. J Exp Biol 214:687–696

Chivers DJ (1984) Feeding and ranging in gibbons. In: Preuschoft H, Chivers DJ, Brockelman WY, Creel N (eds) The lesser apes. Edinburgh University Press, Edinburgh, pp 267–281

Clauss, M (2011) Sauropod Biology and the evolution of gigantism: what do we know. In: Klein N, Remes K, Gee CT, Sander M (eds) Biology of the Sauropod Dinosaurs Indiana University Press, Bloomington, pp. 3–7

Creel N, Preuschoft H (1971) Hominoid taxonomy. A canonical analysis of cranial dimensions. In: Proceeding of the 3rd international congress of primatology, Zürich 1970, vol 1. Karger-Verlag, Basel, pp 79–90

Creel N, Preuschoft H (1976) Cranial morphology of the lesser apes. A multivariate statistical study. In: Rumbaugh DM (ed) Gibbon and Siamang, vol 4. Karger, Basel, pp 219–303

Creel N, Preuschoft H (1984) Systematics of the lesser apes. A quantitative taxonomic analysis of craniometric and other variables. In: Preuschoft H, Chivers DJ, Brockelman WY, Creel N (eds) The lesser apes. Edinburgh University Press, Edinburgh, pp 562–613

Crompton RH, Li Y, Thorpe SK, Wang WJ, Savage R, Payne R et al (2003) The biomechanical evolution of erect bipedality. Courier Forschungs-Institut Senckenberg 243:115–126

Demes B, Preuschoft H (1984) Die biomechanische Bedeutung der Armlänge und der Körpermasse für die hangelnde Fortbewegungsweise. Z Morph Anthropol 74:261–274

Dubbel H (1981) Taschenbuch für den Maschinenbau. Springer, Berlin

Fei H, Ma C, Bartlett TQ, Dai R, Xiao W, Fan P (2015) Feeding postures of Cao Vit gibbons (*Nomascus nasutus*) living in a low-canopy Karst forest. Int J Primatol 36:1036–1054

Fleagle JG (1974) The dynamics of the brachiating siamang (Symphalangus syndactylus). Nature 248:259–260

Fleagle JG (1976) Locomotin and posture of the Malayan siamang and implications for hominid Evolution. Folia primatologica 26:245–269

Fleagle JD (1984) Are there any fossil gibbons? In: Preuschoft H, Chivers DJ, Brockelman WY, Creel N (eds) The lesser apes. Edinurgh University press, Edinburg, pp 4431–4447

Fleagle JG and Mittermaier RA (1980) Locomotor behaviour, body size, and comparative ecology of seven Surinam monkeys. Am J Phys Anthropol 52:301–314

Frisch JE (1973) The hylobatid dentition. In: DM Rumbaugh (ed) Gibbon and Siamang, vol 4, Karger, Basel, pp 56–95

Grand TI (1972) A mechanical interpretation of terminal branch feeding. J Mammal 53:198–201

Günther MM (1989) Funktionsmorphologische Untersuchungen zum Sprungverhalten mehrerer Halbaffen. Dissertation, Freie Universität Berlin

Günther MM, Boesch C (1993) Energetic cost of nut-cracking behaviour in wild chimpanzees. In: Preuschoft H, Chivers DJ (eds) Hands of primates. Springer, Wien, pp 10–132

Harrison T (2016) The fossil record and evolutionary history of hylobatids. In: Reichard UH, Hirohisa H, Barelli C (eds) Evolution of gibbons and siamang. Springer, New York, pp 91–110

Hollihn U (1984) Morphology, selective advantages and phylogeny. In: Preuschoft H, Chivers DJ, Brockelman WY, Creel N (eds) The lesser apes. Edinburgh University Press, Edinburgh, pp 85–95

Ishida H, Kimura T, Okada M, Yamasaki N (1984) Kinesiological aspects of bipedal walking in gibbons. In: Preuschoft H, Chivers DJ, Brockelman WY, Creel N (eds) The lesser apes. Edinburgh University Press, Edinburgh, pp 135–145

Ishida H, Jouffroy FK, Nakano Y (1990) Comparative dynamics of pronograde and upside-down horizontal quadrupedalism in the slow loris (Nycticebus coucang). In: Jouffroy FK, Stack HH, Niemitz C (eds) Gravity, posture and locomotion in primates. Il Sedicesimo, Firenze, pp 209–220

Isler K (2002) Characteristics of vertical climbing in gibbons. Evol Anthropol 11:49–52

Isler K (2003) 3D-Kinematics of vertical climbing in hominoids. Dissertation, Universität Zürich

Jouffroy FK, Petter A (1990) Gravity- related kinematic changes in lorisine horizontal locomotion in relation to position of the body. In: Jouffroy FK, Stack HH, Niemitz C (eds) Gravity, posture and locomotion in primates. Il Sedicesimo, Firenze, pp 199–208

Jouffroy FK, Stern JT (1990) Telemetered EMG-stuy of the antigravity versus propulsive actions of the knee and elbow muscles in the slow loris (Nycticebus coucang). In: Jouffroy FK, Stack HH, Niemitz C (eds) Gravity, posture and locomotion in primates. Il Sedicesimo, Firenze, pp 221–236

Jouffroy FK, Godinot M, Nakano Y (1993) Biometrical characteristics of primate hands. In: Preuschoft H and Chivers DJ (eds.) Hands of primates. Springer, Wien, pp 133–171

Jungers WL (1984)Scaling of the hominoid locomotor skeleton with special reference to lesser apes. In: Preuschoft H, Chivers DJ, Brockelman WY, Creel N (eds.) The lesser apes. Edinburgh University Press, Edinburgh, pp 146–169

Jungers WL (1985) Body size and scaling of limb proportions in primates. In: Jungers WL (ed) Size and scaling in primate biology. Plenum Press, New York, pp 345–381

Jungers WL and Stern JT (1980) Telemetered electromyography of forelimb muscle chains in gibbons (Hylobates lar). Science 208:617–619

Jungers WL and Stern JT (1981) Preliminary electromyographical analysis of brachiation in gibbon and spider monkey. Int J Primatol 2:19–33

Jungers WL, Stern JT (1984) Kinesiological aspects of brachiation in lar gibbons. In: Preuschoft H, Chivers DJ, Brockelman WY, Creel N (eds) The lesser apes. Edinburgh University Press, Edinburgh, pp 119–134

Larson SG (1993) Functional morphology of the shoulder in Primates. In: Gebo DL (ed) Postcranial adaptation in primates. Northern Illinois University Press, DeKalb, pp 45–69

Kimura T (2002) Primate limb bones and locomotor types in arboreal or terrestrial environments. In: Okada M, Preuschoft H (eds) Arboreal locomotor adaptation in primates and its relevance to human evolution. Z Morph Anthropol 83:201–219

Kimura T (2010) Arboreal origin of bipedalism re-examined. Comparative analysis. In: Abstract of the 23rd International Primatological Society Cogress, Kyoto, p 297

Kümmell S (2009) Die Digiti der Synapsida: Anatomie, Evolution und Konstruktionsmorphologie. Shacker Verlag, Aachen

Larson SG, Stern JT (1989) The role of supraspinatus in the quadrupedal locomotion of vervets (Cercopithecus aethiops). Implications for interpretation of humeral morphology. Am J Phys Anthropol 79:369–377

Lehmann T (1974–1977) Elemente der Mechanik, Bände 1–3, Vieweg, Braunschweig

Leutenegger M (1982) Scaling of sexual dimorphism in body weight and canine size. Folia Primatol 37:163–176

Lorenz R (1971) The functional interpretation of the thumb in the Hylobatidae. Proceedings of the 3rd Internat. Congress Primatology, Zürich, Vol 1, 130–136

Lovell NC (1987) Skeletal pathology of pongids. Am J Phys Anthropol 72:227 (Abstract)

Martin R (1928) Lehrbuch der Anthropologie, G.Fischer, Jena

Martin F (2003) Organisationsprinzipien zielgerichteter Bewegungen flexibler Greiforgane. Dissertation, Freie Universität Berlin

Michilsens F, D`Aout K, Aerts P (2011) How pendulum-like are siamangs? Energy exchange during brachiation. Am J Phys Anthropol 154:581–591

Michilsens F (2012) Functional anatomy and biomechanics of brachiating gibbons (Hylobatidae). Ph D-Thesis, Faculteit Wetenschappen, Dept. Biologie, Universiteit Antwerpen

Michilsens F, Vereeke EE, D´Aout K and Aerts P (2010) Muscle moment arms and function of the siamang forelimb during brachiation. J Anat 217:521–535

Mochon S and McMahon TA (1980) Ballistic walking. J Biomech 13:49–57

Mochon S and McMahon TA (1981) Ballistic walking, an improved model. Math Biosciences 52:241–260

Morbeck ME (1979) Forelimb use and positional adaptation in *Colobus guereza*: Integration of behavioural, ecological and anatomical data. In: Morbeck ME, Preuschoft H, Gomberg N (eds) Environment, Behavior and Morphology: Dynamic Interactions in Primates. Georg Fischer, New York, pp 95–117

Napier JR, Napier PH (eds) (1967) A handbook of living primates. Academic Press London, NewYork

Nowak MG, Reichard UH (2016) The torso-orthograde positional behavior of wild white-handed gibbons (*Hylobates lar*). In: Reichard UH, Hirohisa H, Barelli C (eds) Evolution of gibbons and siamang. Springer, New York, pp. 203–225

Nyakatura J, Andrada E (2013) A mechanical link model of two-toed sloths: no pendular mechanics during suspensory locomotion. Acta Theriol 58:83–93

Okada M (1985) Primate bipedal walking. Comparative kinematics. In: Kondo S (ed) Primate morphophysiology, locomotor analysis and human bipedalism. University of Tokyo Press, Tokyo, pp 47–58

Orgeldinger M (1994) Ethologische Untersuchung zur Paarbeziehung beim Siamang (*Hylobates syndactylus*) und deren Beeinflussung durch Jungtiere. Dissertation, Universität Heidelberg

Preuschoft H (1961) Muskeln und Gelenke der Hinterextremität des Gorilla. Morphologisches Jahrbuch 101:432–540

Preuschoft H (1963) Muskelgewichte bei Gorilla, Orang-utan und Mensch. Anthropologischer Anzeiger 26:308–317

Preuschoft, H (1985) On the quality and magnitude of mechanical stresses in the locomotor system during rapid movements. Z Morph Anthrop 75: 245–262

Preuschoft H (1989) Body shape and differences between species. Hum Evol 4:145–156

Preuschoft H (2002) What does 'arboreal locomotion' mean exactly? and what are the relationships between 'climbing', environment and morphology? In: Okada M, Preuschoft H (eds) Arboreal locomotor adaptation in primates and its relevance to human evolution. Z Morph Anthropol 83:171–188

Preuschoft H (2004) Mechanisms for the acquisition of habitual bipedality: are there biomechanical reasons for the acquisition of upright bipedal posture? J Anat 204:363–384

Preuschoft H (2010) The selective value of size and sexual dimorphism in primates. Abstracts of the 23rd Congress of the International Primatological Society in Kyoto, 2010

Preuschoft H, Demes B (1984a) Biomechanics of brachiation. In: Preuschoft H, Chivers DJ, Brockelman WY, Creel N (eds) The lesser apes. Edinburgh University Press, Edinburgh, pp 96–118

Preuschoft H, Demes B (1984b) Biomechanic determinants of arm length and body mass in brachiators. In: Dunker HR, Fleischer G (eds) Vertebrate Morphology, Fortschritte d. Zoologie. Georg Fischer-Verlag, New York, pp 39–44

Preuschoft H, Demes B (1985) Biomechanic determinants of arm length and body mass in brachiators. In: Jungers WL (ed) Size and scaling in primate biology. Plenum Press, New York, pp 383–398

Preuschoft H, Witte H (1991) Biomechanical reasons fort he evolution of hominid body shape. In: Coppens Y, Senut B (eds) Origins of bipedalism in hominids. CNRS, Paris, pp 59–77

Preuschoft H, Witte H (1993) Die Körpergestalt des Menschen als Ergebnis biomechanischer Erfordernise. In: Voland E (ed) Evolution und Anpassung. Warum die Vergangenheit die Gegenwart erklärt. S. Hirzel, Stuttgart, pp 43–74

Preuschoft H, Witzel U (2005) Functional shape of the skull in vertebrates: which forces determine skull morphology in lowe r primates and ancestral synapsids? Anat Rec 283:402–413

Preuschoft H, Chivers DJ, Brockelman WY, Creel N (eds) (1984) The lesser apes. Edinburgh University Press, Edinburgh

Preuschoft H, Witte H, Demes B (1992) Biomechanical factors that influence overall body shape of large apes and humans. In: Matanao S, Tuttle R, Ishida H, Goodman M (eds) Topics in primatology, vol 3., Evolutionary biologyUniversity of Tokyo Press, Tokyo, pp 259–289

Preuschoft H, Godinot M, Beard C, Nieschalk U, Jouffroy FK (1993) Biomechanical considerations to explain important morphological characters of primate hands. In: Preuschoft H, Chivers DJ (eds) Hands of primates. Springer, Wien, pp 245–256

Preuschoft H, Witte H, Christian A, Recknagel S (1994) Körpergestalt und Lokomotion bei großen Säugetieren. Verh Dt Ges Zool 87:147–163

Preuschoft H, Witte H, Christian A, Fischer M (1996) Size influence on primate locomotion and body shape, with special emphasis on the locomotion of 'small mammals'. Folia Primatol 66:93–112

Preuschoft H, Christian A, Günther M (1998) Size dependences in prosimian locomotion and their implications for the distribution of body mass. Folia Primatol 69:60–81

Preuschoft H, Hohn B, Stoinski S, Witzel U (2011) Why so huge? Biomechanical reasons for the acquisition of large size in sauropod and theropod dinosaurs. In: Klein N, Remes K, Sander M (eds) Biology of the Sauropod dinosaurs: understanding the life of giants. Indiana University Press, Bloomington, pp 197–218

Raemaekers J (1984) Large versus small gibbons: relative roles of bioenergetics and competition in their ecological segregation in sympatry. In: Preuschoft H, Chivers DJ, Brockelman WY, Creel N (eds) The lesser apes. Edinburgh University Press, Edinburgh, pp 209–218

Reichard UH, Preuschoft H (2016) Why is the siamang larger than other hylobatids? In: Reichard UH, Hirohisa H, Barelli C (eds.) Evolution of gibbons and siamang. Springer, New York, pp. 167–181. doi: 10.1007/978-1-4939-5614-2_8

Ripley S (1979) Environmental grain, niche diversification, and positional behavior in Neogene primates: an evolutionary hypothesis. In: Morbeck ME, Preuschoft H, Gomberg N (eds) Environment, behavior and morphology: dynamic interactions in primates. Georg Fischer, New York, pp 37–74

Rose MD (1979) Positional behaviour of natural populations: some quantitative results of a field study of *Colobus guereza* and *Cercopithecus aethiops*. In: Morbeck ME, Preuschoft H, Gomberg N (eds) Environment, behavior and morphology: dynamic interactions in primates. Georg Fischer, New York, pp 95–117

Sander PM, Christian A, Clauss M et al (2010) Biology of the sauropod dinosaurs: The evolution of gigantism. Biological Reviews of the Cambridge Philosophical Society. doi: 10.1111/j.1469-185X.2010.00137.x

Schilling D, Preuschoft H (1984) Ererbt oder erlernt? Formspezifische Merkmale der Gesänge von Gibbons. Verh Dtsch Zoolog Ges 77:220

Schultz AH (1930) The skeleton of the trunk and limbs of higher primates. Hum Biol 2:303–438

Schultz AH (1933a) Die Körperproportionen der erwachsenen catarrhinen Primaten. Anthrop Anz 10:154–185

Schultz AH (1933b) Observation on the growth, classification and evolutionary specialisation of gibbons and siamangs. Hum Biol 5:212–255

Schultz AH (1944) Age changes and variability in gibbons. Amer J Phys Anthropol 2:1–129

Schultz AH (1956a) Post-embryonic age changes. In: Hofer H, Schultz AH, Starck D (eds) Primatologia I. Karger, Basel, pp 887–964

Schultz AH (1956b) The occurence and frequency of pathological and teratological conditions and of twinning among non-human primates. In: Hofer H, Schultz AH, Starck D (eds) Primatologia I. Karger, Basel, pp 965–1014

Schultz AH (ed) (1975) Die Primaten, Die Enzyklopädie der Natur. Editions Rencontre, Lausanne. English edition (1969): the life of Primates. Weidenfeld & Nicolson, London

Smith RJ and Jungers WL (1997) Body mass in comparative primatology. J Hum Evol 32:523–559

Stern JT, Wells JP, Vangor AK, Fleagle JG (1977) Electromyography of some muscles of the upper limb in Ateles and Lagotrix. Yrbk Phys Anthropol 20:498–507

Stern JT, Wells JP, Jungers WL, Vangor AK, Fleagle JG (1980) An electromyographic study of the pectoralis major in atelines and Hylobates, with special reference of the pars clavicularis. Am J Phys Anthropol 52:13–25

Tuttle RH(1972) Functional and evolutionary biology of hylobatid hands and feet. In: Rumbaugh DM (ed) Gibbon and Siamang, vol 1, Karger, Basel, pp 136–206

Vereeke EE, Channon AJ (2013) The role of hindlimb tendons in gibbon locomotion: springs or strings? J Exp Biol 216:3971–3980. doi: 10.1242/jeb.083527

Witte H, Preuschoft H, Recknagel S (1991) Human body proportions explained on the basis of biomechanical principles. Z Morph Anthropol 78:407–423

Witte H, Lesch C, Preuschoft H, Loitsch C (1995a) Die Gangarten der Pferde: Sind Schwingungsmechanismen entscheidend? Teil I. Pendelschwingungen der Beine bestimmen den Schritt. Pferdeheilkunde 11:199–206

Witte H, Lesch C, Preuschoft H, Loitsch C (1995b) Die Gangarten der Pferde: Sind Schwingungsmechanismen entscheidend? Teil II. Federschwingungen bestimmen den Trab und den Galopp. Pferdeheilkunde 11:265–272

Yamazaki N (1985) Primate bipedal walking: computer simulation. In: Kondo S (ed) Primate morphophysiology, locomotor analysis and human bipedalism. University of Tokyo Press, Tokyo, pp 105–130

Yamazaki N (1990) The effects of gravity on the interrelationship between body proportions and brchiation in the gibbons. In: Jouffroy FK, Stack HH, Niemitz C (eds) Gravity, posture and locomotion in primates. Il Sedicesimo, Firenze, pp 157–172

Yamazaki N (1992) Biomechanical interrelationshsip among body proportions, posture, and bipedal walking. In: Matana S, Tuttle RH, Ishida H, Goodman M (eds) Topic in Primatology, vol 3, Evolutionary Biology, Reproductive Endocrinology an Virology. University of Tokyo Press, pp 243–257

Yamazaki N, Ishida H (1984) A biomechanical study of vertical climbing and bipedal walking in gibbons. J Hum Evol 13:5673–571

Yamazaki N, Ishida H, Kimura T, Okada M (1979) Biomechanical analysis of primate bipedal walking by computer simulation. J Hum Evol 8:337–349

Yamazaki N, Ishida H, Kimura T, Okada M, Kondo S (1983) Biomechanical evaluation of evolutionary models for prehabitual bipedalism. Ann Sci Nat Zool 5:159–168

Zihlman AL, Mootnick AR, Underwood CE (2011) Anatomical contributions to Hylobatid taxonomy and adaptation. Int J Primatol 32:865–877

Part IV
Gibbon and Siamang Cognition

Chapter 12
Hand Manipulation Skills in Hylobatids

Jacqueline M. Prime and Susan M. Ford

Long-fingered hand of a white-handed gibbon, Khao Yai National Park. Photo credit: Jacqueline Prime

J.M. Prime (✉) · S.M. Ford
Department of Anthropology and Center for Ecology, Southern Illinois University, Carbondale, USA
e-mail: primejm@siu.edu

© Springer Science+Business Media New York 2016   267
U.H. Reichard et al. (eds.), *Evolution of Gibbons and Siamang*,
Developments in Primatology: Progress and Prospects,
DOI 10.1007/978-1-4939-5614-2_12

Introduction

Complex hand use and manual skills of great apes are consistently linked with advanced performance in primate cognitive and behavioral studies, including tool use, manipulative tasks demonstrating problem solving abilities, and sign language communication skills (Wright 1972; Patterson 1978; Savage-Rumbaugh 1978; Byrne and Byrne 1991, 1993; McGrew 1992, 1995). However, information on the manipulative abilities of the small apes is sparse. Although previous researchers have documented that hylobatids possess an ability to grasp objects using their opposable thumb in various positions (Lorenz 1974; Christel 1993), hylobatid hands continue to be habitually referred to as 'hooks' in primate literature (Fleagle 2013: 158). This implies that, due to their specialized morphology with elongated curved phalanges (Tuttle 1969), they are unable to adequately manipulate objects. This seems unreasonable given that, despite differences in hand shape and function across species, all primates use their hands to manipulate objects, particularly to obtain and consume food. Thus, a more detailed understanding of how hylobatids use their hands with their specialized morphology is necessary. This study documents hand use executed by hylobatids (*Hylobates lar* and *Nomascus leucogenys*) when grasping objects, documenting hand manipulation skill and manual dexterity demonstrated by hylobatids through the introduction of inanimate objects to their captive environments.

Few studies have addressed the functional significance of gibbon forelimb structure and manual dexterity beyond relations to brachiation (Jouffroy and Lessertiseur 1960; Beck 1967; Parker 1973; Lorenz 1974). Of all apes, hylobatids appear to be the most unique in their upper limb anatomy, reflective of their highly specialized richochetal brachiation form of locomotion (Larson 1998). However, a closer look at their hand anatomy compared with other apes reveals that gibbon thumbs are not reduced (Tuttle 1969; Schultz 1973). In fact, Tuttle (1969) suggested that gibbon physical features have evolved to allow for highly agile locomotive movement, while concurrently retaining and modifying the morphological requirements necessary for considerable manual dexterity, particularly in the first digit of hylobatid hands. The thumb of hylobatids is unique among all primates with a deep groove separating the thumb from the index finger, which allows for greater thumb mobility when coupled with their unique ball-shaped carpometacarpal joint of the first digit (originally documented by Nuck 1938, and noted by Lewis 1969; Tuttle 1969; Lorenz 1974).

Lorenz (1974) prepared a general overview of thumb use and hand forms characteristic of hylobatids, reporting on how small apes utilized their hands during day-to-day activities (i.e., during travel, grooming, feeding, etc.) with a focus on hand contact with structural objects in their captive environments. This study builds on Lorenz's original ideas by providing additional items to the gibbons to entice manipulation in order to understand how hylobatids use their hands to manipulate objects. The aim of this study is to understand the ways hylobatids use their hands for manipulative purposes of object acquisition, exploration, transfer, and use during nontravel-related activities. Research was centered on one specific question: How do hylobatids utilize their hands to manipulate objects?

Materials and Methods

Study Subjects

Two pairs of captive adult hylobatids, representing the two genera *Nomascus* and *Hylobates*, housed at two separate facilities, were provided with a variety of objects of varying size, shape, color, and texture, in different contexts within their enclosures (i.e., with and without food items to entice manipulation). Data collection took place at the Toronto Zoo, Toronto, Canada, which houses one adult male and one adult female white-handed gibbons (*Hylobates lar*), and the Lincoln Park Zoo, Chicago, U.S.A., which houses a family of white-cheeked gibbons (*Nomascus leucogenys*) consisting of one adult male, one adult female, and their juvenile and infant male offspring at the time of data collection. Unlike other study subjects, the adult *Nomascus* male of the Lincoln Park group was reared from birth by humans; thus he demonstrated a greater amount of interest in human visitors and human objects than did all other study subjects. He also demonstrated stereotypical thumb sucking/chewing behavior (Mootnick and Nadler 1997).

Data Collection

Data were collected at the Toronto Zoo over the course of 28 days for approximately 5 h each day (for a total of approximately 140 h). Data were collected at the Lincoln Park Zoo over the course of 25 days for approximately 5 h each day (for a total of approximately 125 h). Focal animal individuals were followed for 32 min time blocks, alternating every two minutes between the left and right hand of the focal individual, specifically focusing on the use of their hands in all activities except travel without an object. All observed forms of hand contact were recorded and then grouped into broad categories of hand positions for analysis ($N = 24{,}072$), wherein single movement hand manipulation was coded as finger-only, thumb-only, or finger-and-thumb contact. Any contact with an object that was not with the hands (such as mouth contact, foot contact, etc.) was recorded, but grouped into the broad category of non-hand contact for analysis.

Object use was recorded based on broad categories that implied different use of items (Table 12.1). Items were most often manipulated by the hands, mouth, or feet; manipulative contact was distinguished from passive contact based on movement of the body part touching an object (passive contact, manipulate hand, manipulate mouth, manipulate foot). Any object movements caused by the hylobatids' hands were considered to be a manipulation; conversely, if the hand was in contact with an object but not moving it, this was considered passive contact.

Holding onto an object was further distinguished from hand manipulation and passive contact based on whether or not the object moved when pressure was exerted by the hand. If the hand grasped the object and the object did not move this was considered a hold. Generally, all hand contact with structural objects during suspension falls into the 'hold' category, whereas hand contact with introduced objects were considered to be object manipulation to some degree. Consuming a food object was considered an activity ('eating'), and the handling of the food was coded as hand manipulation since the fingers and/or thumb of the individual were frequently moving the food item during consumption.

Objects and Methods

We provided all introduced materials used in this study with the exception of permanent parts of the enclosures and food/drink items that were provided in accordance with zoo feeding routines. A total of 50 different objects of various shapes, sizes, textures, consistencies, and colors were used and grouped into broad categories for data analysis (Table 12.2). These included: (1) structural objects, that were permanent parts of the enclosures, such as ground, fencing, vines, bars, branches, logs, and platform structures; (2) food containers, that were introduced objects of various shapes and sizes made of cardboard, plastic, or metal that contained food for the subjects, these items could be suspended from chains or freely spread around the enclosure; (3) food objects, that comprised any food items distributed by keepers (fruits, vegetables, muffins, peanut butter, nuts, cereals, and/or biscuits), as well as foliage growing in and around the enclosure; and (4) hard and (5) soft nonfood objects such as plush teddy bears and pet toys, tennis balls, rubber balls, balls attached to braided ropes, magazines, newspapers, hats, and t-shirts.

Although the enclosures for each small ape group were different, both enclosures had similar objects that could be combined into eight basic features under the category of structural objects. The structural objects within each enclosure were permanent features that could not be removed, though the orientation of certain items such as ropes or hoses could be altered. Structural objects constitute the majority of features that were touched or utilized by the subjects during all activities and therefore dominate all object contact that occurred for each individual in the study.

Introduced objects included any object given to the subjects for the purposes of this study, and these were further coded into two overlapping categories based on object softness and placement within the enclosure (Table 12.2). All introduced objects with the exception of food (that comprised its own separate category) were distinguished between those made out of hard or soft material. Hard objects consisted of any object that was made of plastic, metal, rubber, or wood material. Soft objects consisted of any object that was made of plush material, paper, fabric, or cardboard. Food containers were also distinguished in a separate category by placement within the enclosure (regardless of object material). All containers that were suspended from branches or bars were grouped as suspended food containers, whereas any containers housing food items that were loosely placed around the enclosure (either on the ground, in trees or on platforms that could easily be picked up and transported around the area) were termed loose food containers. All objects were susceptible to modification based on input and approval of zoo staff members, thus there were some differences between zoos as to what objects were suitable to give to the animals.

Data Analysis

Chi square tests were conducted comparing hand contact with introduced and structural objects divided by species, by sex, and by individuals to account for variability between sexes, species, or zoos; however no significant differences were found. Data across all gibbons were therefore pooled and Chi square tests were used to compare hand position categories in relation to specific structural objects, object material (hard vs. soft), object placement in the enclosure (suspended vs. loose food containers), and food objects. To determine which types of hand contact require thumb use, additional Chi square tests were carried out correlating thumb use with object use. All data analysis was done using JMP 4.0.4 (SAS Institute 2001).

Results

Structural Objects

Hand contact with structural objects reflects hand positions employed by hylobatids in stationary postures, particularly while individuals held on to features in their enclosures to support themselves in standing or suspensory postures or while resting in seated postures. Small apes tended to use only their fingers when handling the majority of structural objects. For example, finger-only contact dominated handling of horizontal bars (75.2 %), horizontal branches (75.3 %), fencing (70.1 %), and vines (74.8 %). While vertical bars were more often contacted with

Table 12.1 Broad categories for objects

Structural objects	Bars, branches, vines, platforms, fencing, ground
Suspended food containers	Hard or soft food containers suspended from branches or bars
Loose food containers	Hard or soft food containers scattered around ground/platforms
Food objects	Branches, leaves, food items distributed by keepers
Hard objects	Plastic, tupperware, rubber, wood—food containers or toys
Soft objects	Plush, paper, fabric, cardboard—food containers or toys

Table 12.2 Object use

Variable	Expression
Carry/Drag	Travel with object from one point to another
Flail/Shake	Flailing hand in air
Hold	Exerting pressure on object but object does not move
Manipulate foot	Moving object with the foot
Manipulate hand	Moving object with the hand
Manipulate mouth	Moving object with the mouth
No contact	No object contact
Passive contact	Contacting object but not moving it in any way
Rub/Rake	Moving fingers or thumb along body surface
Throw/Drop	Tossing or dropping an object intentionally or accidentally
Wear/Cover	Putting the object over any part of the body and leaving it there
Multiple object use	Moving more than one object at once

finger-and-thumb (75.0 %) hand positions ($N = 8824$, Pearson's $Chi^2 = 1256.09$, $r = 0.291$, $df = 35$, $p < 0.0001$; Fig. 12.1).

However, there was an interesting difference evident in the types of hand positions small apes used to hold onto bars versus branches with vertical orientations. As expected, finger-and-thumb contact dominates hand use (75.0 %) on vertical bars, indicating individuals clasp vertical bars tightly with their thumb in opposition to the fingers to combat gravity. But use of vertical branches does not appear to fit with expectations, as finger-only contact actually slightly dominates (55.8 %) over finger-and-thumb contact (42.4 %).

Object Softness (Hard vs. Soft)

The softness or hardness of an object significantly influences how that object is handled by hylobatids. An inverse relationship between finger-only contact and finger-and-thumb contact between hard and soft objects is evident ($N = 1007$, Pearson's $Chi^2 = 229.21$, $r = 0.27$, $df = 10$, $p < 0.0001$; Fig. 12.2). Hylobatids contact hard objects with only their fingers 51.3 % of the time, with finger-and-thumb contact occurring 37.4 % of the time. Soft objects, on the other hand, are most often contacted with finger-and-thumb positions at 46.9 % and finger-only positions only 30.2 % of the time.

Placement of Objects (Loose vs. Suspended Food Containers)

Significant differences were seen when the small apes handled suspended versus loose food containers placed in different locations around their enclosures.

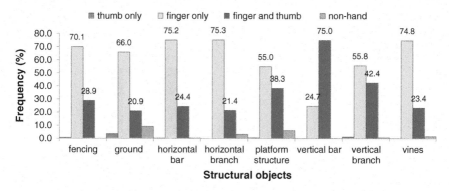

Fig. 12.1 Hand contact with structural objects

Fig. 12.2 Hand contact with introduced objects divided by object softness

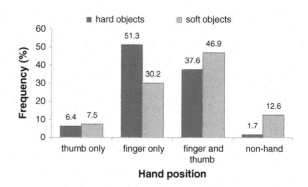

Although finger-only contact dominates in both categories, finger-only contact is statistically significantly used more frequently with suspended food containers at a frequency of 61.8 % of contact than it is with loose food containers at 42.4 % ($N = 498$, Pearson's Chi^2 = 122.11, $r = 0.28$, df = 10, $p < 0.0001$; Fig. 12.3).

Food Objects

Of all hand contact with food objects, 54.6 % is attributed to finger-and-thumb use, with thumb-only contact making up an additional 21.1 % of all contact with food items ($N = 2388$, Table 12.3; Fig. 12.4). This is in stark contrast to the mere 12.8 % of the time finger-only contact occurs with food objects. It is quite apparent, therefore, that when it comes to object manipulation, small apes are consistently using their thumbs much more frequently during food contact than with other objects.

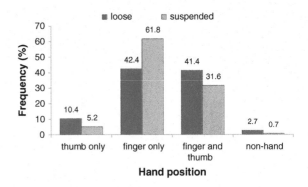

Fig. 12.3 Hand contact with introduced food containers divided by placement

Discussion

All small apes in this study proved to be quite capable of manipulating various objects regardless of the size of the article presented to them. It is important to note that whenever there is a small number of subjects in a study, the individual characteristics of each subject can dramatically influence the results. Thus, our interpretations of the results presented here are valid within the scope of this study and with regard to the individuals studied, and extrapolations to the species or genus level require further study with a larger sample. At the same time, however, it is also reasonable to interpret some results across a wider taxonomic group given the large morphological and behavioral uniformity that generally characterizes the family hylobatidae.

Structural Objects

Gibbons employ a variety of hand positions to clench structural supports. The relatively high frequencies of finger-only contact with all categories of contact with structural objects, except vertical supports, are likely reflective of the typical suspensory postures in which hylobatids engage throughout the majority of their activities (Nowak and Reichard 2016), particularly when grasping horizontal or mesh supports (e.g., fencing). Finger-only contact would be expected, considering horizontal supports would be conducive to the morphologically specialized 'hook-like' hands of gibbons, i.e., horizontal supports would be easy to 'hook' onto, whereas vertical supports should require the force of a griphold to bear weight during suspension.

While a finger-and-thumb grip, wherein the thumb is used in opposition to the fingers to clench the object within the hand, is the predominant hand position on vertical bars, there is a surprising relatively high frequency of finger-only contact with vertical branches over vertical bars. The 'grasp' hand position of clenching an object with just the fingers while the thumb is open and not used in opposition to

Table 12.3 Summary statistics for Chi^2 analyses for all gibbons

Hand use compared to						
	Object	# observers	Chi^2	r	Prob > Chi^2	DF
All gibbons	Structural objects	8824	1256.09	0.29	<0.0001	35
	Material	1007	292.21	0.27	<0.0001	10
	Placement	498	122.11	0.28	<0.0001	10
	Thumb use*	16,414	848.64	0.20	<0.0001	8

*In this summary, thumb use is not compared to hand use; thumb use is compared to object use

Fig. 12.4 Hand contact with food items

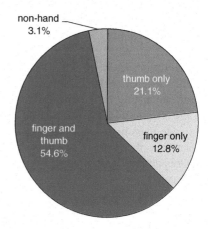

close the grip was often employed to hook the fingers onto vertical branches, and it is likely that the texture of the bark allows hylobatids to maintain a hold on the vertical branches with only their fingers and no thumb.

While gibbons were most often observed to brace themselves with their feet while clinging to vertical supports, their ability to hold onto vertical branches with only their fingers during hand contact demonstrates the tremendous strength of their fingers when clasping objects. It is apparent that regardless of the orientation of the substrate, hylobatids are able to support their entire body weight solely by hanging from their fingers for extended periods of time during suspension. And although they may use their feet to stabilize themselves on vertical supports, which may relieve some of the weight strain on the fingers during vertical clinging, they still employ only their fingers to support their upper bodies and their grip does not appear to be compromised in any way on vertical branches.

Suspensory feeding postures, characteristic of hominoids, have been argued to increase the range of foraging options available to apes by enhancing their abilities to feed in terminal branches, thereby allowing them to reach nutritious foods that are unavailable to quadrupedal monkeys (Grand 1972; Gebo 1996). Unlike apes, monkeys do not suspend below branches from their forelimbs when moving through the tree canopy but instead walk quadrupedally on top of branches. This pronograde movement pattern constrains mobility of their forelimb joints and prevents circumduction of the shoulder (Hunt 1992), and likely strongly impacts how they access and acquire resources in the trees. Grand (1972) hypothesized that primates who could hang from supports with rotating, elongated forelimbs, like gibbons, could theoretically expand their surrounding foraging radius increasing the resources available to them in the trees, while quadrupedal primates would be limited in their foraging radius by their restricted joints and equal limb length. This proposition has been widely accepted and cited in primate literature (Cartmill and Milton 1977; Hunt 1992; Povinelli and Cant 1995; Myatt and Thorpe 2011; McGraw and Daegling 2012). Resource acquisition postures used during food harvesting are central to understanding how different primates have anatomically adapted to their environments, and how those environments have shaped their morphology and behaviors over time. Improved means of exploiting the terminal branches of trees by modern primates has been linked to the evolution of angiosperms in tropical rain forests, noting an adaptive shift in primates to feeding on a milieu of previously unexploited resources (fruits, flowers, young leaves) located with pollinating insects in the flimsy tips of branches in the tree canopy (Sussman 1991; Rasmussen 2001; Sussman et al. 2013). In order to acquire food items, a primate must first be able to stabilize its body on available substrates in a position relative to the resource (Cant and Temerin 1984). Several factors influence a primate's ability to do this, such as body size, limb proportions, joint flexibility, and manual prehensility (Cant and Temerin 1984). Following from this, Cartmill (1985) proposed that the best means for primates to resist tumbling from the top of horizontal branches in the tree canopy is to possess one or more of the following traits: (1) relatively short limbs to keep the body's centre of gravity close to the support, (2) prehensile hands and feet to grip supports and exert a resistive torque against falling, (3) a reduced body size, or (4) a capacity for suspensory postures to safely hang below supports to reduce the risk of falling. Thus, different primates exhibited variations of these traits in order to access and acquire resources in the trees.

In the wild, hylobatids are entirely arboreal feeders whose diet predominantly consists of a variety of fleshy ripe fruits supplemented by leaves and smaller portions of shoots, flowers, insects, and occasionally unripe fruits and they are noted to be terminal branch feeders (Carpenter 1940; Ellefson 1974; Raemaekers 1977; Chivers 1980; Gittins and Raemaekers 1980; Srikosamatara 1984; Stafford et al. 1990; Whitington and Treesucon 1991; Bartlett 2009; McConkey et al. 2002; Elder 2009). Thus, it is reasonable to assume that the strength of hylobatid fingers noted here in their manual use during suspensory postures may have important implications for terminal branch feeding in the wild. Coupled with their smaller body size compared to the larger apes (Roonwal and Mohnot 1977; Marshall and Sugardjito 1986), hylobatids appear to exhibit unique evolutionary adaptations in finger strength and manual dexterity that may allow them to suspend in the tree canopy and extend their reach into the terminal branches to exploit resources providing an advantage over larger apes and quadrupedal monkeys.

Object Softness (Hard vs. Soft Objects)

Object material influences the different ways that small apes are able to handle various objects. Softer objects are more easily manipulated by finger-and-thumb handling because they are malleable and can be gripped or squeezed during manipulation. Hard objects, however, are not so pliable and therefore they cannot conform to the hand during manipulation. If hard objects are larger than the hand or too awkward for the hand to clench, the hylobatids cannot grip these objects with the finger-and-thumb hold. As a result, most often these items were manipulated using only the fingers to lift, push, or flip the objects around during handling. It is important to note, however, that because of the deep groove separating the thumb from the index finger on hylobatid hands, the thumb is widely opposable. Thus, hylobatids are able to grip objects that are much larger than what the average human hand can clutch, using their long fingers and highly opposable thumb.

Hylobatids were often observed to stabilize difficult to open containers with their feet or to bite or gnaw at the edges during manipulation. In addition, larger items were often transported in their legs by tucking the knees up to the chest and cradling the object between the thighs and belly to carry it. The small apes often employed various hand maneuvers in attempts to open different containers, and when their hands could not function appropriately to get containers open, they did attempt to employ other means (using the mouth and feet) to get at food inside.

Placement of Objects

Differences in hand contact related to object placement is reflective of the distinct ways gibbons approach suspended versus loose objects. Loose food containers can be gripped using finger-and-thumb contact because they fit within a gibbon's hand and can be manipulated from any angle if they are transported or moved. Suspended containers, however, are relatively stationary. Although they may swing or swivel during manipulation, suspended containers cannot be picked up and transported to another location. Therefore, when hylobatids were handling suspended objects, individuals were commonly hanging near the objects, suspended from a nearby bar or branch with one hand, using the free hand to manipulate the suspended container or the food inside. In this suspensory position, hylobatids would commonly use finger-only contact to push containers around, attempting to shake items out from inside them. Hylobatids would also stick only their fingers inside the containers to scoop or pull out food items from inside containers until they could grip them more precisely with their thumb.

Hand contact with loose objects tended to provoke greater object manipulation, as hylobatids would turn, squeeze, grip, or lift up food containers, whereas suspended objects would simply be pushed or pulled to reach objects inside them.

Many times, hylobatids would work to spill food items out of suspended containers and then move to the ground to pick up the spilled food items at their leisure.

Food Objects

Food objects are often small and easily manipulated by the hands of hylobatids, both during foraging activities and during consumption, and this is evident in the hand-to-mouth feeding technique characteristic of all hylobatids (Carpenter 1940; Ellefson 1974; Lorenz 1974; Fleagle 2013; pers. obs.). Each of the individual subjects appeared to have his/her own unique ways of moving objects around with their hands, although all exhibited similar hand forms during food object manipulation.

The side grip was the most common form of holding any food item, whereby the piece of food could be tightly clasped with the thumb pressing the object against the palm and index finger. This tight grip on the food object ensures that during travel, foraging, or consumption, the object will not slip out of the hand and fall to the ground, which in the wild would become a lost item, since gibbons live in the highest parts of the forest canopy and do not go down to the ground (Carpenter 1940; Ellefson 1974). Hylobatid diets in zoos primarily consist of monkey chow biscuits, which are extremely hard items that appear to be quite difficult for the small apes to bite into and chew. To eat these food items, small apes would hold the core of a biscuit in the side grip position and break off pieces using the back molars.

Hylobatids generally choose only one food item at a time and consume it entirely before picking up another piece, only rarely collecting several pieces of food together (Carpenter 1940; pers. obs.). During foraging and eating, hylobatids will most often pick up food items from the ground or pluck leaves from trees using a thumb-pick movement that involves sliding the tip of the thumb under the object and pressing it against the side of the index finger. This activity involves exclusively using only the thumb to manipulate food objects, and it is the predominant means of picking up food objects during foraging demonstrated by all hylobatids in this study.

Thumb Use in Relation to Object Softness and Object Use

Hylobatids used their thumbs extensively when handling objects (Fig. 12.5). Lorenz (1974) noted that the skin on the thumb of gibbons and siamangs is much thinner than the skin on the fingers, suggesting that the thumb is more receptive to tactile sensory stimulus than the fingers. In this study, we found that although hylobatids occasionally contact objects using just the index finger, hylobatids more commonly use all four fingers in conjunction when handling objects, while the thumb is considerably more mobile and sensitive. The thumb is used to poke, pick, and delicately explore and maneuver objects, thus it is the thumb which acts as a sensor and precision digit, allowing hylobatids to maintain an essential amount of precision, tactile sensitivity, and control during object manipulation to counterbalance their extremely long and specialized fingers. The relatively equal frequencies of thumb-only contact for both hard (6.4 %) and soft (7.5 %) objects reflect the ways hylobatids delicately manipulate introduced objects during initial contact.

Hylobatids appear to use their thumbs most often when carrying items and when manipulating objects, especially during carry/drag activities, whereby objects are moved from one point to another (especially moving food from the ground to a preferred eating location), but travel between the two locations requires some form of brachiation or climbing. Brachiation demonstrated by hylobatids involves rapid hand-over-hand movements where the fingers hook onto branches and release at very high speeds. Carrying small objects would be extremely difficult without some adaptation to allow hylobatids to tightly grip onto an object while moving quickly. The structure of the thumb allows hylobatids to hold a small food item or soft/pliable object firmly next to the palm, while keeping the long fingers free for travel. This may be beneficial in the wild as hylobatids often move frequently throughout the tree crown during foraging bouts, typically plucking a food item from the terminal branches and moving to an interior location in the tree canopy to consume the item (Prime 2014). Though within group contest competition may be low for hylobatids living in small groups, and therefore they do not need adaptations to collect large quantities of food items quickly to hold for consumption later (as macaques with cheek pouches do, for example), this individual item pluck and carry style with the adaptation to tightly grip on object with the thumb, freeing the fingers for travel, may provide an evolutionary advantage to hylobatids in their suspensory terminal branch feeding style. Thus, the thumb is of paramount importance in the dynamic hand activities of manipulation and transportation of small objects, though the thumb is not predominantly involved in various passive activities, such as resting (passive contact with objects) or scratching (rubbing or raking the fingers over the body).

Intricate Object Manipulation

Layering objects by placing smaller items inside larger containers to provoke manipulation shed light on the dexterous skill of hylobatids when handling objects. Some of these larger containers had holes all over the surfaces of the container, while others had only one specific opening and objects could only be removed from that particular end. Oftentimes when hylobatids manipulated these larger objects, they flipped and turned objects using only their fingers, pushing objects around or turning them over on the ground or on platform surfaces in an attempt to free whatever was inside.

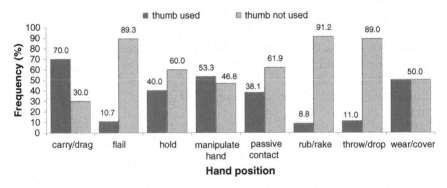

Fig. 12.5 Thumb use in relation to object use/behavior

Hylobatids would frequently attempt to pull the inner objects out of the containers by sticking their long slender fingers into the openings and grasping the smaller inner objects with only their fingers, or probing the thumb into the container and using a thumb grip to remove the object. When this was not sufficient, hylobatids would pick and pull repeatedly at the objects with only their fingers until objects were partially protruding from the containers; items could then be gripped and pulled out completely.

Hylobatids were also observed to grip the entire container and lift it repeatedly, turning it with the open side down until objects would fall out of the opening. Moreover, if the inside object would not fall out of the larger containers, the small apes were observed to clench the larger containers with their hands and feet and bite or gnaw on them in an attempt to tear the containers open. Such attempts, however, were brief, often consisting of one bite or chew on the object before continuing to flip and shake the object.

In one particular instance, the food items placed inside a larger plastic soda pop bottle would not come out of the small opening at the top of the container and the study subjects persistently flipped, chewed, and banged the container around until eventually the container broke open. The hylobatids were observed to work on the item and leave it, returning sporadically to try to open it for approximately 2 h unsuccessfully. When observations commenced the next day, the broken and empty pieces of the container were noted to be in a new location and all the inside material was removed. There was no keeper contact with the hylobatids overnight; therefore, it was evident that only the small apes could have opened this container. However, how and when they finally did get it open was not witnessed.

Hylobatids appear to be very particular about their actions when it comes to small object manipulation, often picking up items or plucking leaves from trees one at a time. In a few instances, hylobatids were observed to employ meticulous efforts to open boxes—peeling off masking tape that held a small box closed with only the thumb and then shaking the masking tape off the thumb, or soaking small cardboard boxes in water until the material softened and could be easily torn apart. However, in most instances hylobatids would simply manipulate and shake containers until the inside objects came out.

Interestingly, if food items were concealed within other objects inside containers (for example, inside crumpled paper, inside a cardboard or plush box; three complex containers were provided on four separate occasions), specific individuals appeared to have trouble finding the food, often opening the box and pulling out the paper but not opening the paper to expose the food. During these instances when one individual would have trouble with a particular container, one of two things would happen. The individual would attempt to take food from other individuals who already had food (4 out of 4 occurrences, 100 %), or the individual would attempt to bring their partner over to their food package and push the food package to them for the partner to open (3 out of 4 occurrences, 75 %). This interaction between individuals was quite interesting among primates as it demonstrates the reliance/cooperation of paired individuals in small hylobatid social groups.

Comparisons of Small Apes with Large Apes

Examinations of large ape hand manipulation skills are almost synonymous with investigations of cognitive abilities in apes. Though humans clearly surpass all other primates in brain, research into the manual dexterity of large ape has revealed that all large apes are capable of more accurate and precise manipulation than monkeys (Marzke and Wullstein 1996), exhibit cognitively complex behaviors in captivity and also in the wild (such as tool use: Galdikas 1982; Tuttle 1986; Boesch and Boesch 1990; McGrew 1992; van Schaik et al. 1996; Hohmann and Fruth 2003), and consistently outperform monkeys in cognitive tests (Anderson 1996; Hare et al. 2006).

The functional morphology of the hands of large apes reflects adaptations for fine motor movements and for morphological compromises between locomotor and manipulative functions. Because all apes are suspensory to some extent (Tuttle 1969; Tuttle and Watts 1985; Hunt 1991, 1992; Doran 1996), their hands have undergone a reduction in thumb length and an increase in finger length. Yet, these apparent adaptive alterations do not appear to have compromised their ability to manipulate objects. Chimpanzees and gorillas appear to be the most capable of manipulatory functions comparable to humans, and despite their diminutive thumbs orangutans also demonstrated a high level of manipulatory proficiency. As Tuttle (1969) explained, it is likely that the more balanced proportions of the thumb and index finger in gorillas allows them to approximate the distal phalanges more completely than other apes, but all large apes have been documented to perform the basic precision grip with slight variation comparable to humans (Napier 1960; Marzke 1997). Furthermore, large apes have been documented to use a scissor grip at times when the finger-thumb opposition, which is the basis of the precision grip, is impaired (Napier 1960; Tuttle 1972).

The small ape behaviors presented here show clear evidence that small apes are capable of using their hands in a variety of manners quite similar to their large ape counterparts, despite their elongated phalanges. Hylobatids are able to form both power and the modified precision grips noted in all other species of apes. In fact, solely looking at the underlying morphology of gibbons and siamangs in comparison to the other apes, it is apparent that with their well-developed pollex musculature and 'ball-and-socket' carpometacarpal joint, the thumb of hylobatids should be (and is) able to flex independently from the other digits, which would allow for opposable precision control, contrary to presumptions that hylobatid hands are rigid hooks. As Tuttle (1969) reasoned, the hands of hylobatids morphologically reconcile the functional use of the hand for richochetal brachiation (which places unique demands on the integrity of the forelimb), while maintaining proficient manipulation abilities most likely for food objects. This study supports this view, since most often the thumb was independently employed by hylobatids during suspension and travel to hold onto food objects while the fingers sustained the weight of the individuals during these activities. It seems this unique ability was a necessary adaptation to a primarily suspensory ape in an arboreal environment. Additionally, because of the independence of the thumb from the other digits, hylobatids also employ the thumb for a variety of activities generally reserved for the index finger in other apes and humans (Christel 1993; Gentile et al. 2011). While other apes frequently use their index and middle fingers to manipulate objects (to pick up objects or probe in an exploratory fashion), hylobatids in this study most frequently utilized their thumbs for tactile sensitivity actions.

It is clear from both behavioral hand use and morphology that the hands of the Hominoidea evolved to allow the last common ancestor of all apes to use their hands to manipulate objects. While our human hands function primarily for manipulative purposes, the hands of apes are concurrently adapted for both positional and manipulation functions. Nonetheless, it is evident that selection for specialized locomotor behavior has not significantly compromised the ability of any ape species, including small apes, to proficiently manipulate objects.

Conclusion

There are considerably fewer similarities between humans and small apes in comparison to those shared by humans and large apes. The uniqueness of hylobatids in comparison with the large apes and humans is reflected in their highly specialized anatomy, richochetal brachiation locomotion, and their well-developed thumbs with a deep cleft on the palm for grasping and manipulation.

This study clearly demonstrates that hylobatids do manipulate objects and are able to execute precision and power grips like all other apes, including humans. Hylobatid hands are very strong and powerful, which is demonstrated by their ability to support their entire body weight on just their fingers during suspension regardless of substrate orientation. They are able to delicately maneuver objects with only their thumbs or use their entire hand to squeeze, lift, or hold objects to examine materials. Furthermore, depending on the location of the object, they are capable of changing their own behaviors/body position/orientation to increase their ability to manually maneuver objects or alter the object (shaking food out from suspended containers) to gain access to desired items.

During manipulation, hylobatids use their thumb extensively in very patterned ways, most often employing the thumb when transporting small objects during travel (carrying food) and when engaging in fine-tuned motor movements, such as picking small objects off of the ground or manipulating objects during handling. In conjunction with this, their frequent use of the thumb alone to pick at, probe, and investigate objects indicates the thumb is of vital importance to hylobatids and is most often employed as a sensory digit. It is apparent that despite their unique morphology, hylobatids are able to use their hands in similar ways to their large ape counterparts. However, instead of prominent use of the index finger as the primary digit involved in tactile and exploratory activities characteristic of large apes, gibbons will use their thumb (this study; Lorenz 1974).

Implications of hand use as a reflection of cognitive abilities require further investigation. However, establishing the underlying basis of hand function and form during object manipulation offers a vital stepping stone for future research. Early research into the cognitive abilities of different primate species based on comparative testing reasoned that hylobatids were less advanced than other apes (and even some species of monkeys) due to their poor performance in captive testing (Yerkes and Yerkes 1929; Klüver 1933; Kohler 1959) and a manipulation-averse hand that was largely fixed into a 'hook'. Beck (1967), however, countered these claims pointing out that design flaws in testing were ill-suited for hylobatids due to their exclusively arboreal lifestyles and their hand morphology. With slight modification to adapt testing apparatus to gibbon preferences (providing an elevated string, rather than a string lying on a flat surface), Beck (1967) demonstrated hylobatids were able to solve the required tasks, demonstrating their cognitive insight with success comparable to the large apes. In addition, more recent investigations into hylobatid cognitive capacity, using a tool-like object (a rake) to gain a food reward out of reach conducted by Cunningham et al. (2006), demonstrated that hylobatids were able to successfully utilize the apparatus to obtain the reward, thereby learning associative rules. By applying this base information of hand manipulation skills toward developing new and modified means of testing cognition in hylobatids, specially designed (or modified) tools can be created to test gibbon and siamang comprehension and neurological skills that control for the potential shortcomings of current research tools. In the research setting, similarities in size and function between human and large ape hands allow investigators to utilize human tools in various forms to test large ape cognition focusing on object identification, manipulation, and recognition. These studies have proven to be successful because large ape hands are easily able to manipulate these objects, or only slight alterations are required to compensate for their unique functional needs (McGrew 1995; Marzke 1997; Susman 1998). While it may be undocumented that hylobatids utilize tools in the wild, research conducted on various species of monkeys and large apes has proven that, regardless of their use of tools in the wild, various primate species are quite capable of utilizing tools in captivity (Torigoe 1985; Visalberghi et al. 1995; Macellini et al. 2012).

With the information provided in this study, it is clear that hylobatids are proficient at using their hands to manipulate objects and they do so in a variety of different ways, demonstrating extraordinary hand control and precision maneuverability, strongly suggesting that the last common hominoid ancestor already exhibited these abilities. Understanding how our earliest ape ancestors and relatives functioned can shed light on the evolution and developments of hominoids and hominids over the course of history. But with the tight connections established in the molecular evidence between humans and the large apes, small apes are often pushed aside. As research on the small apes continues to grow, mounting information on the flexible behaviors and life histories of hylobatids (Barelli et al. 2007; Reichard and Barelli 2008; Asensio et al. 2011; Reichard et al. 2012) is revealing that small apes are by no means less complex or less skilled than any of our large ape relatives.

Acknowledgments We would like to thank the Toronto Zoo, Toronto, Canada, and the Lincoln Park Zoo, Chicago, U.S.A., for permitting observations of their animals. Their assistance and generosity (allowing us to observe the animals 'behind the scenes' on several occasions) was very much appreciated. Thank you to Matthew Nowak for his input during analysis. Thank you to Dr. R. Corruccini and Dr. D. Sutton for their interest, suggestions, and comments on this work. And thank you to Dr. U. Reichard and an anonymous reviewer for their suggestions to improve the manuscript.

References

Anderson J (1996) Chimpanzees and capuchin monkeys: comparative cognition. In: Russon A, Bard KB, Parker S (eds) Reaching into thought: the minds of the great apes. Cambridge University Press, Cambridge, pp 23–56

Asensio N, Brockelman WY, Malaivijitnond S, Reichard UH (2011) Gibbon travel paths are goal oriented. Anim Cogn 14:395–405

Baldwin L, Teleki G (1974) Field research on gibbons, siamangs and orang-utans: an historial, geographical, and bibliographical listing. Primates 15:365–376

Barelli C, Heistermann M, Boesch C, Reichard UH (2007) Sexual swellings in wild white-handed gibbon females (*Hylobates lar*) indicate the probability of ovulation. Horm Behav 51:221–230

Bartlett T (2009) Seasonal home range use and defendability in white-handed gibbons (*Hylobates lar*) in Khao Yai National Park, Thailand. In: Lappan S, Whittaker D (eds) The gibbons: new perspectives on small ape socioecology and population biology. Springer, New York, pp 265–275

Beck BB (1967) A study of problem solving by gibbons. Behavior 28:95–109

Boesch C, Boesch H (1990) Tool use and tool making in wild chimpanzees. Folia Primatol 54:86–99

Byrne RW, Byrne JME (1991) Hand preferences in the skilled gathering tasks of mountain gorillas (*Gorilla g. beringei*). Cortex 27:521–546

Byrne RW, Byrne JM (1993) Complex leaf-gathering skills of mountain gorillas (*Gorilla g. beringei*): variability and standardization. Am J Primatol 31:241–261

Cant J, Temerin LA (1984) A conceptual approach to foraging adaptations in primates. In: Rodman PS, Cant JGH (eds) Adaptations for foraging in nonhuman primates. Columbia University Press, New York, pp 304–342

Carpenter CR (1940) A field study in Siam of the behavior and social relations of the gibbon (*Hylobates lar*). Comp Psychol Monogr 16:1–212

Carpenter CR (ed) (1964) Naturalistic behavior of nonhuman primates. Penn State University Press, University Park

Cartmill M (1985) Climbing. In: Hildebrand D, Bramble M, Liem K, Wake D (eds) Functional vertebrate morphology. Belknap Press, Cambridge, pp 73–88

Cartmill M, Milton K (1977) The lorisiform wrist joint and the evolution of "brachiating" adaptations in the Hominoidea. Am J Phys Anthropol 47:249–272

Chivers D (ed) (1980) Malayan forest primates. Springer, New York

Christel M (1993) Grasping techniques and hand preferences in Hominoidea. In: Preuschoft H, Chivers DJ (eds) Hands of primates. Springer, New York, pp 91–108

Cunningham CL, Anderson JR, Mootnick AR (2006) Object manipulation to obtain a food reward in hoolock gibbons, *Bunopithecus hoolock*. Anim Behav 71:621–629

Doran DM (1996) Comparative positional behavior of the African apes. In: McGrew MC, Marchant LF, Nishida T (eds) Great ape societies. Cambridge University Press, Cambridge, pp 213–224

Elder A (2009) Hylobatid diets revisited: the importance of body mass, fruit availability, and interspecific competition. In: Lappan S, Whittaker D (eds) The Gibbons: new perspectives on small ape socioecology and population biology. Springer, New York, pp 133–159

Ellefson JO (1974) A natural history of white-handed gibbons in the Malayan peninsula. In: Rumbaugh DM (ed) Gibbon and siamang, vol 3: natural history, social behavior, reproduction, vocalization, prehension. Karger, Basel, pp 1–136

Fleagle JG (ed) (2013) Primate adaptation and evolution. Academic Press, San Diego

Fujita K, Kuroshima H, Asia S (2003) How do tufted capuchin monkeys (*Cebus paella*) understand causality involved in tool use? J Exp Psychol Anim Behav Process 29:233–242

Galdikas BMF (1982) Orang-utan tool-use at Tanjung Puting Reserve, Central Indonesian Borneo (Kalimantan Tengah). J Hum Evol 11:19–33

Gebo DL (1996) Climbing, brachiation, and terrestrial quadrupedalism: historical precursors of hominid bipedalism. Am J Phys Anthropol 101:55–92

Gentile G, Petkova VI, Ehrsson HH (2011) Integration of visual and tactile signals from the hand in the human brain: an FMRI study. J Neurophysiol 105:910–922

Gittins SP, Raemaekers JJ (1980) Siamang, lar and agile gibbons. In: Chivers DJ, Raemaekers JJ (eds) Malayan forest primates: ten years' study in tropical rain forest. Springer Press, New York, pp 63–105

Grand TI (1972) A mechanical interpretation of terminal branch feeding. J Mammal 53:198–201

Hare B, Call J, Tomasello M (2006) Chimpanzees deceive a human competitor by hiding. Cognition 101:495–514

Hohmann G, Fruth B (2003) Culture in Bonobos? Between-species and within-species variation in behavior. Curr Anthropol 44:56–571

Hunt KD (1991) Positional behavior in the Hominoidea. Int J Primatol 12:95–118

Hunt KD (1992) Positional behaviour of *Pan troglodytes* in the Mahale Mountains and Gombe Stream national parks. Am J Phys Anthropol 87:83–106

Jouffroy FK, Lessertiseur J (1960) Les specialisations anatomiques de la main chex les singes a pregression suspendue. Mammalia 24:93–151

Klüver H (ed) (1933) Behavior mechanisms in monkeys. The University of Chicago Press, Chicago

Kohler W (ed) (1959) The mentality of apes, 2nd edn. Vintage Books, New York

Larson SG (1998) Parallel evolution in the hominoid trunk and forelimb. Evol Anthropol 6:87–99

Lewis OJ (1969) The hominoid wrist joint. Am J Phys Anthropol 30:251–268

Limongelli L, Boysen ST, Visalberghi E (1995) Comprehension of cause-effect relations in a tool-using task by chimpanzees (*Pan troglodytes*). J Comp Psychol 109:18–26

Lorenz R (1974) On the thumb of the Hylobatidae. In: Rumbaugh DM (ed) Gibbon and siamang, vol 3: natural history, social behavior, reproduction, vocalization, prehension. Karger, Basel, pp 157–175

Macellini S, Maranesi M, Bonini L, Simone L, Rozzi S, Ferrari PF, Fogassi L (2012) Individual and social learning processes involved in the acquisition and generalization of tool use in macaques. Phil Trans R Soc B 367:24–36

Marshall J, Sugardjito J (1986) Gibbon systematics. In: Swindler DR, Erwin J (eds) Comparative primate biology (vol 1): systematics, evolution, and anatomy. Alan R, Liss, New York, pp 137–185

Marzke MW (1997) Precision grips, hand morphology, and tools. Am J Phys Anthropol 102:91–110

Marzke MW, Wullstein KL (1996) Chimpanzee and human grips: a new classification with a focus on evolutionary morphology. Int J Primatol 17:117–139

McConkey K, Aldy F, Ario A, Chivers D (2002) Selection of fruit by gibbons (*Hylobates muelleri* x *agilis*) in the rain forests of central Borneo. Int J Primatol 23:123–145

McGraw WS, Daegling DJ (2012) Primate feeding and foraging: integrating studies of behavior and morphology. Annu Rev Anthropol 41:203–219

McGrew W (ed) (1992) Chimpanzee material culture: implications for human evolution. Cambridge University Press, Cambridge

McGrew W (1995) Thumbs, tools and early humans. Science 268:586–589

Mootnick AR, Nadler RD (1997) Sexual behaviour of maternally separated gibbons (*Hylobates*). Dev Psychobiol 31:149–161

Myatt JP, Thorpe SKS (2011) Postural strategies employed by orangutans (*Pongo abelii*) during feeding in the terminal branch niche. Am J Phys Anthropol 146:73–82

Napier JR (1960) Studies of the hands of living primates. Proc Zool Soc Lond 134:647–657

Napier JR (ed) (1980) Hands. Princeton University Press, Princeton

Nowak MG, Reichard UH (2016) Locomotion and posture in ancestral hominoids prior to the split of hylobatids. In: Reichard UH, Hirohisa H, Barelli C (eds) Evolution of gibbons and siamang. Springer, New York, pp 55–89

Patterson FG (1978) The gestures of a gorilla: language acquisition in another pongid species. Brain Lang 5:72–97

Parker CE (1973) Manipulatory behavior and responsiveness. In: Rumbaugh DM (ed) Gibbon and siamang, vol 2: anatomy, dentition, taxonomy, molecular evolution and behavior. S. Karger, Basel, pp 163–184

Povinelli DJ, Cant JGH (1995) Arboreal clambering and the evolution of self-conception. Q Rev Biol 70:393–421

Prime J (2014) The ape ecological niche: posture and hand use in gibbons and macaques and the influence of manual skill on cognitive development in apes and humans. Dissertation, Southern Illinois University, Carbondale

Raemaekers JJ (1977) Gibbons and trees: comparative ecology of the siamang and lar gibbons. Dissertation, University of Cambridge, Cambridge

Rasmussen DT (2001) Primate origins. In: Hartwig HC (ed) The primate fossil record. Cambridge University Press, London, pp 5–10

Reichard UH, Barelli C (2008) Life history and reproductive strategies of Khao Yai *Hylobates lar*: implications for social evolution in apes. Int J Primatol 28:828–844

Reichard UH, Ganpanakngan M, Barelli C (2012) White-handed gibbons of Khao Yai: social flexibility, complex reproductive strategies, and a slow life history. In: Kappeler PM, Watts D (eds) Long-term field studies of primates. Springer, Berlin, pp 237–258

Roonwal ML, Mohnot SM (eds) (1977) Primates of south Asia: ecology, sociobiology, and behavior. Harvard University Press, Cambridge

Savage-Rumbaugh ES (ed) (1978) Ape language: from conditioned response to symbol. Columbia University Press, New York

Schultz AH (1973) The skeleton of the Hylobatidae and other observations on their morphology. In: Rumbaugh DM (ed) Gibbon and siamang, vol 2: anatomy, dentition, taxonomy, molecular evolution and behaviour. S. Karger, Basel, pp 2–55

Srikosamatara S (1984) Ecology of pileated gibbons in south-east Thailand. In: Preuschoft H, Chivers DJ, Brockelman WY, Creel N (eds) The lesser apes: evolutionary and behavioural biology. Edinburgh University Press, Edinburgh, pp 242–257

Stafford D, Milliken G, Ward J (1990) Lateral bias in feeding and brachiation in *Hylobates*. Primates 31:407–414

Sussman RW (1991) Primate origins and the evolution of angiosperms. Am J Primatol 23:209–223

Susman RL (1998) Hand function and tool behaviour in early hominids. J Hum Evol 35:23–46

Sussman RW, Rasmussen DT, Raven P (2013) Rethinking primate origins again. Am J Primatol 75:95–106

Torigoe T (1985) Comparison of object manipulation among 74 species of non-human primates. Primates 26:182–194

Tuttle RH (1969) Quantitative and functional studies on the hands of Anthropoidea. I. The Hominoidea. J Morphol 128:309–364

Tuttle RH (1972) Functional and evolutionary biology of the hylobatid hands and feet. In: Rumbaugh DM (ed) Gibbons and siamang: a series of volumes on the lesser apes, vol 1: evolution, ecology, behaviour, and captive maintenance. Kargar, Basel, pp 136–206

Tuttle RH (ed) (1986) Apes of the world: their social behaviour, communication, mentality and ecology. William Andrew Publishing, New York

Tuttle RH, Watts DP (eds) (1985) Primate morphphysiology, locomotor analysis and human bipedalism. Tokyo University Press, Tokyo

van Schaik CP, Fox EA, Sitompul AF (1996) Manufacture and use of tools in wild Sumatran orangutans: implications for human evolution. Naturwissenschaften (Hist Arch) 83:186–188

Visalberghi E, Fragaszy DM, Savage-Rumbaugh S (1995) Performance in a tool-using task by common chimpanzees (*Pan troglodytes*), bonobos (*Pan paniscus*), an orangutan (*Pongo pygmaeus*), and capuchin monkeys (*Cebus apella*). J Comp Psychol 109:52–60

Whitington C, Treesucon U (1991) Selection and treatment of food plants by white-handed gibbons (*Hylobates lar*) in Khao Yai National Park, Thailand. Nat Hist Bull SIAM Soc 39:111–122

Wright RVS (1972) Imitative learning of a flaked-tool technology—The case of an orang-utan. Mankind 8:296–306

Yerkes RM, Yerkes AW (eds) (1929) The great apes. Yale University Press, New Haven

Chapter 13
The Evolution of Technical Intelligence: Perspectives from the Hylobatidae

Clare L. Cunningham, James R. Anderson and Alan R. Mootnick

Subadult male agile gibbon's first trial of a rake-in task, Twycross Zoo. Photo credit: Clare Cunningham

C.L. Cunningham (✉)
Department of Psychology, Abertay University, Dundee, UK
e-mail: c.cunningham@abertay.ac.uk

J.R. Anderson
Department of Psychology, University of Stirling, Stirling, UK

A.R. Mootnick
Gibbon Conservation Centre, Saugus, CA, USA

© Springer Science+Business Media New York 2016
U.H. Reichard et al. (eds.), *Evolution of Gibbons and Siamang*,
Developments in Primatology: Progress and Prospects,
DOI 10.1007/978-1-4939-5614-2_13

Introduction

During the early nineteenth century, natural historians began to take an interest in small Asian apes known as gibbons. They became the self-appointed guardians of captive specimens, often taking them into their homes, recording the behavioural repertoires of their charges. Accounts of the intellectual capacities of these small apes varied from those describing the gibbon as being 'most intelligent and very often human-like' (Forbes 1894 cited in Yerkes and Yerkes 1929) and as 'probably the most intelligent of all the apes' (Garner 1900 cited in Yerkes and Yerkes 1929), to those suggesting the siamang (*Symphalangus syndactylus*) 'exhibits an absence of all intellectual qualities' (Geoffroy-Saint-Hilaire and Cuvier 1824).

Despite over 300 years of scientific interest in these hominoid primates (Yerkes and Yerkes 1929), our knowledge of gibbon and siamang ecology and behaviour still lags behind that of other ape genera. In particular, there is a paucity of research investigating the mental capacities of this family, a surprising deficit given their unique phylogenetic position representing evolution's first diversion on the pathway from monkey to ape, and the available diversity of species; they are the only ape family with more than one extant genus. The great apes are considered cognitively advanced, showing some continuity with human intellectual capacities; however, many researchers postulate a significant divide between the mental abilities of apes and monkeys (Povinelli et al. 1992a, b; Byrne 1995; Visalberghi et al. 1995). Gibbons, representing an intermediary divergence between monkeys and great apes, are therefore of interest to the study of cognitive development through evolutionary time.

The aim of this chapter is twofold: to present a brief review of what is currently known about gibbons' understanding of their physical world, and to present data from research investigating one particular aspect of this understanding, the ability to use tools to achieve particular goals. Tool-use is considered a special class of object manipulation indicative of higher-level comprehension of the properties of objects and their relationships to one another. Successful use of tools is also thought to imply knowledge of theoretical notions such as force or gravity that are not directly perceivable (Povinelli 2000), and from an evolutionary perspective, has adaptive significance in that it may allow animals to exploit otherwise inaccessible resources. Many species of primate are capable of using tools in captivity (Tomasello and Call 1997; Anderson 2006; Schmaker et al. 2011); however, in the wild, regular tool-use is restricted to a small number of genera (*Pan troglodytes*: McGrew 1992; *Pongo* sp.: van Schaik et al. 1996, 2003; *Macaca fasicularis*: Gumert et al. 2009, 2011; *Cebus* sp.: Spagnoletti et al. 2011, 2012). Why some species seem cognitively capable of impressive tool-using abilities under the appropriate conditions in captivity, yet do not incorporate tools into their normal behavioural repertoire in the wild has been subject to much speculation. The explanation for the diversity in primate tool-use may come from an unlikely source. Recently, Emery (2013) demonstrated significant tool-using skills in a non-tool-using, large-brained bird (*Corvus frugilegus*), comparing their skills to a habitual tool-user from the same family (*C. moneduloides*) and concluding that understanding properties of objects and how they could be useful in goal attainment must have emerged at least in the common ancestor of the Corvidae. This suggests that tool-using may in part result from a 'domain-general intelligence'; that is from cognitive processes that support general learning and are not specifically adapted to understanding objects as tools. This contrasts with modular accounts of brain evolution (Fodor 1983; Tooby and Cosmides 1992) that suggest tool-use, and indeed all cognitive processes, are supported by encapsulated mental pathways, shaped by evolutionary selection, to solve specific problems. Thus, a general learning mechanism can produce significant cognitive understanding of objects in the absence of environmental pressure and selection for specialised tool-using cognitive modules.

Gibbons and siamang are not habitual tool-users, and reports of their use of objects to achieve goals in captivity are sparse (see review below). They are, however, highly encephalised, having a relative brain size (brain size when body size is controlled for) that is comparable to the great apes and larger than most monkey species (see below for further detail). This suggests cognitive processing power that should endow them with sufficient 'intellect' to achieve a reasonable level of understanding of object properties and relations, even in the absence of specialised modules resulting from evolved adaptations for tool-use.

Gibbon Brains in Comparative Perspective

It has long been known that the primate order is characterised by significantly larger brains than would be predicted for a non-primate mammal of equal body size (Jerison 1973). In absolute terms, gibbon (*Hylobates lar*) brain volume is 83 cc (SD $\pm$ 11.3 cc, $N = 4$) (Rilling and Seligman 2002); thus they have less absolute cortical tissue than the great apes (e.g., orangutan (*Pongo pygmaeus*): 406.9 cc (SD $\pm$ 57.5 cc, $N = 4$); gorilla (*Gorilla gorilla*): 397.3 cc (SD $\pm$ 94.2 cc, $N = 2$); chimpanzee *(P. troglodytes)*: 337.3 cc (SD $\pm$ 38.7 cc, $N = 6$); human (*Homo sapiens*): 1299 cc (SD $\pm$ 127.4 cc, $N = 6$)), and some monkey species (baboon (*Papio cynocephalus*): 143.3 cc (SD $\pm$ 38.7 cc, $N = 2$); mangabey (*Cercocebus atys*): 98.8 cc (SD $\pm$ 3.3 cc, $N = 4$)) (Table 13.1). However, as absolute brain volume is influenced by body size, it is the amount of cortical expansion in excess of that predicted by body size that is assumed critical for advanced cognitive performance. In terms of relative brain size (derived from quantifying species deviations from the least squares regression line of best fit for total brain volume against body weight), hylobatids are more encephalised than expected, as is the case for all apes; however, capuchin monkeys' (*Cebus apella*) relative brain volume exceeds that of the Hylobatidae (Rilling and Insel 1999). Gibbons and siamang are

Table 13.1 Measurements of brain part size in gibbons (*Hylobates* sp.) compared to humans (*Homo sapiens*), chimpanzees (*Pan troglodytes*), capuchins (*Cebus* sp.), baboons (*Papio* sp.) and macaques (*Macaca* sp.)

Measure	Gibbon	Chimpanzee	Human	Capuchin	Baboon	Macaque
Total brain vol. (cc)	83 (11.3)[a]	337.3 (38.7)[a]	1299 (127.4)[a]	66.5 (10.5)[a]	143.3 (38.7)[a]	79.1 (6.8)[a]
Neocortex ratio	1.16[b]	1.03[b]	3.6[b]	1.28[b]	0.8[b]	0.71[b]
Gyrification index	1.9[b]	2.19[b]	2.57[b]	1.6[b]	2.03[b]	1.73[b]
Frontal lobes (percentage of total brain volume)	29.4 (1.98)[c]	35.4 (1.9)[c]	37.7 (0.9)[c]	29.6–31.5[c]	[d]	30.6 (1.5)[c]

Numbers in brackets represent ±1SD
[a]Rilling and Seligman (2002); [b]Rilling and Insel (1999); [c]Semendeferi et al. (2002); [d]No data available at time of writing

therefore highly encephalised primates, but they do not exceed all non-apes in potential processing power.[1]

The increases in brain size that exemplify the Primate order have not been uniform. The neocortex is the outer layer of the mammalian cerebral hemispheres and is proposed to be the seat of higher cognitive functioning; the thinking part of the brain (Dunbar 1998). It is this region that has undergone the most expansion through primate evolution (Finlay and Darlington 1995). Rilling and Insel (1999) measured the neocortical volume of multiple representatives of 11 species of anthropoid primates using magnetic resonance imaging (MRI). This study improved on previous postmortem measurements of brain parts (Stephan et al. 1981; Zilles and Rehkämper 1989) by using living, anaesthetised primates, alleviating problems of shrinkage during fixation and possible brain atrophy due to illness or old age. From this dataset, the proportion of the brain that is given over to neocortex (neocortex ratio, NR), for *H. lar* was 1.16, grouping with the values for the non-human great apes (orangutan: 1.14; gorilla: 1.0; chimpanzee: 1.03). The capuchin monkeys were again shown to be highly encephalised with an NR of 1.28, exceeding the gibbons and non-human great apes in relative neocortical volume (Table 13.1). All other monkey species included in this study had smaller neocortices than gibbons or other apes.

As brain size increases across primate species, the neocortex does not show substantial thickening but an expansion of surface area, resulting in increased folding of the cortical tissue. Generally, larger anthropoid brains are more convoluted than smaller ones (Zilles and Rehkämper 1989; Rilling and Insel 1999) with the level of cortical folding described using a gyrification index (GI) (Zilles and Rehkämper 1989) that determines how much neocortex is buried within the cerebral folds (higher values indicate a greater level of folding). Using this measure, gibbon

---

[1]It should be noted that data on neural architecture of the hylobatid brain are scarce, with most information being derived from a limited number of specimens from a single species, *H. lar*. Applicability to other hylobatid species is presently unknown.

brains are less convoluted than those of the great apes (gibbon GI = 1.9, orangutan GI = 2.29, gorilla GI = 2.07, chimpanzee GI = 2.19, human GI = 2.57) (Rilling and Insel 1999). Their level of gyrification is closer to that of the larger brained monkeys (baboon GI = 2.03, mangabey GI = 1.84, macaque (*Macaca mulatta*) GI = 1.73), although it is noteworthy that the large-brained capuchins have significantly less folding of the neocortex than would be predicted for a primate of their brain size (Rilling and Insel 1999) (Table 13.1). The relevance of cortical folding to cognitive performance, beyond allowing more neural tissue to fit within the confines of the skull is unclear (Jerison 1982; van Essen 1997). One proposal is that it brings areas of the cortex into closer spatial proximity, minimising the length of neural connections between communicating sectors (van Essen 1997; Rilling and Insel 1999); the high metabolic costs of neural tissue may necessitate such conservatism in large-brained primates.

The convoluted neocortex is divided into distinct lobes that are, broadly speaking, functionally specialised. The frontal lobe is involved in creative thinking, planning of future actions, decision making and some aspects of working memory, language and motor control (Semendeferi et al. 1997), and it is this area that is usually considered to be disproportionately enlarged in humans (Semendeferi et al. 2002). Magnetic resonance scans of a range of primate species, including representatives from all great ape genera, one gibbon (*H. lar*) and two monkey species (capuchin (*Cebus* sp.) and macaque (*M. mulatta*)), revealed that the gibbon frontal lobe constitutes 29.4 % (SD ± 1.8 %, N = 4) of total cerebral hemisphere volume. This value groups closely with proportions reported for monkey specimens (capuchin 29.6 and 31.5 % (N = 2); macaque 30.6 % (SD ± 1.5 %, N = 3), rather than those for great apes which are substantially larger (orangutan 37.6 % (SD ± 1.1 %, N = 4); gorilla 35.0 and 36.9 % (N = 2); chimpanzee 35.4 % (SD ± 1.9 %, N = 6); bonobo 34.7 % (SD ± 0.6 %, N = 3); human 37.7 % (SD ± 0.9 %, N = 10)) (Semendeferi et al. 2002). It therefore appears that in some respects the hylobatid brain is clearly ape-like, while in others it more closely resembles that of a monkey. Nothing in the neuroanatomical evidence to date indicates that hylobatids should not be capable of the cognitive operations required to use simple tools; in terms of available cortical processing power, it seems reasonable to suggest that gibbons should be able to learn to use objects to achieve goals, at least to a level mediated by a general intelligence mechanism. However, available data on brain volume and brain part volume is limited for hylobatids. Sample sizes in neuroanatomical studies are often one or two individuals from the genus *Hylobates* (Stephan et al. 1981; Semendeferi et al. 1997, 2002). What is needed to facilitate a more comprehensive understanding of brain evolution in higher primates generally, and in the hylobatid genera specifically, are larger data sets including representatives from all gibbon and siamang species.

Object Manipulation and Tool-Use in the Hylobatidae

In order to successfully navigate the physical environment, animals must have some knowledge of the objects that occupy it. In many species, investigation of objects involves looking at or smelling and tasting items. Manual exploration is less common, although primates in general have evolved hands that support flexible manipulation skills (Passingham 1981), potentially facilitating a greater under-standing of object properties. Hylobatids have been included in a number of comparative studies addressing the reactions of captive primates to presented objects. Bernstein et al. (1963) tested the responses of 11 gibbons (*H. lar, N* = 8; *Hylobates pileatus, N* = 3) and 11 rhesus macaques (*M. mulatta*), ranging from juveniles to adults, to the introduction of novel objects into a cage adjacent to their own. The enclosures were separated by wire mesh that the subjects could reach through to manipulate the test objects. Only a descriptive account of the results was provided, with gibbons reported as quicker to approach and explore objects, leading the authors to postulate that they were more active and curious in unfamiliar sit-uations than monkeys; the latter were often immobile and showed submissive or agonistic displays.

Glickman and Sroges (1966) presented five inanimate objects to 200 zoo-living animals from several different orders, including primates. Reactivity to the objects was measured by latency to approach and the number and types of contact made. Gibbons tested in this study (*H. lar, N* = 2) were less responsive to objects than their phylogenetic position would predict, showing relative indifference to the items and few contacts. However, contrasting data were obtained by Parker (1973) during a comparative study that looked at complexity of manipulations performed as well as overall responsiveness to an aluminium bar with a steel rod through one end that was tethered to the enclosure wall. Parker reported that gibbons (*H. lar*) scored the highest number of responses and spent the most time in contact with the object compared to monkey and prosimian species tested. This was supported by a further study that looked at object manipulation in nine hylobatids (*H. lar, N* = 2; *Hylobates agilis, N* = 2; *Hylobates moloch, N* = 1; *Hylobates klossii, N* = 2; *S. syndactylus, N* = 2) in comparison to 73 other species of primates (Torigoe 1985). The procedure was similar to that in Parker (1973), and involved presentation of a three-stranded nylon rope with knots at each end and a wooden cube. Results showed the great apes, hylobatids and capuchins had the most varied repertoire of manipulations, followed by the Old World monkeys (except leaf eaters). Lemurs, marmosets, spider monkeys and the leaf eaters showed the least diversity of object manipulations (a full list of species is available in Torigoe 1985). Of particular interest is that although gibbons and siamang did engage in several types of ma-nipulation, these were all described as *primary actions* in which the object was moved with no relation to another object (except the global substrate). The gibbons and siamang were never observed to exhibit *secondary actions* whereby the original stimulus object was manipulated in conjunction with another object. For example, an orangutan was observed to wrap the nylon rope around part of the wire mesh, twisting the free ends to make another rope. The chimpanzees floated the wooden cube in their drinking water, pushing it around with their fingers, and a macaque fed the rope in and out of the mesh in a sewing-like action. This type of secondary manipulation may be a precursor to the development of more complex types of object interactions such as tool-use, and so its absence in hylobatids is noteworthy.

Evidence for object-related problem solving is rare in hylobatids. Drescher and Trendelenburg (1927) report observations of a small ape (species not stated) that was confronted with a box containing food, securely fastened with a bolt mechanism. In order to gain entry, the ape had to learn how to slide open the bolt. The small ape was given a number of trials each day, and successfully obtained the reward on day 3. The authors describe the attitude of the subject as extremely alert and interested in the food, but shy and easily distracted. Yerkes and Yerkes (1929) reviewed the work of Drescher and Trendelenburg (1927) and state that the small ape's performance indicates clear inferiority compared to great apes. During studies on captive siamangs (*S. syndactylus*), Fox (1972) observed an instance of object-related problem solving when a young male loosened a tangled rope that was caught around the bars of the enclosure in a series of apparently deliberate actions that freed the entwined end.

The only empirical study to assess object-related problem solving in hylobatids was conducted by Beck (1967), who used patterned string problems similar to those given to chimpanzees by Köhler (1925). Four gibbons (*H. lar*, $N = 3$, *H. pileatus*, $N = 1$) were presented with a series of single string and food configurations designed to assess the apes' understanding of spatial arrangements between component objects. Simple configurations involved a piece of string with a food reward (banana) tied to the far end, out of the subject's direct reach. To obtain the food, the ape had to use the free end of the string, closest to the enclosure, to pull the reward within reach. More complex problems included a 'sham' condition where the string was in place but not attached to the food, and a 'distracter' condition where the string was tied to the food as in the first simple problem, but a second piece of food was closer to the subject, still beyond reach with no string attached. Beck reported that all subjects solved these problems with relative ease, generally adopting the most efficient method of securing the reward. He compared the performance of the gibbons favourably with that of Köhler's chimpanzees, and suggested these small apes matched, if not exceeded great apes in this task.

Few reports exist of hylobatids displaying tool-use behaviours. In the wild, reports are limited to instances of throwing branches at intruders (Beck 1980), while in captivity, Drescher and Trendelenberg (1927) reported that a gibbon used a rake to draw in an out-of-reach food item, and Rumbaugh (1970) observed a captive hylobatid use a cloth to soak up drinking water and make a swing from a piece of rope. One short report discusses hylobatids' understanding of the causal relationships between objects and an environmental feature, using a trap-table paradigm designed by Povinelli and Reaux (2000). A juvenile gibbon (*H. lar*) was given the choice of pulling in one of two rakes placed on a table, one that would lead to retrieval of a reward and one that would not. The ape obtained the incentive on all trials without training, indicating an understanding of the spatial relationships between the food and tool (Inoue and Inoue 2002). A trap was then introduced along the surface of one table. A reward was placed in front of both rakes that could be used to pull the food towards the subject. Selecting the rake on the trap side would result in the food being lost into a hole, while the continuous surface presented an attainable reward. The young gibbon made the correct choice on 26/32 trials. The results of this experiment were assumed to indicate the gibbon was capable of understanding the causal relationships between three factors; the goal object, a tool and an environmental obstruction to goal attainment. Cunningham et al. (2006) presented a similar task to four hoolock gibbons (*Hoolock hoolock*) and found that two individuals spontaneously avoided the trap from first presentation, supporting the suggestion of Inoue and Inoue (2002) that this task was within the cognitive capacity of gibbons.

Similar tasks have been presented to chimpanzees by Povinelli and Reaux (2000) who argued that although the great ape subjects performed well, other explanations that do not imply an understanding of causal relationships could explain their performance. These explanations could also be relevant to results from the small ape studies. For example, subjects could have learned to choose the correct rake in the trap/no-trap condition simply by making an association between the continuous surface and goal attainment, without any consideration of the effects of the trap. Or they could have formed an associative rule such as 'avoid the rake with an obstacle in its path', without necessarily having any concept of the properties of the trap.

Overall, most researchers report hylobatids show a willingness to approach and manipulate objects, with sporadic accounts of reluctance to engage. Their level of manipulatory ability also seems to befit their phyletic status, with most studies indicating a level of complexity in their interactions with objects intermediate between the great apes and monkeys. However, secondary actions that may be a prerequisite for the development of tool-use have not often been observed in hylobatids. Beyond these comparative studies of object manipulation, the data are particularly scant regarding object-mediated problem solving and tool-use in the Hylobatidae. Beck (1967) showed that gibbons and siamang are capable of solving patterned string problems to a level comparable to chimpanzees, and two studies (Inoue and Inoue 2002; Cunningham et al. 2006) reported success with a trap-table paradigm, indicating possible understanding of relationships between the object to be manipulated, the goal object and an environmental obstruction in these small apes.

Development and Understanding of Simple Tool-Use in Small Apes

The remainder of this chapter will focus on a series of experiments that assessed the development and understanding of objects as tools in the Hylobatidae in a more systematic way. Before we can evaluate whether gibbons and siamang can become proficient at tool-using given the relevant opportunity, we must first establish more clearly their basic propensity for tool-use. At this point, it is pertinent to define exactly what is meant by 'tool-use'. The most widely accepted definition for many years was proposed by Beck (1980). The relevant points are (1) that the tool must be an unattached object used to change the form, position or condition of another object, another organism or the user itself, (2) it must be held or carried by the user during or immediately prior to use and (3) the user must be responsible for creating the appropriate orientation of the tool for use. Although recently revised by Schmaker et al. (2011) to more clearly distinguish tool-use as a special category of behaviour (see Hunt et al. 2013 for a more detailed discussion of the changes), the main points of definition have remained constant for over 30 years.

The research described here assessed whether hylobatids could spontaneously learn to manipulate a rake-shaped object to gain a food reward. The type of manipulation required in this task is classified as *zero-order manipulation* (action on one object results in an action on a second object by default) as the subjects were not responsible for producing a relationship between the two objects involved, but simply made use of a pre-existing relationship set up by the experimenter (Fragaszy et al. 2004). Although this does not qualify as tool-use according to Beck's strict definition, for ease of expression the object used to retrieve food will be referred to as a tool.

Using a rake, placed with the handle oriented towards the subject, to draw in a food item situated directly in line (thus requiring no adjustment of the tool as in zero-order manipulation), is within the capacities of a number of primate species, including some that do not habitually use tools (*Saguinus oedipus*: Santos et al. 2005a; *C. paella*: Cummins-Sebree and Fragaszy 2005; *Cercopithecus aethiops*: Santos et al. 2005a; *Prosimian sp.*: Santos et al. 2005b; *P. troglodytes*: Povinelli and Reaux 2000). This research recorded the development of skills and understanding needed to reach solution in a similar zero-order raking-in task in a sample of hylobatids with representatives from all four extant genera, some that were given prior exposure to the objects before testing and others that were exposed to the potential tools for the first time on test.

Experimental Design

Twenty-nine hylobatids served as subjects (Table 13.2); 22 were housed at the Gibbon Conservation Center (GCC), a conservation and behavioural research establishment in California (USA), and the remainder ($N = 7$) at Twycross Zoological Park (TZ), West Midlands (UK). At GCC, hylobatids were housed in outdoor enclosures 10×3 4 m, with an adjacent smaller area $4 \times 3 \times 2.5$ m that was generally available at all times but could be closed off to separate individuals as required. All cages were a minimum of 5.5 m apart and visual barriers in the form of solid tarpaulin sheets and planted vegetation obstructed direct views between adjacent enclosures (see Mootnick 1997b for more details of enclosure design). The feeding regime at GCC varied with season and was tailored to individuals. Generally, the gibbons and siamang were fed four times a day, beginning with a breakfast of fruits and primate biscuits, a main feed of fruits and vegetables and two further feeds of apples, bananas and greens (Mootnick 1997a). A proportion of this food was handed to individuals; the design of enclosures allowed the apes to extend their arms through the fencing to accept food from caregivers. Water was available ad libitum.

Hylobatids at TZ were housed in similar-sized enclosures except the smaller area was an indoor space that was available at all times (except during cleaning). Enclosures were made of wire mesh, as at GCC; however, they were organised in two adjacent blocks of six placed back to back, with indoor areas facing inwards onto a walk-through corridor for zoo visitors. This resulted in each cage sharing at least one adjoining fence with another. This may have allowed gibbons (or siamang) to view others taking part in the tasks. To avoid this as far as possible, only apes in every other enclosure were selected as subjects, making visible access negligible in all but two small apes that were housed together but tested separately. No order effects were evident for this pair. The cages were furnished with branches, ropes and several free objects including bags, buckets, and infant or pet toys. Hylobatids were given one main feed of fruit, vegetables and primate biscuits at the end of the day. This was presented in the indoor space to encourage the small apes inside at night. Another smaller feed was provided in the morning when the gibbons and siamang were let out into their outdoor area, including bread, eggs, celery and primate biscuits, a proportion of which was handed to individuals through the wire mesh and the remainder scattered onto the enclosure floor. Water was available ad libitum.

The task involved pulling in a rake-like tool to obtain a food reward. A wooden table (110 × 27 × 12 cm) was placed outside the main enclosure adjacent to the feeding platform. The table had a 2 cm lip along three edges (not the edge aligned with the cage) to prevent the rake and food item sliding off. The rake consisted of an aluminium rectangle (wooden for tests at TZ) (25 × 12 cm) fixed to one end of a 115 cm handle with the free end protruding through the chain link fence approximately 5 cm into the enclosure (Fig. 13.1). This elevated the end of the rake to facilitate grasping by the hylobatids' elongated hands (Beck 1967, for a similar arrangement with a string pulling task).

Subjects at GCC were separated into two groups, no prior exposure (NE) and prior exposure (PE) (Table 13.2). The PE group was exposed to the apparatus (table and rake) in situ for seven consecutive days prior to testing. The table and rake were placed as they would be during the test situation, although no reward was used. The tool could be manipulated in its location; however, the plate attached to the end prevented it being pulled completely into the enclosure. Each morning, the rake was reset onto the table in its original starting position if necessary. No other interaction with the apparatus by the experimenter occurred. The NE group was exposed to the apparatus for the first time on presentation of the first test trial. Each test trial commenced when a raisin (a food item not usually included in the normal diet but highly palatable) was placed, in view of the subject when they were attending to the experimenter, on the end of the table a few centimetres beyond direct reach and in front of the end plate of the rake. The only way to obtain the reward was by pulling the rake in toward the cage, thus moving the raisin along the table until it was within reach (Cunningham et al. 2006, 2011 for further details of apparatus and method).

No training was given. Each subject was given a maximum of 30 min from placement of the food item to gain the reward in each trial. Additional single raisins were added at 0, 10 and 20-min intervals to encourage the small apes to return to the task if they had moved out of the target area (designated as a 1.5 m^2 area surrounding the apparatus). If they did not obtain the reward during the first 30-min session, the apparatus were removed and the next trial commenced the following day. If the subject did not reach a solution in any of the first 10 trials, testing was discontinued with that individual. If a reward was obtained within these 10 trials, the subject was presented with a further 9 trials, each time for a maximum of 30 min. Thus, successful individuals were those that used the tool to retrieve the food once within 10 trials, and then proceeded to complete a further 9 trials (10 trials in total). All trials were videotaped to allow later coding of behaviours. Time taken to first solution, taken as time within the target area surrounding the apparatus and attending to the task, was measured from the recorded footage and compared across genera and PE and NE groups. Subjects that went on to successfully obtain the reward on 9 consecutive trials after first solution were considered to fully understand the task demands and were classified as 'successful', while those that only completed some trials within the 10 presented may have understood the task demands but were less motivated to participate; these apes were therefore referred to as 'partially successful'.

Results

Of the 29 gibbons tested, nearly ¾ (21 individuals) showed clear signs of understanding the task. Seventeen individuals (58.6 %) obtained the reward on 10 consecutive trials (successful), a further four (13.8 %) used the tool to retrieve the reward on some trials but did not perform consistently (partially successful), and the remaining eight (27.6 %) did not reach solution on any trial. One female gibbon (*H. moloch*) was successful on every trial but was excluded from the analyses due to the method adopted; she repeatedly jumped on the rake handle causing vibrations through the table that moved the food within reach. Motivation, as measured by percentage of time in target area and attending to the task, differed between individuals that were successful in 10 consecutive trials compared to those that did not maintain successful performance and those that were unsuccessful (one-way GLM (log transformed): $F_{(2, 25)} = 72.22$, $p < 0.001$) (Fig. 13.2). Post hoc tests revealed that the significant differences were between the successful small apes and the two other groups (Bonferroni pairwise comparisons: $p < 0.001$); however, there was no difference in percentage time in target area between the partially successful and unsuccessful subjects (Bonferroni pairwise comparisons: $p = 1.00$) (Fig. 13.2).

Table 13.2 Subject details for hylobatids used in the zero-order (experiment 1) and true tool-use (experiment 2) tasks

Subject	Genus	Species	Sex	Age (yers)	Group	Housing	Institution	Experiment
Maung	*Hoolock*	*hoolock*	M	4	NE	Solitary	GCC	1, 2
Chester	*Hoolock*	*hoolock*	M	5	NE	M/F pair	GCC	1,2
Betty	*Hoolock*	*hoolock*	F	5	PE	M/F pair	GCC	1
Arthur	*Hoolock*	*hoolock*	M	9	PE	M/F pair	GCC	1
Sasha	*Nomascus*	*leucogenys*	M	27	PE	Solitary	GCC	1
Ricky	*Nomascus*	*leucogenys*	F	15	NE	Family group	GCC	1, 2
Vok	*Nomascus*	*leucogenys*	M	17	NE	Family group	GCC	1, 2
Clara	*Nomascus*	*leucogenys*	F	29	–	M/F pair	TZ	1
Fred	*Nomascus*	*leucogenys*	M	29	–	M/F pair	TZ	1
Kino	*Symphalangus*	*syndactylus*	M	20	NE	Solitary	GCC	1
Dudlee	*Symphalangus*	*syndactylus*	F	9	PE	F/F sib pair	GCC	1
Kimbo	*Symphalangus*	*syndactylus*	F	5	PE	F/F sib pair	GCC	1
Chloe	*Hylobates*	*moloch*	F	13	NE	Family group	GCC	1, 2
Ivan	*Hylobates*	*moloch*	M	30	PE	Solitary	GCC	1
Chillibi	*Hylobates*	*moloch*	M	16	PE	Solitary	GCC	1
Khusus[a]	*Hylobates*	*moloch*	F	9	NE	Family group	GCC	1
Tuk	*Hylobates*	*pileatus*	F	12	PE	Solitary	GCC	1, 2
Valentina	*Hylobates*	*pileatus*	F	7	PE	Family group	GCC	1
Birute	*Hylobates*	*pileatus*	M	22	NE	Family group	GCC	1
JR	*Hylobates*	*pileatus*	F	15	NE	Family group	GCC	1
Kanako	*Hylobates*	*pileatus*	F	4	NE	Family group	GCC	1
Jason	*Hylobates*	*pileatus*	M	33	–	Family group	TZ	1
Jay	*Hylobates*	*pileatus*	M	2	–	M/M sib pair	TZ	1
Ruby	*Hylobates*	*agilis*	F	18	NE	Mother/infant	GCC	1

(continued)

Table 13.2 (continued)

Subject	Genus	Species	Sex	Age (yers)	Group	Housing	Institution	Experiment
Bebop	Hylobates	agilis	M	15	PE	Father/daughter	GCC	1
Lilleth	Hylobates	agilis	F	4	PE	Father/daughter	GCC	1
Sirikit	Hylobates	agilis	F	11	–	Family group	TZ	1
Charlie	Hylobates	agilis	M	25	–	Family group	TZ	1
Chloe	Hylobates	agilis	F	4	–	Family group	TZ	1

Only those housed at GCC were given prior exposure to the tools before testing and therefore the NE group for comparison of the effects of experience only included gibbons housed at the same institution

M male; F female; NE No prior exposure; PE Prior exposure; GCC Gibbon Conservation Center, California; TZ Twycross Zoological Park, West Midlands, UK; [a]Excluded from analyses (see text for explanation)

Only small apes that were successful on 10 consecutive trials ($N = 16$) were considered in this analysis. Time to first solution varied significantly between genera (one-way GLM (log transformed): $F_{(3, 16)} = 3.68$, $p = 0.047$), with *Hoolock* obtaining the reward in the shortest time (mean 57.25 s, SE = 13.71 s, $N = 4$) (Fig. 13.3). The greatest difference was between the mean time to first solution in *Hoolock* individuals, who were the most proficient genus, and the much slower *Hylobates* individuals (mean 233.40 s, SE = 75.17 s, $N = 5$), although post hoc tests were marginally nonsignificant (Bonferroni pairwise comparisons: $p = 0.06$). The differences between other genera did not approach significance.

Prior exposure to the rake did not influence the likelihood of solving the task in the GCC hylobatids: 7/11 and 6/10 hylobatids obtained the reward in NE and PE groups, respectively. However, there were significant differences in how males and females responded to prior exposure to the apparatus when genera were pooled; sample sizes were not sufficient to allow for sex difference evaluation between genera (Fig. 13.4). PE females required less time to solution than their NE counterparts (NE mean = 194.00 s (SE = 26.00); PE mean = 67.50 s (SE = 18.50); $F_{(1,5)} = 7.298$, $p = 0.043$). No significant difference was found for males in time to solution (NE mean = 66.25 s (SE = 27.762); PE mean = 67.50 s (SE = 18.50), $F_{(1,4)} = 0.125$, $p = 0.74$).

Fig. 13.1 Pileated gibbon (*H. pileatus*) pulling in reward in a zero-order manipulation task where food was placed in direct alignment with rake that the subject then draws in to bring the goal object within reach. Reproduced from Cunningham et al. 2011, Fig. 13.1. Copyright © 2011 by Springer

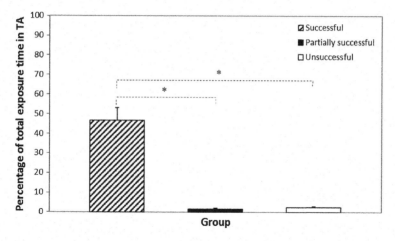

Fig. 13.2 Mean percentage of total exposure time engaged with task for three groups of hyloabtids in the zero-order manipulation task; successful (obtaining the reward on 10 consecutive trials), partially successful (obtaining the reward within the first 10 trials but failing to complete 10 consecutive trials) and unsuccessful (did not reach solution within the first 10 trials). *significant at $p < 0.01$ level

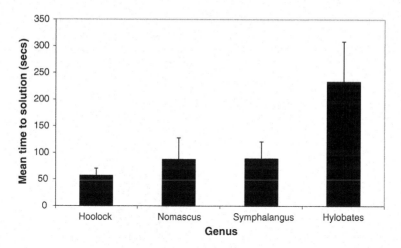

Fig. 13.3 Mean time to first solution (seconds) in zero-order raking-in task for successful gibbons by genera. *Error bars* represent +1SE

These results demonstrate that in general, hylobatids are able to manipulate a pulling tool in order to gain a reward, without explicit training. However, the effects of prior object-relevant experience varied between sexes. For males, previous exposure did not increase the likelihood of successful object use. In contrast, females exhibited a significant decrease in time to solution with previous experience. While the small sample size makes conclusions tentative, it may be that females exhibit caution in their initial responses to a novel object that can be offset by prior exposure to that object, enabling them to learn first-hand of its neutrality and affordances (see Cunningham et al. 2011 for further discussion of this result).

True Tool Use in Gibbons

Zero-order object manipulation to obtain a food reward is within the cognitive capacities of many species, and given their level of cortical development and phylogenetic position, it is not surprising that hylobatids were also capable of this level of object-mediated problem solving. Further research modified the raking-in task to necessitate a lateral movement of the rake head towards the incentive by the subject, thus elevating the manipulation to true tool-use. Six hylobatids housed at GCC that were previously successful at using the rake to pull in an out of reach food item took part in this experiment (Table 13.2). The task required the small apes to use the same T-shaped rake to pull in an out-of-reach food item placed in one of five positions: in direct alignment with the rake head as in the previous zero-order manipulation, offset by approximately 5 cm to the left or right of the rake head (offset), or positioned 30 cm to the left or right of the handle's midpoint (far). Subjects completed 50 trials with the reward appearing in each position for ten presentations randomly assigned across each block of ten trials.

All subjects were successful on all zero-order trials, as expected given their previous success with this tool-reward configuration. However, performance was

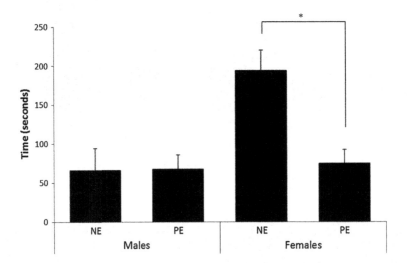

Fig. 13.4 Mean (+1SE) time to solution in those gibbons that reached solution in the zero-order task by gender (* = $p < 0.05$). Adopted from Cunningham et al. 2011, Fig. 13.2. Copyright © 2011 by Springer

generally poor when a reorientation of the tool was necessary for reward attainment, with a significant drop in successful outcomes in the two offset conditions (χ^2 (6) = 2322.11, $p < 0.001$). In the offset position, one hoolock gibbon (Maung—*H. hoolock*) obtained 55 % of the rewards when a slight movement to the left or right of centre was required. The female pileated gibbon (Tuk—*H. pileatus*) pulled in 45 % of the offset rewards, with the remaining subjects obtaining between 20 and 30 % of rewards in this position. All hylobatids found the 'far' condition difficult. Only 15 % of rewards presented in this position were retrieved, two by the same hoolock gibbon with the best success on the offset condition, and one by the female moloch gibbon (Chloe—*H. moloch*).

These results strongly suggest that gibbons do not immediately comprehend the relationships between the tool and goal object sufficiently well to allow systematic success when the necessary relationship for goal attainment was not directly perceivable. The successes that did occur appeared to be the result of chance manipulation, resulting from a general exploration of the objects (rather than directed, goal-led manipulation) that inadvertently produced the correct orientation for retrieval. This process might have eventually facilitated learning of the required action had more trials been presented. Ishibashi et al. (2000) reported that extensive exposure was required for Japanese macaques (*M. fuscata*) to learn to use a rake to pull in a reward located in an offset position. Learning occurred after several fortuitous manipulations of the tool that led to unintentional food retrieval in over 1100 trials. It would therefore seem that, unlike reports of spontaneous use of sticks to pull in out-of-reach items by great apes (Schmaker et al. 2011), the hylobatids are more monkey-like in their capacities for tool manipulation. However, this research presented relatively few trials; with more exposure to the objects, it is possible that they could have reached levels close to those demonstrated by great apes.

Concluding Remarks

The data from this series of experiments suggest a divide between the types of goal-directed object manipulations that are readily acquired by small apes and those that are not. For successful tool-mediated problem solving to develop within the exposure time in these studies, the necessary relationships between the object to be manipulated and the goal object had to be directly perceptible. Despite their level of cortical development and phylogenetic position, gibbons do not seem to be able to spontaneously use a rake tool that must be repositioned to become functional as in a true tool-use task. They were however, capable of learning to use an object to secure a reward in a zero-order manipulation without extensive exposure. This could indicate a cognitive limitation in the Hylobatidae compared to great apes that spontaneously reorient a tool to make it functional (Köhler 1925; Jordan 1982; Lethmate 1982; Fontaine et al. 1995), aligning hylobatids more with monkeys who have difficulty in tasks that require positioning of the tool before use (Ishibashi et al. 2000).

Such a limitation in level of understanding of objects as tools could be the result of these cognitions being underpinned by a domain general intelligence module in non-habitual-tool-using species such as the hylobatids (as suggested by Emery 2013), rather than a specialised evolved module for specific tool-use activities associated with food acquisition strategies in the wild state that may be present in habitual tool-users. Such a domain general mechanism may allow tool-using species to learn new skills beyond those adaptations evolved in a very specific context. For example, Japanese macaques could be considered habitual tool-users as they use stone tools in a foraging context (Tan et al. 2015). However, they take a considerable number of trials (approximately 1100) to learn systematic use of a rake tool to reach rewards out of direct reach (Ishibashi et al. 2000), implicating a more general learning mechanism in acquisition of novel tool-use skills.

Although some non-tool-using species have demonstrated impressive tool-use skills, it is likely that such a domain general mechanism requires more input from the environment to become efficient in the use and manipulation to tool-related information. The small apes in these studies were given very limited exposure to the tasks, and to the author's knowledge, had very little previous experience with objects that could be used as tools. Providing additional exposure to objects that could be functional in certain problem-solving situations, and more opportunities to learn about the properties of objects and how to use them during critical developmental windows particularly (Kummer and Goodall 1985), may have improved the performance of the hylobatids, giving them more time to develop the necessary cognitive schemas about 'objects as tools'. There is much research still needed to elucidate the cognitive differences between the primate groups. Hylobatids provide an exciting opportunity to differentiate the cognitive changes that have occurred during the transition from monkey to ape.

References

Anderson JR (2006) Tool use: a discussion of diversity. In: Fujita K, Itakura S (eds) Diversity of cognition. Kyoto University Press, Kyoto, pp 367–400

Beck BB (1967) A study of problem solving by gibbons. Behaviour 29:95–109

Beck BB (ed) (1980) Animal tool behaviour: the use and manufacture of tools by animals. Garland STPM Press, New York

Bernstein IS, Schusterman RJ, Sharpe LG (1963) A comparison of rhesus monkey and gibbon responses to unfamiliar situations. J Comp Psychol 5:914–916

Byrne R (ed) (1995) The thinking ape. Oxford University Press, Oxford, Evolutionary origins of intelligence

Cummins-Sebree SE, Fragaszy DM (2005) Choosing and using tools: capuchins (*Cebus apella*) use a different metric than tamarins (*Saguinus Oedipus*). J Comp Psychol 119:210–219

Cunningham CL, Anderson JR, Mootnick AR (2006) Object manipulation to gain a reward in hoolock gibbons (*Bunopithecus hoolock*). Anim Behav 71:621–629

Cunningham CL, Anderson JR, Mootnick AR (2011) A sex difference in effect of relevant object-mediated problem-solving in gibbons (Hylobatidae). Anim Cogn 14:599–605

Drescher K, Trendelenburg W (1927) Weiterer Beitrag zur Intelligenzprüfung an Affen (Einschließlich Anthropoiden). Zeitschrift für vergleichende Physiologie 5:613–642

Dunbar RIM (1998) The social brain hypothesis. Evol Anthropol 6:178–190

Emery NJ (2013) Insight, imagination and invention: tool understanding in a non-tool-using corvid. In: Sanz CM, Call J, Boesch C (eds) Tool use in animals: cognition and ecology. Cambridge University Press, Cambridge, pp 67–88

Finlay BL, Darlington RB (1995) Linked regularities in the development and evolution of mammalian brains. Science 268:1578–1584

Fodor J (ed) (1983) The modularity of mind. MIT Press, Cambridge

Fontaine B, Moisson PY, Wickings EJ (1995) Observations of spontaneous tool making and tool use in a captive group of western lowland gorillas (*Gorilla gorilla gorilla*). Folia Primatol 65:219–223

Forbes HO (ed) (1894) A hand-book to the primates, vol 2. Allen's Naturalist's Library, London

Fox GJ (1972) Some comparisons between siamang and gibbon behaviour. Folia Primatol 18:122–139

Fragaszy DM, Visalberghi E, Fedigan LM (eds) (2004) The complete capuchin. Cambridge University Press, Cambridge, The biology of the genus Cebus

Geoffroy-Saint-Hilaire E, Cuvier F (1824) Cited in Yerkes RM, Yerkes AW (eds) (1929) The great apes: a study of anthropoid life. Yale University Press, New Haven

Glickman S, Sroges R (1966) Curiosity in zoo animals. Behaviour 26:151–187

Gumert MD, Kluck M, Malaivijitnond S (2009) The physical characteristics and usage patterns of stone axe and pounding hammers used by long-tailed macaques in the Andaman Sea region of Thailand. Am J Primatol 71:594–608

Gumert MD, Hoong LK, Malaivijitnond S (2011) Sex differences in the stone tool-use behavior of a wild population of burmese long-tailed macaques (Macaca fascicularis aurea). Am J Primatology 73:1239–1249

Hunt GR, Gray RD, Taylor AH (2013) Why is tool use rare in animals? In: Sanz CM, Call J, Boesch C (eds) Tool use in animals: cognition and ecology. Cambridge University Press, Cambridge, pp 89–118

Inoue Y, Inoue E (2002) The trap-table problem with a young white-handed gibbon (*Hylobates lar*). In: Abstract of the joint symposium of COE2/SAGA5; evolution of apes and the origins of human beings. Inuyama, Japan

Ishibashi H, Hihara S, Iriki A (2000) Acquisition and development of monkey tool-use: behavioural and kinematic analyses. Can J Physiol Pharmacol 78:958–966

Jerison H (ed) (1973) Evolution of brain and intelligence. Academic Press, New York

Jerison H (1982) Allometry, brain size, cortical surface, and convolutedness. In: Armstrong E, Falk D (eds) Primate brain evolution. Plenum Press, New York, pp 77–84

Jordan C (1982) Object manipulation and tool-use in captive pygmy chimpanzee (*Pan paniscus*). J Hum Evol 11:35–39

Köhler W (ed) (1925) The mentalities of apes. Routledge and Kegan Paul, London

Kummer H, Goodall J (1985) Conditions of innovative behaviour in primates. Phil Trans R Soc Lond B 308:203–214

Lethmate J (1982) Tool-using skills of orang-utans. J Hum Evol 11:49–64

McGrew WC (ed) (1992) Chimpanzee material culture: implications for human evolution. Cambridge University Press, Cambridge

Mootnick AR (1997a) Management of gibbons Hylobates spp. at the international center for gibbon studies, California: with a special note on Pileated gibbons Hylobates pileatus. Int Zoo Yearbook 35:271–279

Mootnick AR (1997b) Nutrition, health and sanitation standards used at the International Center for Gibbon Studies which could be applied at a Javan Gibbon Rescue. In: Proceedings of the international workshop on rescue and rehabilitation, pp 20–24

Parker CE (1973) Manipulatory behavior and responsiveness. In: Rumbaugh DM (ed) Gibbon and siamang. Anatyomy, dentition, taxonomy, molecular evolution and behavior, vol 2. Karger, Basel, pp 185–207

Passingham RE (1981) Primate specialization in brain and intelligence. Symp Zool Soc Lond 46:361–388

Povinelli DJ (2000) Folk physics for apes: the chimpanzee's theory of how the world works. Oxford University Press, Oxford

Povinelli DJ, Reaux JE (2000) The trap-table problem. In: Povinelli DJ (ed) Folk physics for apes: the chimpanzee's theory of how the world works. Oxford University Press, Oxford, pp 132–148

Povinelli DJ, Parks KA, Novak MA (1992a) Role reversal by rhesus macaques, but no evidence of empathy. Anim Behav 44:269–281

Povinelli DJ, Parks KA, Novak MA (1992b) Comprehension of role reversal in chimpanzees—evidence of empathy. Anim Behav 43:633–640

Rilling JK, Insel TR (1999) The primate neocortex in comparative perspective using magnetic imaging. J Hum Evol 37:191–223

Rilling JK, Seligman RA (2002) A quantitative morphometric comparative analysis of the primate temporal lobe. J Hum Evol 42:505–533

Rumbaugh DM (1970) Learning skills of anthropoids. In: Rosenblum LA (ed) Primate behavior: developments in field and laboratory research. Academic Press, New York, pp 1–70

Santos LR, Pearson H, Spaepen G, Tzsao F, Hauser MD (2005a) Probing the limits of tool competence: experiments with two non-tool-using species (*Cercopithecus aethiops* and *Saguinus oedipus*). An Cogn 9:94–109

Santos LR, Mahajan N, Barnes JL (2005b) How prosimian primates represent tools: experiments with two lemur species (*Eulemur fulvus* and *Lemur catta*). J Comp Psychol 119:394–403

Schmaker RW, Walkup KR, Beck BB (eds) (2011) Animal tool behavior; the use and manufacture of tools by animals. John Hopkins University Press, Baltimore

Semendeferi K, Damasio H, Frank R (1997) The evolution of the frontal lobes: a volumetric analysis based on three-dimensional reconstructions of magnetic resonance scans of human and ape brains. J Hum Evol 32:375–388

Semendeferi K, Lu A, Schenker N, Damasio H (2002) Humans and great apes share a large frontal cortex. Nat Neurosci 5:272–276

Spagnoletti N, Visalberghi E, Ottoni E, Izar P, Fragaszy D (2011) Stone tool use by adult wild bearded capuchin monkeys (*Cebus libidinosus*). Frequency, efficiency and tool selectivity. J Hum Evol 61:97–107

Spagnoletti N, Visalberghi E, Verderane MP, Ottoni E, Izar P, Fragaszy D (2012) Stone tool use in wild bearded capuchin monkeys, Cebus libidinosus. Is it a strategy to overcome food scarcity? Anim Behav 83:1285–1294

Stephan H, Frahm H, Baron G (1981) New and revised data on volumes of brain structures in primates and insectivores. Folia Primatol 35:1–29

Tan A, Tan SH, Vyas D, Malaivijitnond S, Gumert MD (2015) There is more than one way to crack an oyster: identifying variation in Burmese long-tailed macaque (*Macaca fascicularis aurea*) stone-tool use. PLoS One 10(5):e0124733

Tomasello M, Call J (eds) (1997) Primate cognition. Oxford Press, Oxford

Tooby J, Cosmides L (1992) The psychological foundations of culture. In: Barkow JH, Cosmides L, Tooby J (eds) The adapted mind: evolutionary psychology and the generation of culture. Oxford University Press, Oxford, pp 19–137

Torigoe T (1985) Comparison of object manipulation among 74 species of nonhuman primates. Primates 26:182–194

van Essen DC (1997) A tension-based theory of morphogenesis and compact wiring in the central nervous system. Nature 385:313–318

van Schaik CP, Fox EA, Sitompul AF (1996) Manufacture and use of tools in wild Sumatran orangutans. Naturwissenschaften 83:186–188

van Schaik CP, Fox E, Fechtman LT (2003) Individual variation in the rate of tree-hole tools among wild orang-utans: implications for hominin evolution. J Hum Evol 44:11–23

Visalberghi E, Savage-Rumbaugh S, Fragaszy DM (1995) Performance in a tool-using task by common chimpanzees (*Pan troglodytes*), bonobos (*Pan paniscus*), an orang-utan (*Pongo pygmaeus*) and capuchin monkeys (*Cebus apella*). J Comp Psychol 109:52–60

Yerkes RM, Yerkes AW (eds) (1929) The great apes: a study of anthropoid life. Yale University Press, New Haven

Zilles K, Rehkämper G (1989) The brain, with special reference to the telencephalon. In: Schwartz JH (ed) Orang-utan biology. Oxford University Press, Oxford, pp 157–176

Chapter 14
Communication and Cognition
of Small Apes

Katja Liebal

Juvenile siamang. Photo credit: Manuela Lembeck

K. Liebal (✉)
Department of Education and Psychology, Comparative Developmental Psychology,
Freie Universität Berlin, 14195 Berlin, Germany
e-mail: katja.liebal@fu-berlin.de

© Springer Science+Business Media New York 2016
U.H. Reichard et al. (eds.), *Evolution of Gibbons and Siamang*,
Developments in Primatology: Progress and Prospects,
DOI 10.1007/978-1-4939-5614-2_14

Introduction

One of the distinguishing features of primates is their tendency to be highly social throughout all life stages. They realize a variety of different social systems with varying degrees of complexity (Kappeler and van Schaik 2002) that range from solitary species that only temporarily aggregate with other conspecifics to group-living species with groups comprised of up to several hundred individuals. Within groups, individuals form different social relationships (e.g., egalitarian or despotic social organization), and different sexual relationships (e.g., monogamous or polygamous mating systems).

This characterizing feature of primates has two important implications. First, across these different types of primate social systems, communication *within* a group, as well as *between* groups, is of central importance to maintain a group and to negotiate this variety of relationships between individuals (Altmann 1967). Groups that include more individuals with varying social roles and/or varying composition are suggested to require a more complex communicative system than groups with few individuals that have stable roles and little variation in group composition (Freeberg et al. 2012). Second, living in social groups also comes with certain costs, such as competition over mating partners and food resources and therefore creates situations that require 'social problem-solving' (Humphrey 1976). The complex and variable nature of social relationships within primate groups has thus provided the selective pressure for the evolution of more sophisticated socio-cognitive skills that are essential to compete or cooperate with other group members (Jolly 1966; Byrne and Whiten 1988; Kappeler and van Schaik 2006).

Based on these two important implications of primates' sociality, the purpose of this chapter is to review existing literature on the communicative and socio-cognitive skills of small apes. The next section describes the social and ecological features of small apes, followed by a discussion of the influence of these factors on the communication and ending with a section about the cognitive skills of gibbons and small apes.

Ecological and Social Characteristics of Small Apes

Hylobatids are closely related to both great apes and Old World monkeys and therefore have a unique phylogenetic position within the primates (Groves 2001). The phylogenetic relationships of small apes, as well as the characterizing features in regard to their morphology, are discussed elsewhere in this book (Parts II and III). In this section, special attention is paid to the ecology and social systems of gibbons to demonstrate how these characteristics might have shaped their communicative and cognitive skills.

In regard to their ecology, small apes are—together with orangutans—the only arboreal apes. They live in the canopy of the densely foliated rain forests of South-East Asia and have long limbs and digits which enable them to perform their unique brachiating style of locomotion (Leighton 1987; Fleagle 1999). They are highly territorial and defend their territories against other hylobatid groups (Marshall and Marshall 1976; Leighton 1987; Palombit 1993). Similar to orangutans, gibbons are primarily frugivorous (Harrison and Marshall 2011) and are highly selective eaters (McConkey et al. 2002). Siamangs (*Symphalangus syndactylus*) mostly feed on leaves (Gittins and Raemaekers 1980; MacKinnon and MacKinnon 1980), although there seems to be considerable variation in feeding habits between different siamang populations (Chivers 1974; MacKinnon 1977).

In terms of their social system, small apes are unique among apes since their groups only comprise an adult breeding pair and their offspring (Leighton 1987). As a consequence, group size is drastically smaller than in great apes, with close social bonds between the two partners, lack of a pronounced dominance hierarchy, and a shared territory (Carpenter 1940; Chivers 1974; Ellefson 1974; Brockelman et al. 1998). However, the traditional notion of mandatory nuclear families with the adult male and female maintaining lifelong, sexually monogamous relationships is increasingly challenged (Fuentes 2000; Barelli et al. 2008b; Reichard 2009; Reichard et al. 2012). Across hylobatid genera, groups with more than two adult individuals have been reported (Malone et al. 2012), with multi-male/single female groups representing the most frequent type of social organization besides pair-living (Barelli et al. 2008a, b; Lappan 2007; Reichard et al. 2012). For example, in white-handed gibbons (*Hylobates lar*) at Khao Yai National Park, Thailand, two-thirds of the males and almost half of the females lived in at least one other type of group structure in addition to pair-living, which was still the most frequent type of social organization (Reichard et al. 2012, p. 242). Furthermore, there is increasing evidence that hylobatid females are sexually polyandrous (Barelli et al. 2008b), since they engage in extra-pair copulations while living in pairs (Palombit 1994; Reichard 1995) or copulate with both males in multi-male groups (Barelli et al. 2008b). These findings, together with the fact that extra-pair copulations can result in fertilization (Kenyon et al. 2011; Barelli et al. 2013a), require a new perspective on the social organization and mating behavior of small apes, characterized by a considerable degree of social flexibility and the active role of females in pursuing their reproductive interests (Sommer and Reichard 2000; Reichard et al. 2012).

Within their group, siamangs (Chivers 1976) and Kloss's gibbons (*Hylobates klossii*) (Whitten 1980) synchronize up to 75 % of their daily activities in a way that most, if not all, group members simultaneously engage in the same activity at the same place. According to Chivers (1976), synchronized activity is mostly achieved by perceiving activities of others and then reacting accordingly (e.g., to start feeding when others are feeding) rather than by explicit communication between group members. At the same time, wild siamangs and other small apes spend surprisingly little time socializing with others, since social behaviors including singing, grooming or social play take up only a small proportion of their activity period (see Bartlett 2003, for a summary of several studies). Siamangs differ from gibbons because of their direct paternal care for the infant (Lappan 2008), closer pair bonds (Palombit 1996), and a 'greater harmony of group life' (Chivers 1976, p. 132). However, at least in captive settings, intra-group interactions of siamangs do not differ from those of white-handed gibbons as they show similar frequencies of aggression, percentages of successful food transfer, and occurrences of social grooming (Fischer and Geissmann 1990). There is no strict dominance hierarchy between the two pair-partners; however, several studies mention that pair partners might take different roles with females usually leading the group while traveling, while males are more dominant in encounters with other groups (Chivers 1976; Reichard and Sommer 1997; Barelli et al. 2008a). Intergroup interactions can account up to 9 % of the daily activities in white-handed gibbons at Khao Yai (Reichard and Sommer 1997), with the majority representing chases between males, but only little contact aggression (Reichard and Sommer 1997; Bartlett 2003). During such encounters, individuals of different groups might also engage in calling. Different groups also feed, travel, and rest together for a considerable amount of their daily activity and immatures might engage in play bouts with members of other groups (Reichard 1995; Bartlett 2003).

Based on these ecological and social factors characterizing hylobatids, the next sections will point out the predictions that can be made for their communication and cognition, respectively, followed by an overview of the current knowledge in regard to their communicative and cognitive skills.

Communication

What Are the Predictions for Hylobatids' Communicative Skills?

Because small apes live in arboreal habitats with dense vegetation that hinders the extensive use of visual signals, they should rely more on vocalizations than on visual signals such as facial expressions or gestures (Marler 1965; Maestripieri 1999). This should be particularly true for communication between groups, because of their territorial lifestyle and the corresponding distance between different groups (Chivers 1976; Brockelman et al. 1998). Within groups, however, both tactile and visual signals might still be expected as efficient signals for short-distance communication, because of the strong group cohesion and the close bond between pair partners (Chivers 1976; Orgeldinger 1999).

Regarding the influence of the social system, it has been proposed that more complex social groups are characterized by more complex communicative systems (Freeberg et al. 2012). To capture social complexity, it is not sufficient to only consider the number of individuals living in a group, but also the types of relationships between group members, as well as interactions with other groups (Freeberg et al. 2012). Considering the number of individuals, an increase of group size is correlated with an increase in facial and vocal repertoires (McComb and Semple 2005; Dobson 2009). Therefore, because hylobatids live in comparably small groups with the majority including only two adult individuals, they should use a limited repertoire consisting of few visual and tactile signals for close proximity interactions within their group (Chivers 1976; Orgeldinger 1999). Furthermore, considering the relationship between group members, Maestripieri (1999) proposed, based on the analysis of gestural signals of three macaque species with different social systems, that primate species in which most social interactions are predictable—i.e., based on rigid dominance and important kinship structures—should use comparably few communicative signals. On the other hand, primate species in which individuals must negotiate relationships frequently—for example, in more egalitarian species with relatively weaker dominance structures and less important kinship ties—should use a wider repertoire of communicative signals. Applying this to small apes, their lack of a strict dominance hierarchy, resulting in a more egalitarian group structure, might require the frequent negotiation of their relationships within their groups. However, given that these groups normally comprise only two adult individuals, a smaller signal repertoire is expected in hylobatids compared to great ape and monkey species that live in larger groups with more complex social relationships.

Hylobatids also frequently interact with other groups (Reichard and Sommer 1997; Bartlett 2003). Therefore, vocal communication in the form of loud, species- and sex-specific calls, which travel large distances through the dense vegetation, is expected to play an important role in defending their territories, but also to attract potential mates. However, because intergroup encounters also include interactions in close proximity, hylobatids might use a variety of signals to negotiate relationships between groups, including more subtle vocalizations as well as tactile and visual signals, which need to be adjusted to the communicative situation, such as the recipient's attentional state and the context of interaction.

What Is Known About Hylobatid Communication?

Primate communication research differentiates between several signal modalities. While some researchers use the term 'modality' to refer to the specific sensory mode of the stimulus (Partan and Marler 1999), a broader definition is used for the purpose of this chapter, because of the classic scientific distinctions made between different signal types and the different cognitive mechanisms that might underlie their production (Liebal et al. 2013). Thus, in contrast to some studies on monkeys, which refer to facial expressions as 'facial gestures' (Maestripieri 1996a, b), this chapter discusses research on facial expressions separately from gestures, which include visual, tactile, and auditory gestures. Auditory gestures produce sounds, but unlike vocalizations, these sounds are not produced with the vocal cords, but—for example—by slapping on the ground or on the own body.

1. *Vocalizations of small apes*

Vocalizations are produced from vibrations of the vocal folds within the larynx. However, it is still fiercely debated whether primate vocalizations are voluntarily produced, and thus represent intentionally used signals (Liebal et al. 2013). While many researchers working on primate gestures would argue that they are not (Tomasello 2008), there is increasing evidence for intentional use of vocalizations, particularly in chimpanzees (Crockford et al. 2012; Schel et al. 2013). Vocalizations are commonly separated into signals used for communication within a group (intra-group vocalizations) and those used for communication between groups (intergroup vocalizations).

Intergroup vocalizations are often loud calls, which are required to travel over longer distances, often in dense vegetation (Mitani and Stuht 1998). In hylobatids, they take the form of singing behavior, which is one of their distinguishing features separating them from the great apes (Geissmann 2000; Koda 2016). These loud, long-distance vocalizations are by far the most intensively studied form of hylobatid communication, with a considerable amount of research conducted in the wild (e.g., Marshall and Marshall 1976; Raemaekers et al. 1984; Haimoff and Gittins 1985; Clarke et al. 2006, 2015; Barelli et al. 2013a, b; Koda et al. 2013).

Small apes mostly sing in the early morning hours, with species-specific preferences for hours before, around, or after dawn (Geissmann 2000). Although some gibbons, such as Kloss's-gibbons (*Hylobates klossii*) and Silvery gibbons (*H. moloch*), produce solo songs even if they are mated in pairs, most species produce species- and sex-specific duets (Geissmann 1993, 2002) (Fig. 14.1). Thus, females produce species-specific great calls that remain largely unchanged throughout a song bout, which can last up to 30 min (Marshall and Marshall 1976; Marler and Tenaza 1977; Haimoff 1984; Geissmann 1993). Males start with single notes and gradually build up their phrases over the course of several minutes (Raemaekers et al. 1984; Mitani 1988).

While mated males of some gibbon species also produce solo song bouts, mated males of siamangs (*Symphalangus syndactylus*), crested gibbons (*Nomascus ssp.*), and hoolocks (*Hoolock ssp.*) only sing in duets with their females. In duets, each of the pair partners contributes their song to a complex, but rather stereotyped vocal interaction (Marshall and Marshall 1976; Haimoff 1984; Geissmann 1993). For example, in siamangs, both the male and female start with *booms* (with an inflated throat sac) (Fig. 14.2a) and collective barks, referred to as *chatters*. This is then followed by several *bark* series of the female, and each series is terminated by a *scream* of the male, accompanied by vigorous movements including branch shaking of both sexes (Chivers 1974, 1976) (Fig. 14.2b).

Since differences in song structure can be used to identify and distinguish different species or subspecies (Haimoff 1984; Geissmann 2002; Thinh et al. 2011), it has been repeatedly suggested that song repertoires of gibbons are largely genetically determined (Carpenter 1940; Marshall and Marshall 1976; Marler and Tenaza 1977; Brockelman and Schilling 1984). Studies on hybrids of different gibbon species in both captive and natural habitats seem to confirm this notion, since their songs combine elements of both parent species (Geissmann 1984; Tenaza 1985). Furthermore, hylobatid songs might not only be species- and sex-specific, but also reveal the identity of the caller, as found in female songs of white-handed gibbons (Terleph et al. 2015), agile gibbons (*Hylobates agilis*) (Haimoff and Gittins 1985; Oyakawa et al. 2007), and Silvery gibbons (Dallmann and Geissmann 2001), as well as in male songs of black crested gibbons (*Nomascus concolor*) (Sun et al. 2011) and Bornean southern gibbons (*Hylobates albibarbis*) (Wanelik et al. 2013).

Interestingly, these studies also highlight the *variability* of calls within and between individuals, with the latter exceeding intra-individual variation, therefore enabling the discrimination of individuals (Terleph et al. 2015). Furthermore, the comparison of different populations of the same species found that variation in call structure *between* populations can exceed variation between individuals *within* a population, indicating the geographical variability of calls (Dallmann and Geissmann 2009). As a result, it is possible to assign individuals to a specific location based on their call (Keith et al. 2009).

Together these studies show that on the one hand, hylobatid songs seem to be rather stereotyped, since they can be reliably used to identify species, subspecies, sex, and even the identity of the individual caller. On the other hand there is considerable evidence for intra- and interindividual variation in call structure. However, whether the variability between individuals is based on genetic differences or whether individuals within a certain population produce similar calls because they are able to modify their call structure to adjust to other group members' calls, is an open question. A study with wild and captive agile gibbons suggests a genetic component of female songs, as the females still produced their species-specific song when raised alone, despite some flexibility in their song patterns (Cheyne et al. 2007). Interestingly, Barelli et al. (2013b), who studied interindividual variation in the songs of white-handed gibbon males, found that their call structures varied as a function of their androgen levels, in a way that calls of males with higher androgen levels were characterized by a higher pitch. Barelli et al. (2013b) suggested that this enables receivers of these calls to retrieve information about the singing male's attributes. This study also shows, however, that many factors might influence the acoustic structure of gibbon vocalizations. Therefore, at the current state of knowledge, it is too early to conclude that interindividual variation is the result of *either* genetic differences between individuals *or* social factors shaping individual song patterns.

Finally, to convincingly show that receivers identify individuals based on their calls and are able to extract specific information from their calls, it is necessary to conduct playback experiments to study the recipient's response to calls of different individuals. Until now, only very few of such studies have been conducted (Mitani 1984, 1987; Raemaekers and Raemaekers 1985a), but currently there is no convincing evidence that recipients identify individuals based on their songs (Mitani 1985).

Different functions have been suggested for hylobatid duet songs, such as the advertisement of a territory, strengthening of pair bonds, and attracting new mates (Chivers 1976; Brockelman and Srikosamatara 1984; Raemaekers and Raemaekers 1985b; Cowlishaw 1992). However, although intensively studied, research into the function of hylobatid songs has produced inconsistent results (Fan et al. 2009), most likely because different selection pressures have acted on males' and females' contributions to their duets (Marshall and Marshall 1976). For example, Cowlishaw (1992) compared existing studies on nine hylobatid species and suggested that

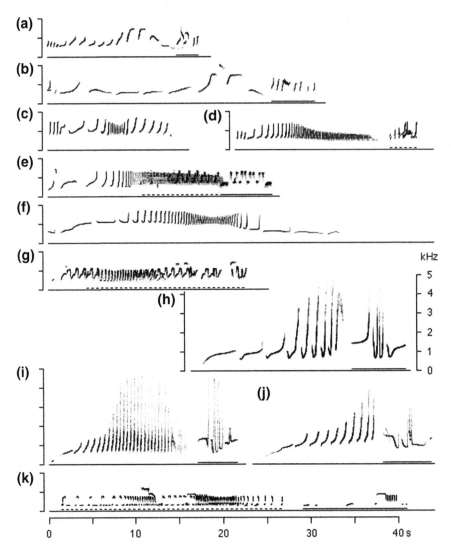

Fig. 14.1 Sonogram of female gibbons' great call sequences. Sonograms **c** and **f** are excerpts from female solo song bouts; all other sonograms show duets. Male solo contributions to duets are underlined with a *solid line*, synchronous male and female vocalizations are underlined with a *dashed line*. **a** *Hylobates agilis* (Asson Zoo); **b** *H. lar* (Paignton Zoo); **c** *H. moloch* (Munich Zoo); **d** *H. muelleri* (Paignton Zoo); **e** *H. pileatus* (Zürich Zoo); **f** *H. klossii* (South Pagai, recording: R.R. Tenaza); **g** *H. hoolock* (Kunming Zoo); **h** *H. concolor* (Xujiaba, Ailao Mountains); **i** *H. leucogenys* (Paris, Ménagerie); **j** *H. l. gabriellae* (Mulhouse Zoo); **k** *H. syndactylus* (Metro Zoo, Miami). Figure from Geissmann (2000)

different information might be encoded in the songs of males and females. While male songs act as advertisement calls that signal the potential to defend the female and offspring, female calls probably signal the ability to defend the territory.

However, there was no evidence that the production of coordinated duets is related to pair-bonding (Cowlishaw 1992).

Others have suggested that duet songs indeed affect the pair bond by increasing the cohesiveness between male and female (Wickler 1980). Because the coordination of duets requires both partners' contributions particularly at the beginning of the formation of a new pair, this could represent a mechanism to reduce the probability of one partner leaving the other as this would be costly, since a new coordination process would become necessary with every new partner. Indeed, Geissmann (1999) found that although siamang pairs had stable song patterns, after a forced partner exchange, the new pair produced a higher number of atypical great call sequences indicating difficulties in synchronizing their duets. Furthermore, songs of both new pair partners underwent changes over time, interpreted as 'partner-directed learning effort' to increase the synchronization of the duets in these newly established pairs (Geissmann 1999, p. 1007). Since duetting in siamangs correlates with grooming activity and behavioral synchronization, Geissmann and Orgeldinger (2000) propose that the production of duets is related to pair-bonding. However, the benefits for pairs with better synchronized duets and/or closer pair bonds have not been convincingly demonstrated yet, as there are no studies confirming that pairs with more synchronized duets have more reproductive success, have better access to resources, or are more successful in defending their territory.

Fig. 14.2 a A duetting pair of siamangs with throat sacks inflated. **b** Locomotor display of a male siamang. Both photos are kindly granted by Thomas Geissmann

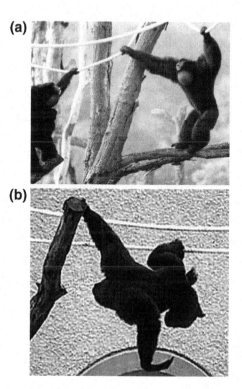

The increasing synchronization of duets in newly formed siamang pairs also suggests that despite the stereotyped nature of siamang songs, there seems to be at least some flexibility in their vocal production (Haraway and Maples 1998). This is supported by recent evidence from vocal interactions of mother–daughter pairs of wild agile gibbons (Koda et al. 2013). During co-singing, songs of mothers and daughters matched in their acoustic properties indicating that gibbons are able to converge their songs over a short timescale. Furthermore, the finding that mothers used a more stereotyped pattern while co-singing compared to when singing alone was interpreted as a mean to enhance the development of the species-specific female song in their daughters (Koda 2016).

Clarke et al. (2006) also found evidence for some flexibility in the song production of white-handed gibbons. Thus, although gibbons usually produce duet vocalizations without any external stimuli (Geissmann 2000), they do occasionally vocalize in response to alarming situations, such as in encounters with predators (Uhde and Sommer 2002). Clarke et al. (2006) investigated whether songs produced in response to predators differed in their acoustic properties from those usually uttered in the morning. Although gibbons only responded to terrestrial, but not aerial predators, these songs were longer, were introduced by more *hoo* notes and differed in several other structural and temporal measures compared to their 'normal' duets. This seems to indicate that white-handed gibbons can combine a finite number of call components into more complex sequences depending on the contextual situation (Clarke et al. 2006).

Compared to these loud calls, characterized by at least some flexibility and mostly used in intergroup communication, much less is known about vocal communication *within* hylobatid groups, which usually includes much softer sounds. According to Marler (1965), intra-group vocalizations are characterized by lower amplitude and are acoustically more variable than intergroup calls. They are used in a variety of contexts such as play, grooming, and sexual or aggressive behavior. For siamangs, Chivers (1976) described only five of such vocalizations and concluded that '[…] the variety and frequency of calls within a siamang group are […] less than in the smaller gibbons' (Chivers 1976, p. 119). For white-handed gibbons, Ellefson (1974) reported seven within-group vocalizations, while others refer to nine vocalizations, with some of them being used both for within- and between group communication (Carpenter 1940; Baldwin and Teleki 1976).

Chivers (1976) described *squeals* in siamangs used in situations when an individual had been lunged by an older animal, while Fox (1977) mentioned that young siamangs utter these calls usually during play. These vocalizations often cause an adult to approach the two play partners or result in the termination of the play bout (Orgeldinger 1999). Fox (1977) mentions *gacks*, which—in contrast to *squeals*—do not cause an adult's intervention. *Chirps* and *squeals* of white-handed gibbons occur during the initiation of play, but they also function as greeting signals (Carpenter 1940; Ellefson 1967). *Bleating distress calls* are uttered by infant siamangs that get isolated from their parents (Chivers 1974). *Soft grunts* are produced in early stages of an alarm response to potential danger, while *glunks* occur during feeding. However, the latter most likely represent physical noise caused by swallowing rather than a vocalization (Chivers 1976).

These examples demonstrate that little is known about intra-group vocalizations, which are usually mentioned as part of more general descriptions of a species' behavioral repertoire. Thus, apart from these qualitative studies, mostly on siamangs and white-handed gibbons, there are virtually no quantitative studies, which analyze the acoustic structure of intra-group vocalizations and how they are used in different contexts. A recent study investigated the *hoo* call of white-handed gibbons (Clarke et al. 2015), which is one of their most common close-range vocalization. The analysis of six acoustic parameters showed that these gibbons produce context-dependent *hoo* calls, which significantly differ as a function of the context they are used in. Interestingly, raptor *hoo*s seem to represent a distinct call class, which may point to the special importance of avian raptors for this gibbon species (Clarke et al. 2015).

2. *Facial expressions of small apes*

Primate facial expressions are movements of the cartilage of the external ear and the alar cartilage of the nose, the skin of the face, and the lips, eyelids and nares, which are produced by the facial (or mimetic) musculature. Dissections of the facial masks of different hylobatid species identified 22 facial muscles (Burrows et al. 2011) that resemble the number of muscles found in rhesus macaques and chimpanzees. However, hylobatids' facial muscles are generally gracile and less complex compared to facial muscles of the other two species (Burrows et al. 2006, 2009).

For hylobatids, there are only a few descriptive studies that mention a limited repertoire of facial expressions, mostly in the context of more general descriptions of a species' behavior (Fox 1972, 1977; Baldwin and Teleki 1976; Orgeldinger 1999) (Table 14.1). The most frequently described facial expression is the *open mouth face* or *threat face* (Fig. 14.3) and its different variants, although different terms are used across studies (Table 14.1). For siamangs, Fox (1977) described a *mild open-mouth threat* (mouth only slightly opened and corners of the mouth somewhat withdrawn) and an *intense open-mouth threat* (mouth widely opened and canines visible), which most likely correspond to the *mouth open half* and *mouth open full* expressions described by Liebal and colleagues (Liebal et al. 2004b). This expression is also present in white-handed gibbons (Ellefson 1974; Baldwin and Teleki 1976) and agile gibbons (Gittins 1979). It is often combined with a stare (Chivers 1976; Gittins 1979) and is then used as threat in agonistic contexts, or to terminate an ongoing interaction such as intense play (Liebal et al. 2004b). The *open mouth face* can also precede *chewing mouth movements* in aggressive encounters (also described as *rapid-open-and-close mouth threat, mouth clamping*) (Carpenter 1940; Baldwin and Teleki 1976). It is currently not known whether these variants of the *open mouth face* systematically change depending on the intensity of an interaction, and/or whether each variant has a distinct social function, as has been found for different variants of play faces in gorillas (Waller and Cherry

Table 14.1 Comparison of gestures and facial expressions in siamangs and gibbons

Species	Gesture/facial expression	Signal[a]	Functional context[a]	Signal [translated]	Functional context
Symphalangus syndactylus	Tactile gestures	Embrace	Affiliation, grooming	Frontal embrace[b]	Nursing/carrying infant
				Embrace[c]	Affiliation
				[Embrace][d]	Reassurance, greeting
		Formal bite	Agonism grooming	[Inhibited bite][d]	Play, nursing, sexual
		Gentle touch	Play, affiliation	[Short touch][d]	Play, aggression, affiliation
		Hold tight	Play, agonism	[Hold][d]	Play
		Pull	Play, grooming	[Pull][d]	Play
		Push	Grooming, agonism	[Push][d]	Play
		Shake body	Play	Shaking[b]	Play
		Slap	Play, agonism	Slap, kick, jerk, hit[b]	Play, agonism
				[Slap][d]	Play, aggression
	Visual gestures	Jerking body movements	Sexual behavior, play	Upward thrusts[c]	Sexual behavior
				[Jerking body movements][d]	Play, sexual behavior, follow
		Offer body part	Grooming	[Offer body part][d]	Grooming
				Present[c]	Grooming
		Present genitals	Sexual behavior, submission	[Present genitals][d]	Sexual behavior, reassurance
				Present genitals[b]	Aggression
		Throwback head	Play	[Throwback head][d]	Follow, sexual
	Facial expressions	Grin	Sexual behavior, play	Grimace[b]	Reassurance
				Grimace/bared-teeth[c]	Appeasement
				[Grin][d]	Play, submission, arousal

(continued)

Table 14.1 (continued)

Species	Gesture/facial expression	Signal[a]	Functional context[a]	Signal [translated]	Functional context
		Mouth-open full	Agonism, grooming	Intense open-mouth threat[b]	
				Staring open-mouth face[e]	
				Open-mouth threat[c]	Aggression
				[Directed threat][d]	Aggression
		Mouth-open half	Agonism, grooming, play	Mild open-mouth threat[b]	
				Relaxed open mouth face[e]	
		Pull a face	Grooming, sexual behavior	[Pull a face][d]	Sexual behavior
Hylobates lar	Tactile gestures			Embrace[f]	Greeting
				Embrace[g]	Submission
				Embrace[h]	Vocalizing, arousal
				Embrace[c]	Affiliation
				Touch[h]	Play
				Holding[g]	Play
				Push[g]	Play
				Slap, cuff[h]	Play
				Hit, kick, jerk[g]	Play
				Strike[h]	Agonism
	Visual gestures			Lowering head[g]	Grooming
				Body present[h]	Grooming

(continued)

Table 14.1 (continued)

Species	Gesture/facial expression	Signal[a]	Functional context[a]	Signal [translated]	Functional context
				Present[c]	Grooming, reassurance
				Posterior present[h]	Grooming, reassurance
				Genital present[h]	Sexual behavior
				Sexual crouch[h]	Sexual behavior, submission
				Sexual crouch[g]	Sexual behavior
				Head tilt[h]	Play, travel
				Throwing back head[f]	Antagonistic
	Facial expressions			Grimace[h]	Fear, uncertainty
				Grimace[g]	Submission
				Grimace/Bare-teeth[c]	Appeasement
				Open-mouth toward[g]	
				Open-mouth threat[c]	
					Aggression
Hylobates agilis	Visual gesture			Present[i]	Grooming
	Facial expression			Grimace[i]	Appeasement
				Open mouth stare[i]	Aggression

To demonstrate the different terms used to label communicative behaviors, the findings of Liebal et al. (2004a, b) are compared with other studies. For each signal, the corresponding reference and its functional contexts are shown. Underlined signal indicate studies in natural habitats

[a]Data from Liebal et al. (2004b)
[b]Fox (1977)
[c]Palombit (1992)
[d]Orgeldinger (1999)
[e]Chivers (1976)
[f]Carpenter (1940)
[g]Ellefson (1967)
[h]Baldwin and Teleki (1976)
[i]Gittins (1979)

2012). Several studies specifically mention a *relaxed open mouth face* or *play face* in different hylobatid species (Chivers 1976; Ellefson 1974; Fox 1977; Gittins 1979; Liebal et al. 2004b). Although most studies do not differentiate between this expression and the *open mouth facial expression*, there is some evidence that the *play face* is structurally different, because in addition to the opened mouth, the lip corners are pulled backwards (Waller et al. 2012).

The *grin* facial expression, or *grimace*, corresponds to the *silent-bared teeth face* described in a variety of primate species (van Hooff 1962, 1967), which—depending on the species' social system—can serve different functions (Preuschoft 2004). Thus, in primate species with strict, linear dominance hierarchies, this facial expression is used in only two contexts, appeasement and submission, and is expressed asymmetrically (i.e., from a subordinate toward a dominant individual). In species with more relaxed dominance styles, this facial expression is flexibly used in a greater variety of contexts such as affiliation, reconciliation, and reassurance (Preuschoft 2004). White-handed gibbons use the *grin* as an expression of fear or uncertainty (Baldwin and Teleki 1976) or submissive behavior (Ellefson 1967), while siamangs use it as a signal of reassurance and during play, but males also show it in the sexual context (*copulation face*) (Fox 1977; Liebal et al. 2004b). In siamang females, the *pull a face* expression seems to represent the female version of the male's *copulation face*, since they show it during sexual behavior, but occasionally also in the context of grooming (Orgeldinger 1999; Liebal et al. 2004b).

This overview of facial repertoires demonstrates that the terms used to label facial expressions differ across studies (Table 14.1). While some authors describe facial expressions based on their social function or the context in which they are used (e.g., *play face, open mouth threat*), others differentiate them based on their structural properties or form (e.g., *relaxed open mouth face, bared teeth face*). Therefore, comparisons between species and studies are difficult. The GibbonFACS, which is a modified version of the human Facial Action Coding System (FACS) (Ekman and Friesen 1978), has recently been developed based on observations of seven hylobatid species (siamangs, pileated gibbons (*Hylobates pileatus*), silvery gibbons, three species of crested gibbons (*Nomascus siki, N. gabriellae, N. leucogenys*), and Müller's gibbons (*Hylobates muelleri*) (Waller et al. 2012). Instead of differentiating facial expressions based on their social function or context in which they occur, they are described as combinations of defined action units, which are based on specific muscle movements that underlie facial movements. Combinations of different action units then constitute a facial expression, such as different variants of the *play face* (e.g., action units 12 + 25 + 26 or action units 12 + 26 + 27, described for chimpanzees, Parr et al. 2007). The GibbonFACS was successfully applied to describe facial movements in seven hylobatid species, and 18 action units were identified across these species (Waller et al. 2012). This variety of facial movements in hylobatids bears great similarity to that of primate species living in more complex social groups. This is surprising, given that hylobatids live in comparatively small and stable groups and thus should use a relatively small repertoire of facial movements (Dobson 2009).

Fig. 14.3 Open-mouth facial
expression of a siamang.
Photo Manuela Lembeck
(from GibbonFACS Manual,
see www.gibbonfacs.com and
Waller et al. 2012)

The unexpected variety of facial movements in hylobatids thus points to several
important issues. First, the social and therefore communicative complexity of small
apes might have been simply underestimated (*see* section 'Ecological and social
characteristics of small apes'). Second, it is not sufficient to only consider sizes of
facial repertoires, but necessary to also investigate the frequency and contexts of
their use to capture the diversity of facial communication in hylobatids. However,
such studies are still very rare. For example, the low number of facial expressions
siamangs use in their social interactions with others (Liebal et al. 2004b) may seem
surprising given the rather high variety of facial movements described for this
species and other hylobatids (Waller et al. 2012). However, it is important to differentiate between single *facial movements* (action units) and *facial expressions*,
which are combinations of several of such action units. Furthermore, not all facial
movements are necessarily used for communicative purposes (Scheider et al. 2016)
and therefore were not considered by Liebal et al. (2004b). However, despite the
low number of *types* of facial expressions, in terms of *frequency*, they accounted
for approximately one third of *all* observed communicative signals, which also
included visual and tactile gestures (Liebal et al. 2004b). For example, the *mouth
open full expression* (Fig. 14.3) was not only the most frequently used facial
expression (67 % of all facial expressions), but also the most frequent signal out of
the total signals observed in this study (24 % of 2782 signals including facial

expressions and gestures). Scheider et al. (2014) compared facial movements of five hylobatid species of three genera (*Symphalangus, Hylobates, Nomascus*) (Scheider et al. 2014) and specifically differentiated between the repertoire of facial movements and their frequency of use. While siamangs differed from gibbons in their higher rate of using facial expressions, there were no significant differences in the repertoires of facial movements across the five species. Furthermore, no correlation was found between group size and either repertoire size or frequency of use. However, it is important to point out that this study was conducted on captive individuals and the observation time was rather limited compared to studies conducted on wild populations. Therefore, it is possible that the complexity of facial communication in these species was not fully captured.

3. *Gestural communication of small apes*

Gestures of nonhuman primates are defined as intentional, voluntarily controlled movements of limbs or the head and body postures, which are directed toward a particular recipient and are motorically ineffective (Call and Tomasello 2007; Liebal and Call 2012). They include visual gestures (e.g., *extend arm*), tactile gestures (e.g., *slap*), and auditory gestures when a sound is created without using the vocal cords (e.g., *chest beats* of gorillas). Most gesture research so far has been conducted on great apes, while little is known about gestures and their use in small apes or monkeys (Slocombe et al. 2011). No unified method has been developed yet resembling the FACS to identify and describe gestures based on structural properties (but *see* Roberts et al. 2012). As a consequence, there is a variety of terms used to describe the same gesture, hindering a comparison across studies and species (see Table 14.1).

In regard to tactile gestures, captive siamangs use *slaps*, *kicks* and *jerks*, mostly in agonistic and play contexts (Fox 1977; Liebal et al. 2004b), which are also described for white-handed gibbons (Baldwin and Teleki 1976) (Table 14.1). *Embrace* gestures are reported for both siamangs and white-handed gibbons, which function as reassurance or greeting gestures (Ellefson 1974; Palombit 1992; Orgeldinger 1999), or occur in the context of nursing or carrying between the infant and an adult siamang (Orgeldinger 1999).

For visual gestures, *offer body part* is the most frequent visual gesture of captive siamangs that mostly present their back to initiate grooming (Liebal et al. 2004b), while white-handed gibbons take up a sitting position and then lower their head (Ellefson 1967). Other visual gestures, such as *present genitals, jerking body movements* and *throw-back head*, are used by adult female siamangs and white-handed gibbons to initiate a copulation (Palombit 1992; Orgeldinger 1999; Liebal et al. 2004b). However, several studies mention that females also *present genitals* to appease males in aggressive interactions (Ellefson 1974; Baldwin and Teleki 1976; Orgeldinger 1999; Liebal et al. 2004b). There is one report on a *wrist offer* gesture used by the subadult daughter of a captive siamang group (Liebal et al. 2004b). Since her peripheralization (which naturally results in the emigration of maturing individuals from their natal group) already started, she was increasingly isolated from interactions with other group members and thus used this submissive gesture in interactions with her parents, similar to chimpanzees who use this gesture when approaching dominant individuals (Goodall 1986).

Taken together, the little we know about hylobatid gestural communication is from studies with white-handed gibbons and siamang, who use a variety of tactile and visual gestures, while much less is known about the gesture use of other hylobatid species. It is difficult to provide estimates of repertoire sizes for each species, because terms and definitions vary across studies (Table 14.1). Furthermore, it is important to note that individual repertoires might vary depending on group composition, the presence or absence of offspring, or the reproductive status of a female. Individual repertoires also vary depending on age, with immature siamangs using larger repertoires than adults, mostly because they use a variety of different gestures to initiate play bouts. Adults in general play less often and as a consequence, also use fewer gesture types (Liebal et al. 2004b). On the other hand, adult individuals may use some gestures in sexual or aggressive contexts not present in younger individuals. Finally, although a gesture is found in both immature and adult individuals, both age groups might use it in different contexts. For example, immature siamangs perform the *throw-back head* gesture to initiate play or to get the attention of a parent, while adult siamang females use this gesture to initiate copulation (Liebal et al. 2004b).

Cognition

The focus of this chapter is specifically on those socio-cognitive skills which are related to the communication of hylobatids, while other cognitive skills—such as the understanding of object relationships and tool use—are discussed elsewhere (Prime and Ford 2016).

What Are the Predictions for Gibbons' Cognitive Skills?

Life in complex social groups requires sophisticated cognitive skills because of the amount and variable nature of relationships between group members (Byrne and Whiten 1988). According to Dunbar (2003), larger brains and here specifically neocortex volume and sophisticated socio-cognitive skills coevolved as adaptations to social complexity and the need to interact with an increasing number of individuals. To capture social complexity, Freeberg et al. (2012) suggested additional measures and proposed that instead of only considering group size (with larger groups being socially more complex) to also include the roles and relationships of individual group members as well as their spatial distribution and interactions with neighboring groups (*see* section '*What are the predictions for hylobatids' communicative skills?*').

Hylobatids live primarily in small, stable groups with no strict dominance hierarchy between pair partners and relatively limited opportunities to interact with larger numbers of individuals within and between groups (Chivers 1976). Therefore, there is no obvious need for qualitatively complex interactions related to the negotiation of social relationships and to cooperate with or to outcompete other group members for access to resources. Therefore, several studies predict that hylobatids' cognitive skills should be less sophisticated than those of great apes or monkeys that live in socially more complex groups (Liebal et al. 2004b; Liebal and Kaminski 2012). However, it is important to consider that there is increasing evidence that social complexity of small apes is not as limited as previously assumed (see section 'Ecological and social characteristics of small apes'). Contrary to the classic approach to hylobatid social organization of socially and sexually monogamous pairs, research has shown that 'individuals are rarely exclusively pair-living with one partner' (Reichard and Barelli 2008, p. 837), but have serial relationships with different partners or even live in groups with more than two adult pair partners, such as multi-male groups (Lappan 2007; Barelli et al. 2008b; Reichard et al. 2012). This raises the question whether the potential variability in group size and the resulting varying relationships between group members may predict more complex socio-cognitive skills than previously assumed.

Considering life history, hylobatids seem to resemble great apes more than monkeys. For example, white-handed gibbons have long inter-birth intervals, with females starting to reproduce late around 11 years of age and a long juvenile period of more than 9 years (Reichard and Barelli 2008). For great apes, it has been proposed that their extended life history is closely related to their complex cognitive skills, which represent strategies to respond to socio-ecological challenges (van Schaik and Deaner 2003; van Schaik et al. 2004). Reichard and Barelli (2008) therefore suggested that it might be worth investigating the cognitive skills of gibbons in the light of their extended life history.

What Is Known About Cognitive Skills of Small Apes?

In 1976, Abordo published a review on the learning skills of gibbons and concluded that they are largely understudied in comparison to other primate species, particularly in regard to their socio-cognitive skills. Almost three decades later, this situation has not substantially changed, since only a few new studies on hylobatid cognitive skills have since been published.

1. *General cognitive skills*

To look in the direction others are looking and thus to follow another individual's gaze is one of the most basic socio-cognitive skills (Butterworth and Jarrett 1991), which has been investigated in a variety of non-primate (Miklósi et al. 1998; Bugnyar et al. 2004; Kaminski et al. 2005) and primate species (Rosati and Hare 2009).

In its simplest form, gaze following represents a reflexive shift of gaze in response to another individual's gaze shift to search for something interesting by following their line of sight, referred to as visual co-orientation (Gómez 2005). A cognitively more complex form related to the interpretation of another individual's visual perspective is geometrical gaze following, which is the ability to follow others' gaze around barriers which block the direct line of sight (Tomasello et al. 1999). It requires an understanding of the visual perspective of another individual and thus what others can and cannot see, which is an important skill for primates that live in complex social groups (Hare et al. 2000, 2001). For hylobatids, only two studies have investigated gaze following in response to a gaze shift. One study tested two pileated, hand-raised gibbons and found a tendency for visual co-orientation with the gaze direction of both conspecifics and humans depicted in photographic images (Horton and Caldwell 2006). A second study used a different paradigm and investigated the small apes' ability to follow a human's gaze shift to the ceiling in a larger sample comprised of 24 individuals including pileated gibbons, silvery gibbons, white-handed gibbons, and siamangs (Liebal and Kaminski 2012). Although the small apes co-oriented with the gaze of a human and looked up more to the ceiling when she was looking upward, they did not show evidence for habituation and no repeated looks upward to check where the human was gazing, as found for great apes (Bräuer et al. 2005). Furthermore, unlike great apes, hylobatids also gazed more in other directions apart from looking up, indicating that they were generally more vigilant and checked the surroundings for relevant events (Liebal and Kaminski 2012). These findings indicate that gibbons do not take the visual perspective of humans, but seem to rather co-orient their gaze to the human's gaze shifts. Their ability to follow others' gaze around barriers, however, has not been studied yet.

The ability to recognize oneself in mirrors has also been extensively studied in a variety of primate and non-primate species (Lethmate and Duecker 1973; Suarez and Gallup 1981; Parker et al. 1994; Patterson and Cohn 1994; Bard et al. 2006; Plotnik et al. 2006; Prior et al. 2008) since Gallup's initial study with chimpanzees (Gallup 1970). It has been suggested that individuals that recognize themselves in the mirror have—at the minimum—a mental model of what they look like (Nielsen et al. 2006), while others propose that mirror-self recognition is evidence for self-awareness (Gallup 1998). The few studies with small apes so far produced inconsistent findings. For example, one white-handed gibbon and one siamang hybrid showed mirror-mediated and self-directed behavior, e.g., the exploration of different body parts (Hyatt 1998). Similar to monkeys (Anderson 1984), however, they did not pass the crucial mark test, since they did not show any reactions to some dye put on their foreheads when looking at the mirror afterwards (Lethmate and Duecker 1973). However, Ujhelyi and colleagues concluded that 'the variety and nature of mirror-mediated behavior in two of our gibbons goes beyond what has previously been reported for gibbons exposed to mirrors' (Ujhelyi et al. 2000, p. 260), because one of their three subjects showed mirror-mediated, self-directed behaviors in front of the mirror and was able to find food hidden out of their view by using the mirror to locate the hidden item. However, two of the three gibbons failed the mark test. An extensive study testing 17 individuals from three different genera (*Hylobates, Symphalangus* and *Nomascus*) found that although all gibbons showed different kinds of behavior directed toward the mirror, they produced very little self-directed behaviors (Suddendorf and Collier-Baker 2009). Interestingly, all but two individuals reached or looked behind the mirror, 'as if the subject was searching for 'the other gibbon'' (Suddendorf and Collier-Baker 2009, p. 1674). Most importantly, however, none of the subjects showed any interest in the mark placed on their head, although all of them were interested in the mark when visibly placed on their limbs. This pre-test was conducted to investigate whether they would touch these marks at all (Suddendorf and Collier-Baker 2009). Together these studies seem to indicate that hylobatids—despite their interest in mirrors—are not able to recognize themselves in mirrors and thus most likely have no mental image of themselves. However, based on the current state of knowledge, it seems too early to draw this conclusion. Thus, even in great apes, there is considerable interindividual variability in regard to their ability to recognize themselves in mirrors. For example, while chimpanzees and orangutans usually pass this test (Gallup et al. 1971; Lethmate and Duecker 1973), results for gorillas are less consistent (Suarez and Gallup 1981; Patterson and Cohn 1994), most likely because different social and ecological factors affect their behavior in this test (Shillito et al. 1999; Swartz and Evans 1994). Furthermore, the methodological approach has been repeatedly criticized (Heyes 1994, 1995; De Veer and van den Boos 1999), partly because in the initial studies, subjects were anaesthetized to place the dye on their heads, which might have influence their rates of face-touching behavior (Heyes 1995; for a response, see Gallup et al. 1995). Given this debate about methodological issues, the limited number of studies available for hylobatids, and the variability between individuals across many studies with nonhuman primates, it seems not justified to conclude that small apes are not able to recognize themselves in a mirror.

2. *Cognitive skills related to communication*

To know what others can see and cannot see by taking their visual perspective is also an important ability in the context of communication. As the previous section on gaze following demonstrated, the few existing studies on hylobatids indicate that they do not take the visual perspective of others (Horton and Caldwell 2006; Liebal and Kaminski 2012). However, in communicative interactions with conspecifics, siamangs use visual gestures and facial expressions only if their partner is visually attending to them resembling patterns of gesture use in great apes (Call and Tomasello 2007). This seems to indicate that siamangs know that others need to be attentive for their visual gestures to be successfully perceived (Liebal et al. 2004b). However, this does not necessarily mean that siamangs understand the visual perspective of others, since the producer of a visual gesture could have simply learnt that potential recipients only respond if they face the producer of this gesture. Furthermore, siamangs also use tactile gestures more often to an attending than non-attending recipient, although this type of gesture does not require the visual attention of others to be successfully perceived. The use of attention-getting gestures, which manipulate the attention of others and are frequently studied in great apes (Tomasello et al. 1994; Hostetter et al. 2001; Liebal et al. 2004a), might provide more insight into hylobatids' communicative and cognitive skills; however, no such study has yet been conducted with small apes.

This adjustment of signal use depending on the recipient's attentional state is often highlighted as evidence for flexible usage of a given repertoire (Liebal et al. 2013). However, how flexibility is operationalized may differ substantially across studies and signal modalities (Tomasello and Zuberbühler 2002; Seyfarth and Cheney 2010). In general, flexibility in the *production* of a signal implies that the sender has some control over when and how they produce a signal and therefore can modify its structure, while flexibility in *usage* refers to the use of one signal in different contexts, and vice versa, the use of several signals within one particular context. Finally, another measure of flexibility is the *combination* of single signals into longer, meaningful sequences, which are then used in other or new contexts than their single-signal components (e.g., Zuberbuehler 2002; Liebal et al. 2004a; Hobaiter and Byrne 2011).

In regard to flexibility in production, most studies in hylobatids consider their vocalizations to examine their degree of intra- and interindividual variability (Dallmann and Geissmann 2009; Sun et al. 2011). As pointed out earlier (section 'Vocalizations of small apes'), small ape songs are rather stereotyped vocalizations and are most likely genetically determined (Geissmann 1984; Brockelman and Schilling 1984). However, there is increasing evidence that at least to some extent, individuals are able to change the structure of calls depending on the contextual situation, which is most compatible with the notion of flexibility (Clarke et al. 2006; Koda et al. 2013). No such studies on the production of gestures exist yet for hylobatids, since studies on gestural communication of nonhuman primates generally focus on the usage of gestures, but not on their structural properties (Liebal et al. 2004b). Similarly, there are virtually no studies on the facial communication in small apes. However, this is currently changing, since the GibbonFACS provides a new method to study structural properties of hylobatids' facial movements (Waller et al. 2012; Scheider et al. 2014, 2016).

Considering flexibility in signal use, a study on siamangs found that they produce the majority of their visual and tactile gestures in more than one context and they also use different gestures to achieve the same goal (Liebal et al. 2004b). For example, they utilize the tactile gestures *nudge* and *slap* in a variety of contexts, including play, aggression, grooming, or sexual behavior. Similarly, they use one of their facial expressions (*mouth open full*) across the majority of social contexts. Thus, instead of using the same behavior to obtain the same goal, siamangs flexibly used some of their gestures and at least one of their facial expressions flexibly in different contexts (Liebal et al. 2004b). How this compares to flexibility in signal usage in great apes or monkeys is difficult to conclude, since different definitions (e.g., differentiating manual gestures from facial expressions) or varying methodological approaches (e.g., how is flexibility operationalized, how many contexts have been differentiated) have been used across studies (but see Call and Tomasello 2007).

In regard to signal combinations, Liebal et al. (2004b) only briefly mention that siamangs combine approximately one-quarter of their signals into sequences, mostly consisting of two signals, which in about 40 % of the time represent repetitions of the same signal, resembling findings for captive orangutans and chimpanzees (Liebal et al. 2004a; Tempelmann and Liebal 2012). How these sequences emerge, whether they are meaningful combinations, or if they reflect strategies to adjust the communication to the behavior of the recipient as demonstrated for great apes (Genty and Byrne 2010; Hobaiter and Byrne 2011) is currently not known.

All studies described so far referred to the cognitive skills relevant for the *production* of signals. There is one study, however, that investigates the hylobatids' *comprehension* of communicative behaviors, produced by a human partner (Inoue et al. 2004). Thus, one human-raised, three-year-old female white-handed gibbon was able to use a human experimenter's directional cues (e.g., pointing, eye gaze, body orientation) to find food hidden under one of two containers. This is surprising, since great apes usually fail to use humans' communicative behaviors to locate hidden food in such object-choice tasks (Miklósi and Soproni 2006). Whether her superior performance in this test can be generalized more broadly and is thus representative for small apes or whether it can be explained by her special raising history, is currently unknown (Inoue et al. 2004).

Discussion of the Communicative and Cognitive Skills of Gibbons

The starting point of this chapter was that the small groups of hylobatids require only limited communicative repertoires with little variation between hylobatid species, because they mostly share the same basic set of social and ecological characteristics. Thus, vocal signals should dominate communication between groups, because small apes are territorial and need to communicate with other groups over larger distances in habitats with mostly dense vegetation, while only a small repertoire of facial expressions and gestures should be used for communication within groups, because of the limited number of group members. Similarly, hylobatid cognitive skills should be only modest compared to other primates that live in socially more complex groups.

First, an important conclusion from reviewing the current literature is that the increasing evidence for social flexibility across hylobatid species (Lappan 2007; Malone et al. 2012; Reichard et al. 2012) has not been fully acknowledged by research interested in their communicative skills. Thus, particularly studies examining their gestural and facial communication usually predict to observe only small repertoires and refer to the traditional notion that hylobatids live in small and stable groups compared to other ape species (Liebal et al. 2004b; Scheider et al. 2014). Indeed, from the little that is known about gestural communication of hylobatids and here mostly siamangs, they seem to use fewer gestures than great apes (Liebal et al. 2004b; Call and Tomasello 2007). However, it is important to highlight that most of the few existing studies on hylobatids' gestural and visual communication have been conducted in captive settings, with groups always consisting of one adult pair and their offspring, and thus with no possibility for variability in their social organization (Orgeldinger 1999; Liebal et al. 2004b). In other words, the little that is currently known about hylobatids' facial and gestural communication is from captive groups and might therefore not be representative for their communication in natural environments. On the other hand, a study with captive small apes revealed an unexpected variety of facial movements across different hylobatid species (Scheider et al. 2014), resembling those of primate species that live in more complex social groups. However, it is important to note that it was not differentiated between facial movements used in social interactions and those performed when individuals were on their own. Thus, apart from some qualitative studies (Baldwin and Teleki 1976; Chivers 1976; Fox 1977), very little is known about how hylobatids use their facial communicative repertoire in interactions with others and across different social contexts, in both captive and wild settings, and how this compares to facial communication of other primates (Parr et al. 2005; Pollick and de Waal 2007).

Second, most research on hylobatid communication concerns their vocalizations, particularly their songs. Although they are species- and sex-specific and seem to be largely genetically determined (Brockelman and Schilling 1984; Geissmann 1993), there is increasing evidence for both intra- and interindividual variation, particularly in wild settings (Dallmann and Geissmann 2009; Sun et al. 2011; Wanelik et al. 2013; Terleph et al. 2015) (see Koda 2016). Furthermore, more subtle calls used for intra-group communication have been shown to systematically vary in their acoustic features as a function of the social context they are used for (Clarke et al. 2015). Thus, particularly when studied in wild populations, hylobatid vocal communication seems to be characterized by a considerable degree of variability. However, more research is needed to investigate whether this variation within call types is also meaningful for the recipients of these vocalizations.

The fact that all hylobatid species produce their characterizing songs seems to support the prediction that they should mostly rely on loud vocalizations, because they live in an arboreal habitat (Marler 1965; Maestripieri 1999). However, it is too early to conclude that this is their predominant mode of communication, because softer vocalizations have not been studied extensively yet (Clarke et al. 2015). Furthermore, there are no systematic studies, which use a multimodal approach and compare hylobatid vocal repertoires with those of gestures and facial expressions, which would be essential to reveal the predominant use of one modality over the others. Finally, because the current knowledge about facial and gestural communication stems from captive settings and thus from interactions *within* groups, virtually nothing is known about the role of these communicative means during encounters *between* groups. Since variability in communicative repertoires and their flexible use is particularly expected if individuals have to constantly negotiate their relationships with others (Maestripieri 1999), communication between groups might reveal a much richer repertoire of signals, particularly gestures and facial expressions, than has been previously observed.

Similar to the predictions made for the communicative repertoires of hylobatids, their socio-cognitive skills are expected to be comparably modest in comparison to primate species that live in socially more complex groups. Thus, their social system primarily consisting of only two adults seems not particularly conducive to complex cognitive abilities (Humphrey 1976; Byrne and Whiten 1988). However, throughout this chapter this common notion has been challenged, because several hylobatid species have been shown to realize a complex social system—despite living in pairs (Fuentes 2000; Reichard et al. 2012). Thus, rather than perceiving hylobatid groups as single, isolated units, it is important to consider them as complex web of single units, which frequently interact and might compete over access to food or mating partners. Although competition between hylobatid groups has not been systematically studied yet, there is evidence that the density of gibbon populations is constrained by the availability of certain food types (Marshall and Leighton 2006; Marshall et al. 2009).

The consequence of this changed perspective on the complexity of the hylobatids' social system is to revise the predictions usually made regarding their socio-cognitive skills. Thus, they might not be as limited as previously assumed and might thus be more similar to those of species that live in complex social groups, further supported by the fact that life histories of hylobatids seem to resemble those of great apes more than those of monkeys (Reichard and Barelli 2008). However, the few existing studies do not support this assumption of more sophisticated socio-cognitive skills of hylobatids, since they produced inconsistent or negative findings. Thus, hylobatids seem to only co-orient with others' gazes rather than to take the visual perspective of others (Horton and Caldwell 2006; Liebal and Kaminski 2012) and they have largely failed to recognize themselves in mirrors (Ujhelyi et al. 2000; Suddendorf and Collier-Baker 2009). This seems to differentiate them from great apes, although it should be acknowledged that on an individual level, great apes' performance also varies considerably, since not all individuals successfully passed the mirror test either (Swartz and Evans 1994). In communicative interactions with conspecifics, however, siamangs have been shown to use their gestures as flexibly as great apes: they adjust gesture use to the attentional state of the recipient and only use visual gestures if the other is visually attending (Liebal et al. 2004b; Call and Tomasello 2007). Furthermore, like great apes, they use most of their gestures flexibly across different social contexts.

Apart from these studies, however, very little is known about hylobatids' socio-cognitive skills. Different reasons are discussed to explain the lack of research on small apes, such as their low levels of motivation to participate in experimental studies (Thompson et al. 1965; Fedor et al. 2008), the use of inappropriate experimental designs that have not considered the morphological features of hylobatid hands (Beck 1967), difficulties in obtaining large samples of hylobatids species, and problems in temporarily separating individuals from other group members for the purpose of such experiments (Liebal and Kaminski 2012). Finally, a common assumption might be that they are not an interesting group to study, because of their 'simple' social system and correspondingly 'simple' communicative and cognitive skills. However, it is important to highlight our current knowledge on socio-cognitive skills of small apes is exclusively based on research on captive groups, which live in comparably poor environmental conditions with no possibilities to interact with other groups.

The Way Forward

Communicative and cognitive skills of hylobatids—in contrast to great apes and monkeys—have not yet been studied in sufficient detail. This final section will point to some suggestions to address this lack of knowledge.

Integrated Approach to Social and Communicative Complexity

The increasing body of research demonstrating a large degree of flexibility in hylobatids' social system has not been incorporated yet into theories regarding the complexity of their communicative and cognitive skills. Furthermore, since most of hylobatid visual and tactile communication has been studied in captive settings, the focus has been limited to within-group communication, while a much richer repertoire might be expected during interactions between groups. Therefore, to fully capture the complexity of hylobatid communication, it is necessary to consider their social flexibility and to compare their communication both within and between groups.

Multimodal Approach

To enable systematic comparisons across modalities, researchers are encouraged to study more than one modality at once (Slocombe et al. 2011). The joint investigation of facial expressions, vocalizations, and gestures and their usage will enable researchers to capture the diversity of hylobatid communication that so far might have been missed by an exclusive focus on one modality (Slocombe et al. 2011; Liebal et al. 2013).

Method Development

It is necessary to develop methods that allow gesture classification based on structural properties, similar to methods that measure acoustic properties of vocalizations or discrete units of facial movements. This will enable more objective measurements based on form and standardized comparisons across different studies and species.

Usage of Signals

Although the size of signal repertoires correlates with group size as demonstrated for facial and vocal signals (McComb and Semple 2005; Dobson 2009), it is necessary to also investigate the frequency of occurrence of signals and their usage across different social contexts. This is important because despite a large repertoire, only a small proportion of signals may be used regularly. Furthermore, investigations into the frequency of occurrences across modalities allow conclusions about which signal types are predominantly used, for example, as a function of context or age.

Mechanisms Underlying Signal Acquisition and Use

For vocalizations, there are some studies that investigate the mechanisms underlying the emergence and development of gibbon songs (Geissmann 1984; Koda et al. 2013). However, nothing is yet known about developmental aspects of facial expressions or gestures in hylobatids. Understanding the ontogeny of hylobatid communication will provide important insights into the mechanisms underlying the acquisition of different signal types and the cognitive foundations of their use. For example, although the structure of a particular signal may be largely genetically determined, interactions with other group members might still influence certain structural properties as shown for the acquisition of gibbon songs (Koda et al. 2013). Similarly, individuals may still need to learn the appropriate use of a signal in the corresponding context and/or the appropriate reaction to it, as has been demonstrated for monkey vocalizations (Seyfarth and Cheney 1980; Fischer 2003).

Comparison Across Species and Habitats

Most knowledge on hylobatid communication is limited to not even a handful of species, which is particularly true for their facial and gestural communication.

While vocalizations are often studied in natural habitats, facial expressions and gestures are largely studied in captive settings. Although this can be partly explained by methodological constraints (e.g., better visibility of visual signals in captivity), it is important to highlight that it is currently not clear whether the knowledge obtained from captive populations also applies to wild hylobatids, particularly because it is not possible to study intergroup interactions in captive settings. Although Fox (1977) mentioned that there are no considerable differences between wild and captive siamangs, much more research is needed which systematically compares the communication of wild and captive populations.

Concluding, hylobatids are a largely understudied group of primates and the little that is known about their communication and cognition comes from few studies and species only. Although hylobatids—in comparison to primate species living in larger social groups—seem to have a more limited communicative repertoire, particularly of gestures and facial expressions, and only modest cognitive skills, it is important to highlight that many aspects of their communication and cognition are currently unknown.

References

Abordo EJ (1976) The learning skills of gibbons. In: Rumbaugh DM (ed) Gibbon and siamang, vol 4. Karger, Basel, pp 106–134

Altmann SA (1967) The structure of primate social communication. In: Altmann SA (ed) Social communication among primates. University of Chicago Press, Chicago, pp 325–362

Anderson JR (1984) Monkeys with mirrors: some questions for primate psychology. Int J Primatol 1:81–98

Baldwin LA, Teleki G (1976) Patterns of gibbon behavior on Hall's Island, Bermuda: a preliminary ethogram for *Hylobates lar*. In: Rumbaugh DM (ed) Gibbon and siamang, vol 4. Karger, Basel, pp 21–105

Bard KA, Todd BK, Bernier C, Love J, Leavens DA (2006) Self-awareness in human and chimpanzee infants: what is measured and what is meant by the mark and mirror test? Infancy 9:191–219

Barelli C, Boesch C, Heistermann M, Reichard UH (2008a) Female white-handed gibbons (*Hylobates lar*) lead group movements and have priority of access to food resources. Behaviour 145:965–981

Barelli C, Heistermann M, Boesch C, Reichard UH (2008b) Mating patterns and sexual swellings in pair-living and multimale groups of wild white-handed gibbons, *Hylobates lar*. Anim Behav 75:991–1001

Barelli C, Matsudaira K, Wolf T, Roos C, Heistermann M, Hodges K, Ishida T, Malaivijitnond S, Reichard UH (2013a) Extra-pair paternity confirmed in wild white-handed gibbons. Am J Primatol 75:1185–1195

Barelli C, Mundry R, Heistermann M, Hammerschmidt K (2013b) Cues to androgens and quality in male gibbon songs. PLoS ONE 8(12):e82748

Bartlett TQ (2003) Intragroup and intergroup social interactions in white-handed gibbons. Int J Primatol 24:239–259

Beck B (1967) A study of problem solving by gibbons. Behaviour 28:95–109

Bräuer J, Call J, Tomasello M (2005) All great ape species follow gaze to distant locations and around barriers. J Comp Psychol 119:145–154

Brockelman WY, Schilling D (1984) Inheritance of stereotyped gibbon calls. Nature 312:634–636

Brockelman WY, Srikosamatara S (1984) Maintenance and evolution of social structure in gibbons. In: Preuschoft H, Chivers DJ, Brockelman WY, Creel N (eds) The lesser apes. Evolutionary and behavioural biology. Edinburgh University Press, Edinburgh, pp 298–323

Brockelman WY, Reichard UH, Treesucon U, Raemaekers JJ (1998) Dispersal, pair formation and social structure in gibbons (*Hylobates lar*). Behav Ecol Sociobiol 42:329–339

Bugnyar T, Stowe M, Heinrich B (2004) Ravens, *Corvus corax*, follow gaze direction of humans around obstacles. Proc R Soc B 271:1331–1336

Burrows AM, Waller BM, Parr LA, Bonar CJ (2006) Muscles of facial expression in the chimpanzee (*Pan troglodytes*): descriptive, comparative and phylogenetic contexts. J Anat 208:153

Burrows AM, Waller BM, Parr LA (2009) Facial musculature in the rhesus macaque (*Macaca mulatta*): evolutionary and functional contexts with comparisons to chimpanzees and humans. J Anat 215:320–334

Burrows AM, Diogo R, Waller BM, Bonar CJ, Liebal K (2011) Evolution of the muscles of facial expression in a monogamous ape: evaluating the relative influences of ecological and phylogenetic factors in hylobatids. Anat Rec 294:645–663

Butterworth G, Jarrett N (1991) What minds have in common is space: spatial mechanisms serving joint visual attention in infancy. Brit J Dev Psychol 9:55–72

Byrne RW, Whiten A (eds) (1988) Machiavellian intelligence: social expertise and the evolution of intellect in monkeys, apes and humans. Clarendon Press, Oxford

Call J, Tomasello M (eds) (2007) The gestural communication of apes and monkeys. Lawrence Erlbaum Associates, Mahwah, New Jersey

Carpenter CR (ed) (1940) A field study in Siam of the behavior and social relations of the gibbon (*Hylobates lar*). In: Comparative psychology monographs, vol 16. The John Hopkins Press, Baltimore, Maryland

Cheyne SM, Chivers DJ, Sugardjito J (2007) Covarvariation in the great calls of rehabilitant and wild gibbons (*Hylobates albibarbis*). Raffles Bull Zool 55:201–207

Chivers DJ (1974) The siamang in Malaya: a field study of a primate in tropical rain forest, vol 4. In: Contribution to primatology. Karger, Basel

Chivers DJ (1976) Communication within and between family groups of siamang (*Symphalangus syndactylus*). Behavior 57:116–135

Clarke E, Reichard UH, Zuberbühler K (2006) The syntax and meaning of wild gibbon songs. PLoS ONE 1(1):e73

Clarke E, Reichard UH, Zuberbühler K (2015) Context-specific close-range "hoo" calls in wild gibbons (*Hylobates lar*). BMC Evol Biol 15:56

Cowlishaw G (1992) Song function in gibbons. Behaviour 121:131–153

Crockford C, Wittig RM, Mundry R, Zuberbühler K (2012) Wild chimpanzees inform ignorant group members of danger. Curr Biol 22:142–146

Dallmann R, Geissmann T (2001) Individuality in the female songs of wild Silvery Gibbons (*Hylobates moloch*) on Java, Indonesia. Contrib Zool 70:41–50

Dallmann R, Geissmann T (2009) Individual and geographical variability in the songs of wild Silvery gibbons (*Hylobates moloch*) on Java, Indonesia. In: Lappan S, Whittaker DJ (eds) The gibbons: new perspectives on small ape socioecology and population biology. Springer, Berlin, pp 91–110

De Veer MW, van den Boos R (1999) A critical review of methodology and interpretation of mirror self-recognition research in nonhuman primates. Anim Behav 58:459–468

Dobson SD (2009) Socioecological correlates of facial mobility in nonhuman anthropoids. Am J Phys Anthropol 139:413–420

Dunbar RIM (2003) The social brain: mind, language, and society in evolutionary perspective. Annu Rev Anthropol 32:163–181

Ekman P, Friesen WV (eds) (1978) Facial action coding system. Consulting Psychologists Press, Palo Alto

Ellefson JO (1967) A natural history of gibbons in the Malay Peninsula. Disserattion, University of California, Berkeley

Ellefson JO (1974) A natural history of white-handed gibbons in the Malayan peninsula. In: Rumbaugh DM (ed) Gibbon and siamang, vol 3. Karger, Basel, pp 1–136

Fan PF, Xiao W, Huo S, Jiang XL (2009) Singing behavior and singing functions of black-crested gibbons (*Nomascus concolor jingdongensis*) at Mt. Wuliang, central Yunnan, China. Am J Primatol 71:539–547

Fedor A, Skollár G, Szerencsy N, Ujhelyi M (2008) Object permanence tests on gibbons (Hylobatidae). J Comp Psychol 122:403–417

Fischer J (2003) Developmental modifications in the vocal behavior of non-human primates. In: Ghazanfar AA (ed) Primate audition ethology and neurobiology. CRC Press, Boca Raton

Fischer JO, Geissmann T (1990) Group harmony in gibbons: comparison between white-handed gibbon (*Hylobates lar*) and siamang (*Hylobates syndactylus*). Primates 31:481–494

Fleagle JG (ed) (1999) Primate adaptation and evolution. Academic Press, New York

Fox GJ (1972) Some comparisons between siamang and gibbon behaviour. Folia Primatol 18:122–139

Fox GJ (1977) Social dynamics in siamang. Dissertation, The University of Wisconsin, Milwaukee

Freeberg TM, Dunbar RIM, Ord TJ (2012) Social complexity as a proximate and ultimate factor in communicative complexity. Phil Trans R Soc B 367:1785–1801

Fuentes A (2000) Hylobatid communities: changing views on pair bonding and social organization in hominoids. Am J Phys Anthropol 113:33–60

Gallup G (1970) Chimpanzees: self-recognition. Science 167:86–87

Gallup GG (1998) Self-awareness and the evolution of social intelligence. Behav Process 42:239–247

Gallup G, McClure M, Hill S, Bundy R (1971) Capacity for self recognition in differentially reared chimpanzees. Psychol Rec 21:69–74

Gallup G, Povinelli D, Suarez S, Anderson J, Lethmate J, Menzel E (1995) Further reflections on self-recognition in primates. Anim Behav 50:1525–1532

Geissmann T (1984) Inheritance of song parameters in the gibbon song analysed in two hybrid gibbons (*Hylobates pileatus* x *Hylobates lar*). Folia Primatol 42:216–235

Geissmann T (1993) Evolution of communication in gibbons (*Hylobates spp.*). Dissertation, University of Zürich, Zürich

Geissmann T (1999) Duetting songs of the siamang, *Hylobates syndactylus*: II. Testing the pair-bonding hypothesis during a partner exchange. Behaviour 136:1005–1039

Geissmann T (2000) Gibbon songs and human music in an evolutionary perspective. In: Wallin NL, Merker B, Brown S (eds) The origins of music. MIT Press, Cambridge, MA, pp 103–123

Geissmann T (2002) Duet-splitting and the evolution of gibbon songs. Biol Rev 77:57–76

Geissmann T, Orgeldinger M (2000) The relationship between duet songs and pair bonds in siamangs, *Hylobates syndactylus*. Anim Behav 60:805–809

Genty E, Byrne R (2010) Why do gorillas make sequences of gestures? Anim Cogn 13:287–301

Gittins SP (1979) The behaviour and ecology of the agile gibbon (*Hylobates agilis*). Dissertation, University of Cambridge, Cambridge

Gittins SP, Raemaekers JJ (1980) Siamang, lar and agile gibbons. In: Chivers DJ (ed) Malayan forest primates. Ten years' study in tropical rain forest. Plenum Press, New York, pp 63–105

Gómez JC (2005) Species comparative studies and cognitive development. Trends Cogn Sci 9:118–125

Goodall J (ed) (1986) The chimpanzees of Gombe: patterns of behavior. Harvard University Press, Cambridge

Groves CP (ed) (2001) Primate taxonomy. Smithsonian Institution Press, Washington, London

Haimoff EH (1984) Acoustic and organizational features of gibbon songs. In: Preuschoft H, Chivers DJ, Brockelman WY, Creel N (eds) The lesser apes. Evolutionary and behavioural biology. Edinburgh University Press, Edinburgh, pp 333–353

Haimoff EH, Gittins SP (1985) Individuality in the songs of wild agile gibbons (*Hylobates agilis*) of Peninsular Malaysia. Am J Primatol 8:239–247

Haraway MM, Maples EG (1998) Flexibility in the species-typical songs of gibbons. Primates 39:1–12

Hare B, Call J, Agnetta B, Tomasello M (2000) Chimpanzees know what conspecifics do and do not see. Anim Behav 59:1–15

Hare B, Call J, Tomasello M (2001) Do chimpanzees know what conspecifics know? Anim Behav 61:139–151

Harrison ME, Marshall AJ (2011) Strategies for the use of fallback foods in apes. Int J Primatol 32:531–565

Heyes CM (1994) Reflections on self-recognition in primates. Anim Behav 46:177–188

Heyes CM (1995) Self-recognition in primates: further reflections create a hall of mirrors. Anim Behav 50:1533–1542

Hobaiter C, Byrne RW (2011) Serial gesturing by wild chimpanzees: its nature and function for communication. Anim Cogn 14:827–838

Horton K, Caldwell C (2006) Visual co-orientation and expectations about attentional orientation in pileated gibbons (*Hylobates pileatus*). Behav Process 72:65–73

Hostetter AB, Cantero M, Hopkins WD (2001) Differential use of vocal and gestural communication by chimpanzees (*Pan troglodytes*) in response to the attentional status of a human (*Homo sapiens*). J Comp Psychol 115:337–343

Humphrey NK (1976) The social function of intellect. In: Bateson PPG, Hinde RA (eds) Growing points in ethology. Cambridge University Press, Cambridge, pp 303–317

Hyatt CW (1998) Responses of gibbons (*Hylobates lar*) to their mirror images. Am J Primatol 45:307–311

Inoue Y, Inoue E, Itakura S (2004) Use of experimenter-given directional cues by a young white-handed gibbon (*Hylobates lar*). Jpn Psychol Res 46:262–267

Jolly A (1966) Lemur social behaviour and primate intelligence. Science 153:501–506

Kaminski J, Riedel J, Call J, Tomasello M (2005) Domestic goats, *Capra hircus*, follow gaze direction and use social cues in an object choice task. Anim Behav 69:11–18

Kappeler PM, van Schaik CP (2002) Evolution of primate social systems. Int J Primatol 23:707–740

Kappeler PM, van Schaik CP (eds) (2006) Cooperation in primates and humans. Springer, Berlin

Keith SA, Waller MS, Geissmann T (2009) Vocal diversity of Kloss's gibbons (*Hylobates klossii*) in the Mentawai Islands, Indonesia. In: Lappan S, Whittaker DJ (eds) The Gibbons: new perspectives on small ape socioecology and population biology. Springer, Berlin, pp 51–71

Kenyon M, Roos C, Binh VT, Chivers D (2011) Extrapair paternity in golden-cheeked gibbons (*Nomascus gabriellae*) in the secondary lowland forest of Cat Tien National Park, Vietnam. Folia Primatol 82:154–164

Koda H (2016) Gibbon songs: Understanding the evolution and development of this unique form of vocal communication. In: Reichard UH, Hirohisa H, Barelli C, (eds) Evolution of gibbons and siamang. Springer, New York, pp. 347–357

Koda H, Oyakawa C, Kato A, Masataka N (2007) Experimental evidence for the volitional control of vocal production in an immature gibbon. Behaviour 144:681–692

Koda H, Lemasson A, Oyakawa C, Pamungkas J, Masataka N (2013) Possible role of mother-daughter vocal interactions on the development of species-specific song in gibbons. PLoS ONE 8(8):e71432

Lappan S (2007) Social relationships among males in multimale siamang groups. Int J Primatol 28:369–387

Lappan S (2008) Male care of infants in a siamang (*Symphalangus syndactylus*) population including socially monogamous and polyandrous groups. Behav Ecol Sociobiol 62:1307–1317

Leighton DR (1987) Gibbons: territoriality and monogamy. In: Smuts BB, Cheney DL, Seyfarth RM, Wrangham RW, Struhsaker TT (eds) Primate societies. The University of Chicago Press, Chicago, pp 135–145

Lethmate J, Duecker G (1973) Untersuchungen zum Selbsterkennen im Spiegel bei Orang-Utans und einigen anderen Affenarten. Z Tierpsychol 33:248–269

Liebal K, Call J (2012) The origins of non-human primates' manual gestures. Phil Trans R Soc B 367:118–128

Liebal K, Kaminski J (2012) Gibbons (*Hylobates pileatus, H. moloch, H. lar, Symphalangus syndactylus*) follow human gaze, but do not take the visual perspective of others. Anim Cogn 5:1211–1216

Liebal K, Call J, Tomasello M (2004a) Use of gesture sequences in chimpanzees (*Pan troglodytes*). Am J Primatol 64:377–396

Liebal K, Pika S, Tomasello M (2004b) Social communication in siamangs (*Symphalangus syndactylus*): use of gestures and facial expressions. Primates 45:41–57

Liebal K, Waller BM, Burrows A, Slocombe K (eds) (2013) Primate communication: a multimodal approach. Cambridge University Press, Cambridge

MacKinnon JR (1977) A comparative ecology of Asian apes. Primates 18:747–772

MacKinnon JR, MacKinnon KS (1980) Niche differentiation in a primate community. In: Chivers DJ (ed) Malayan forest primates: ten years' study in tropical rain forest. Plenum Press, New York, pp 167–190

Maestripieri D (1996a) Gestural communication and its cognitive implications in pigtail macaques (*Macaca nemestrina*). Behaviour 133:997–1022

Maestripieri D (1996b) Social communication among captive stump-tailed macaques (*Macaca arctoides*). Int J Primatol 17:785–802

Maestripieri D (1999) Primate social organization, gestural repetoire size, and communication dynamics. In: King B (ed) The origins of language: what nonhuman primates can tell us. School of American Research Press, Santa Fe, pp 55–77

Malone N, Fuentes A, White FJ (2012) Variation in the social systems of extant hominoids: comparative insight into the social behavior of early hominins. Int J Primatol 33:1251–1277

Marler P (1965) Communication in monkeys and apes. In: DeVore I (ed) Primate behaviour. Field studies of monkeys and apes. Holt, Rinehart and Winston, New York, pp 544–584

Marler P, Tenaza R (1977) Signaling behavior of apes with special reference to vocalizations. In: Sebeok TA (ed) How animals communicate, vol II. Indiana University Press, London, pp 965–1033

Marshall AJ, Leighton M (2006) How does food availability limit the population density of white-bearded gibbons? In: Hohmann G, Robbins MM, Boesch C (eds) Feeding ecology in apes and other primates. Ecological, physical and behavioral aspects. Cambridge University Press, Cambridge, pp 311–333

Marshall JT, Marshall ER (1976) Gibbons and their territorial songs. Science 193:235–237

Marshall AJ, Boyko CM, Feilen KL, Boyko RH, Leighton M (2009) Defining fallback foods and assessing their importance in primate ecology and evolution. Am J Phys Anthropol 140:603–614

McComb K, Semple S (2005) Coevolution of vocal communication and sociality in primates. Biol Lett 1:381–385

McConkey KR, Aldy F, Ario A, Chivers DJ (2002) Selection of fruit by gibbons (*Hylobates muelleri* × *agilis*) in the rain forests of central Borneo. Int J Primatol 23:123–145

Miklósi Á, Soproni K (2006) A comparative analysis of animals' understanding of the human pointing gesture. Anim Cogn 9:81–93

Miklósi Á, Polgárdi R, Topál J, Csányi V (1998) Use of experimenter-given cues in dogs. Anim Cogn 1:113–121

Mitani JC (1984) The behavioural regulation of monogamy in gibbons (*Hylobates muelleri*). Behav Ecol Sociobiol 15:225–229

Mitani JC (1985) Responses of gibbons (*Hylobates muelleri*) to self, neighbor, and stranger song duets. Int J Primatol 6:193–200

Mitani JC (1987) Territoriality and monogamy among agile gibbons (*Hylobates agilis*). Behav Ecol Sociobiol 20:265–269

Mitani JC (1988) Male gibbon (*Hylobates agilis*) singing behavior: natural history, song variations and function. Ethology 79:177–194

Mitani JC, Stuht J (1998) The evolution of nonhuman primate loud calls: acoustic adaptation for long-distance transmission. Primates 39:171–182

Nielsen M, Suddendorf T, Slaughter V (2006) Mirror self-recognition beyond the face. Child Dev 77:176–185

Orgeldinger M (ed) (1999) Paarbeziehungen beim Siamang-Gibbon (*Hylobates syndactylus*) im Zoo: Untersuchungen über den Einfluß von Jungtieren auf die Paarbindung. Schüling Verlag, Münster

Oyakawa C, Koda H, Sugiura H (2007) Acoustic features contributing to the individuality of wild agile gibbon (*Hylobates agilis agilis*) songs. Am J Primatol 69:777–790

Palombit RA (1992) Pair bonds and monogamy in wild siamang (*Hylobates syndactylus*) and white-handed gibbon (*Hylobates lar*). Dissertation, University of California, Davis

Palombit RA (1993) Lethal territorial aggression in a white-handed gibbon. Am J Primatol 31:311–318

Palombit RA (1994) Extra-pair copulations in a monogamous ape. Anim Behav 47:721–723

Palombit RA (1996) Pair bonds in monogamous apes: a comparison of the siamang *Hylobates syndactylus* and the white-handed gibbon *Hylobates lar*. Behaviour 133:321–356

Parker ST, Mitchell RW, Boccia ML (eds) (1994) Self-awareness in animals and human. Cambridge University Press, Cambridge

Parr LA, Cohen M, de Waal FBM (2005) Influence of social context on the use of blended and graded facial displays in chimpanzees. Int J Primatol 26:73–103

Parr LA, Waller BM, Vick SJ, Bard KA (2007) Classifying chimpanzee facial expressions using muscle action. Emotion 7:172–181

Partan S, Marler P (1999) Communication goes multimodal. Science 283:1272–1273

Patterson FG, Cohn R (1994) Self-recognition and self-awareness in lowland gorillas. In: Parker ST, Boccia ML, Mitchell R (eds) Self awareness in animals and humans. University Press, Cambridge, pp 273–290

Plotnik J, de Waal F, Reiss D (2006) Self-recognition in an Asian elephant. Proc Natl Acad Sci 103:17053–17057

Pollick AS, de Waal FBM (2007) Ape gestures and language evolution. Proc Natl Acad Sci 104:8184–8189

Preuschoft S (2004) Power and communication. In: Thierry B, Singh M, Kaumanns W (eds) Macaque societies: a model for the study of social organization. Cambridge University Press, Cambridge, pp 56–60

Prime JM, Ford SM (2016) Hand Manipulation Skills in Hylobatids. In: Reichard UH, Hirohisa H, Barelli C, (eds) Reichard evolution of gibbons and siamang. Springer, New York, pp. 267–287

Prior H, Schwarz A, Güntürkün O (2008) Mirror-induced behavior in the magpie (*Pica pica*): evidence of self-recognition. PLoS Biol 6:e202

Raemaekers JJ, Raemaekers PM (1985a) Field playback of loud calls to gibbons (*Hylobates lar*)— territorial, sex-specific and species-specific responses. Anim Behav 33:481–493

Raemaekers PM, Raemaekers JJ (1985b) Long-range vocal interactions between groups of gibbons (*Hylobates lar*). Behaviour 95:26–44

Raemaekers JJ, Raemaekers PM, Haimoff EH (1984) Loud calls of the gibbon (*Hylobates lar*): Repertoire, organisation, and context. Behaviour 91:146–189

Reichard UH (1995) Extra-pair copulations in a monogamous gibbon (*Hylobates lar*). Ethology 100:99–112

Reichard UH (2009) The social organization and mating system of Khao Yai white-handed gibbons: 1992–2006. In: Lappan S, Whittaker DJ (eds) The gibbons: new perspectives on small ape socioecology and population biology. Springer-Verlag, Berlin, pp 347–384

Reichard UH, Barelli C (2008) Life history and reproductive strategies of Khao Yai *Hylobates lar*: implications for social evolution in apes. Int J Primatol 29:823–844

Reichard U, Sommer V (1997) Group encounters in wild gibbons (*Hylobates lar*): agonism, affiliation, and the concept of infanticide. Behaviour 134:1135–1174

Reichard UH, Ganpanakngan M, Barelli C (2012) White-handed gibbons of Khao Yai: social flexibility, complex reproductive strategies, and a slow life history. In: Kappeler P, Watts DP (eds) Long-term field studies of primates. Springer, Berlin, pp 237–258

Roberts AI, Vick S-J, Roberts SGB, Buchanan-Smith HM, Zuberbühler K (2012) A structure-based repertoire of manual gestures in wild chimpanzees: statistical analyses of a graded communication system. Evol Hum Behav 33:578–589

Rosati AG, Hare B (2009) Looking past the model species: diversity in gaze-following skills across primates. Curr Opin Neurobiol 19:45–51

Scheider L, Waller BM, Oña L, Burrows A, Liebal K (2014) A comparison of facial expression properties in five hylobatid species. Am J Primatol 76:618–628

Scheider L, Waller BM, Oña L, Burrows AM, Liebal K (2016) Social use of facial expressions inhylobatids. PLoS ONE 11:e0151733

Schel AM, Townsend SW, Machanda Z, Zuberbühler K, Slocombe KE (2013) Chimpanzee alarm call production meets key criteria for intentionality. PLoS ONE 8(10):e76674

Seyfarth R, Cheney D (1980) The ontogeny of vervet monkey alarm calling behaviour: a preliminary report. Z Tierpsychol 54:37–56

Seyfarth RM, Cheney DL (2010) Production, usage, and comprehension in animal vocalizations. Brain Lang 115:92–100

Shillito DJ, Gallup GG, Beck B (1999) Factors affecting mirror behaviour in western lowland gorillas, *Gorilla gorilla*. Anim Behav 57:999–1004

Shultz S, Dunbar RI (2007) The evolution of the social brain: anthropoid primates contrast with other vertebrates. Proc R Soc B 274:2429–2436

Shultz S, Opie C, Atkinson QD (2011) Stepwise evolution of stable sociality in primates. Nature 479:219–222

Slocombe KE, Waller BM, Liebal K (2011) The language void: the need for multimodality in primate communication research. Anim Behav 81:919–924

Sommer V, Reichard UH (2000) Rethinking monogamy: the gibbon case. In: Kappeler PM (ed) Primate males: causes and consequences of variation in group composition. Cambridge University Press, Cambridge, pp 159–168

Suarez S, Gallup GG (1981) Self-recognition in chimpanzees and orangutans, but not gorillas. J Hum Evol 10:175–188

Suddendorf T, Collier-Baker E (2009) The evolution of primate visual self-recognition: evidence of absence in lesser apes. Proc R Soc B 276:1671–1677

Sun GZ, Huang B, Guan ZH, Geissmann T, Jiang XL (2011) Individuality in male songs of wild black crested gibbons (*Nomascus concolor*). Am J Primatol 73:431–438

Swartz KB, Evans S (1994) Social and cognitive factors in chimpanzee and gorilla mirror behaviour and self-recognition. In: Parker ST, Boccia ML, RM (eds) Self awareness in animals and humans. University Press, Cambridge, pp 189–206

Tempelmann S, Liebal K (2012) Spontaneous use of gesture sequences in orangutans: a case for strategy? In: Pika S, Liebal K (eds) Recent developments in primate gesture research. John Benjamins Publishing Company, pp 73–92

Tenaza RR (1985) Songs of hybrid gibbons (*Hylobates lar* x *H. muelleri*). Am J Primatol 8:249–253

Terleph TA, Malaivijitnond S, Reichard UH (2015) Lar gibbon (*Hylobates lar*) great call reveals individual caller identity. Am J Primatol 77:811–821

Thinh VN, Hallam C, Roos C, Hammerschmidt K (2011) Concordance between vocal and genetic diversity in crested gibbons. BMC Evol Biol 11:36

Thompson WD, Kirk RE, Koestler AG, Bourgeois AE (1965) A comparison of operant conditioning responses in the baboon, gibbon, and chimpanzee. In: Vagtborg H (ed) The baboon in medical research. Proceedings of the first international symposium on the baboon and its use as an experimental animal. University of Texas Press, Austin, pp 81–93

Tomasello M (ed) (2008) Origins of human communication. The MIT Press, Cambridge

Tomasello M, Zuberbühler K (2002) Primate vocal and gestural communication. In: Bekoff M, Allen CS, Burghardt G (eds) The cognitive animal: empirical and theoretical perspecitives on animal cognition. MIT Press, Cambridge, pp 293–329

Tomasello M, Call J, Nagell K, Olguin R, Carpenter M (1994) The learning and use of gestural signals by young chimpanzees: a trans-generational study. Primates 35:137–154

Tomasello M, Hare B, Agnetta B (1999) Chimpanzees, *Pan troglodytes*, follow gaze direction geometrically. Anim Behav 58:769–777

Uhde N, Sommer V (2002) Antipredatory behavior in gibbons (*Hylobates lar*, Khao Yai/Thailand). In: Miller LE (ed) Eat or be eaten: predator sensitive foraging among primates. Cambridge University Press, Cambridge, pp 268–291

Ujhelyi M, Merker B, Buk P, Geissmann T (2000) Observations on the behavior of gibbons (*Hylobates leucogenys, H. gabriellae, and H. lar*) in the presence of mirrors. J Comp Psychol 114:253–262

van Hooff J (1962) Facial expressions in higher primates. Symp Zool Soc Lond 8:7–68

van Hooff J (1967) The facial displays of the catarrhine monkeys and apes. In: Morris D (ed) Primate ethology. Weidenfels and Nicolson, London, pp 7–68

van Schaik CP, Deaner RO (2003) Life history and cognitive evolution in primates. In: de Waal FBM, Tyack PL (eds) Animal social complexity. Harvard University Press, Cambridge, pp 5–25

van Schaik CP, Preuschoft S, Watts DP (2004) Great ape social systems. In: Russon AE, Begun DR (eds) The evolution of thought: evolutionary origins of great ape intelligence. Cambridge University Press, Cambridge, pp 190–209

Waller BM, Cherry L (2012) Facilitating play through communication: significance of teeth exposure in the gorilla play face. Am J Primatol 74:157–164

Waller BM, Lembeck M, Kuchenbuch P, Burrows AM, Liebal K (2012) GibbonFACS: a muscle-based facial movement coding system for hylobatids. Int J Primatol 33:809–821

Wanelik KM, Azis A, Cheyne SM (2013) Note-, phrase- and song-specific acoustic variables contributing to the individuality of male duet song in the Bornean southern gibbon (*Hylobates albibarbis*). Primates 54:159–170

Whitten AJ (1980) The Kloss gibbon in Siberut. Dissertation. University of Cambridge, Cambridge

Wickler W (1980) Vocal dueting and the pairbond: I. Coyness and partner commitment. A hypothesis. Z Tierpsychol 52:201–209

Zuberbuehler K (2002) A syntactic rule in forest monkey communication. Anim Behav 63:293–299

Chapter 15
Gibbon Songs: Understanding the Evolution and Development of This Unique Form of Vocal Communication

Hiroki Koda

White-handed gibbon female singing. Photo credit: Aru Toyoda

H. Koda (✉)
Primate Research Institute, Kyoto University, 41 Kanrin, Inuyama, Aichi 484-8506, Japan
e-mail: koda.hiroki.7a@kyoto-u.ac.jp

© Springer Science+Business Media New York 2016   347
U.H. Reichard et al. (eds.), *Evolution of Gibbons and Siamang*,
Developments in Primatology: Progress and Prospects,
DOI 10.1007/978-1-4939-5614-2_15

Uniqueness of Gibbon Songs

Gibbon songs have particularly unique characteristics. They are composed of temporally organized stereotypic consecutive notes, and are mainly identified as two song types: solos and duets (for comprehensive reviews, see Marshall and Marshall 1976; Haimoff 1984; Preuschoft et al. 1984; Geissmann 2002).

Solos are a serious of phrases, composing a song bout, voiced by either sex. Male solos may be uttered from the night trees at dawn and can last from a few minutes up to 4 h (Raemeakers and Raemeakers 1985), while female solos occur primarily in daylight and produce a single great call, or occasionally more than one, commonly without preamble or follow-up vocalizations. By contrast, a song bout composed of both male and female contributions are called duet. During duets, males and females exchange their own song parts with temporal antiphonal structures, which is behaviourally comparable to bird duets, defined just by a stable temporal alternation of male-female singing (Farabaugh 1982). Most gibbons sing species- and sex-specific songs, which means that songs are distinctive among species and between males and females (Marshall and Marshall 1976; Geissmann 2000). However, although hylobatid songs are in general a behaviourally unique form of communication in regards to both acoustic structure and coordination style, not all gibbon species behave in the same way. Moloch gibbons (*Hylobates moloch*) and Kloss's gibbons (*H. klossi*), for example, lack duet songs, and males never reply to a partner female's great calls (Geissmann 2002). Gibbon songs are extremely loud vocalizations (Todd and Merker 2004), reaching up to 1 or 2 km throughout the rainforest, suggesting gibbon communication would mainly rely on auditory rather than visual signals, possibly resulting from ecological adaptations to their arboreal life in the canopy. Previous studies dealing with gibbon song function have hypothesized at least three main functions of solos and duets: territorial defence, mate attraction and pair-bond strengthening (Mitani 1985; Cowlishaw 1992; Geissmann and Orgeldinger 2000). However, none of these studies have addressed the contribution of female gibbons in the duet, which are clearly unique as the female parts are more elaborate compared to those of males and because they show remarkable acoustic features (Haimoff 1984).

Despite the broadly recognized uniqueness of gibbon songs, there are still no sufficient studies addressing the fundamental question: why and how gibbon songs have evolved? So far, previous studies have accumulated 'subtle' evidence about behavioural and ecological functions of songs (Gittins 1984; Haimoff 1984; Mitani 1985; Cowlishaw 1992; Geissmann 1999, 2000, 2002; Geissmann and Orgeldinger 2000; Clarke et al. 2006; Oyakawa et al. 2007), yet they do not provide findings addressing the potential causes leading to the evolution of gibbon songs. To my knowledge, only two pioneering studies have addressed such a question, one by Marshall and Marshall (1976) on gibbon song species-specificity and the other on inheritance of species-specific features by Brockelman and Schilling (1984). The former study was the first to report species- and sex-specific characteristics in gibbon songs (Marshall and Marshall 1976), and the latter showed that species-specific patterns in gibbon songs are primarily genetically determined and rarely influenced by auditory experience with mothers (Brockelman and Schilling 1984). Both studies clearly suggested that gibbon songs are directly linked to gene expressions, and that species-specific patterns diverged in accordance with the complicated speciation process caused by the biogeographical background in Sundaland. A quarter of a century after the latter breakthrough, new findings have updated the hypotheses and have accumulated similar findings of song character inheritance in gibbons (Geissmann 1984; Tenaza 1985). Unfortunately, no further findings on the evolution of gibbon songs have been discovered. Here, I attempt to discuss a possible scenario of the evolution of gibbon songs in light of recent findings on gibbon song features (Koda et al. 2012, 2013).

Mother-Daughter Co-singing: 'Triggering' and 'Tutorial' Roles

Although sex specificity in gibbon songs is found in most gibbon species, our knowledge of the ontogeny of such sexual differentiation remains limited. Great calls, the stereotyped female vocal contribution, would be a good candidate model to trace the gibbon vocal developmental pathway. Although great calls present easily distinguishable acoustic structures, they are rarely studied. Predictably, great calls develop in females in accordance with social and physical maturation, since new-born and very young females do not produce great calls (Brockelman and Schilling 1984; Merker and Cox 1999). However, at a later age, gibbon researchers have reported curious and unique mother-daughter vocal interactions during duets, where a juvenile or subadult daughter would simultaneously sing along with the mother, whereas this does not occur with male offspring (Brockelman and Schilling 1984; Merker and Cox 1999). Brockelman and Schilling (1984) hypothesized that co-singing interactions could provide a form of 'learning' or 'practice' for young daughters before reaching adulthood. A comprehensive contribution on the life history of wild lar gibbons also included co-singing as an indication of a females' stages of maturation (Table 13.2 in Reichard 2003). Reichard (2003) reported for white-handed gibbons (*Hylobates lar*) that the co-singing rate changes depending on a females' developmental maturity, observing that adolescent females (4–6 years of age, Reichard 2003 for definition) occasionally co-sing with their mothers, subadult females (6-8 years) regularly co-sing, but fully adult females (>8 years; full adult body size) still residing in their natal group again only occasionally or rarely co-sing with their mother. Co-singing may reflect changes in social bonding between mother and daughter, however, such observations were only based on qualitative, not quantitative, data (Reichard 2003) without specifically testing whether mother-daughter co-singing plays any role in the development of female great calls. Indeed, a recent study by Koda et al. (2013) attempted to investigate such relationship and discuss possible implications of co-singing for the acquisition of mature great calls in gibbons. To address this question, Koda et al. (2013) collected data on six mother-daughter pairs of free-ranging agile gibbons (*Hylobates agilis agilis*) on Sumatra, Indonesia and analysed co-singing interactions in terms of both co-singing rates (the number of mother-daughter co-singing great calls divided by the total number of the mother's great calls) and similarities in acoustic structures between the calls of mothers and daughters. The similarity index is an acoustic score which compares how similar great calls are, generally represented as a degree of overlapping portions of two sonogram bitmaps (Fig. 15.1).

First, our Koda et al. (2013) findings revealed that co-singing rates strongly varied between the six studied groups and were negatively correlated with the temporal precision of the songs' synchronization and with the degree of mother-daughter acoustic resemblance. Daughters, who co-sang less often with their mothers, presented acoustically much more similar maternal songs than with other adult females' song. It was also found that mothers adjusted the songs they sang to a more stereotyped pattern when co-singing with their daughters occurred compared to when they were singing alone, which was particularly evident in groups with low duet rates (Koda et al. 2013). These results suggested that

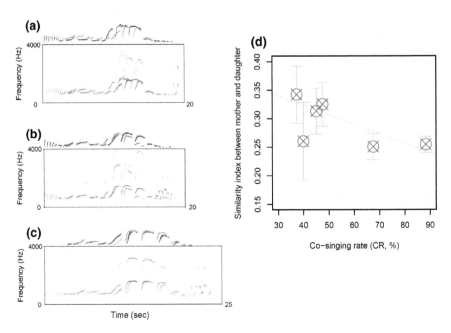

Fig. 15.1 a, b, c Sonograms of mother-daughter co-singing. Mother and daughter pitch contours are illustrated above the sonograms in *blue* and *orange*, respectively. The examples show the typical cases of less similar/less overlapped sonogram (**a**), and the more similar/more overlapped one (**b, c**). In (**d**) are plotted mother-daughter similarity indices ($MD_{co-singing}$) with co-singing rates (CRs) of the six pairs. *Open circles with crosses* indicate the group means, and *error bars* represent 95 % confidence intervals. The regression line is plotted using coefficients of estimated generalized linear mixed models. Adapted from Koda et al. (2013, Figs. 2 and 3). Copyright © 2013 by PLOS. Reprinted by permission of PLOS

mother-daughter co-singing may play an important role in the development of vocal control (temporal synchronization) and acoustic features (great call acoustic structures) of subadult females. Moreover, Koda et al. (2013) suggested that the decrease of co-singing rate at a daughter's older age would reflect the progress of daughters' independency from the mother's singing, while an increase in acoustic similarity would be a result of daughters' refinement of singing great calls. This interpretation is consistent with the notion of previous studies describing a decrease in co-singing from subadult to fully mature, adult females who were still residing in their natal group (Reichard 2003).

Koda et al. (2013) observations were based on daughters at the estimated age of 8 years old, at their advanced stage of social independence just prior to emigration. This may imply that the complete acquisition of great calls has already been achieved, which may explain the decrease in co-singing rates. Longitudinal observations would be ideal to investigate the complete sequence of acquisition and maturation processes of female great calling activities, however, data are currently limited and the only other study available on co-singing is on a captive yellow-cheeked gibbon (*Nomascus gabrielle*). The study revealed that co-singing rates between mother and immature daughter seemed to increase gradually after the onset of the performance of great calling, at the age of three years, until acquisition of the whole great call pattern (Merker and Cox 1999). The combination of Merker and Cox's (1999) and Koda et al. (2013) findings imply a possible inverse U-shaped process in co-singing rate (Fig. 15.2). Specifically, immature daughters will start to co-sing with their mothers when daughters' great calls are immature and incomplete, as suggested by Merker and Cox (1999). As the female matures, daughters acquire the species-specific song type and the co-singing rate increases in accordance with acoustic refinement of the entire great call pattern. Later, as suggested by the Koda et al. (2013) study, after a peak of co-singing, co-singing rates decrease, reflecting the maturation of daughter's sociality to be ready to emigrate from the natal group. In the latter process, the temporal regulation of song production control would become more precise as a result of the daughter's maturation (Fig. 15.2).

The above scenario may provide novel insights for debates on great call ontogeny in gibbons, and propose an innovative function. Great calls are female-specific songs, and in most gibbon species a male partner replies to a female's great calls with a male coda. Accordingly, great calls are believed to serve as female vocal displays communicating mainly with her partner. However, it is now suggested that gibbon mothers' songs may have 'triggering' and 'tutorial' roles in the development of daughter's great calls. This seems likely, because great calls of daughters always occurred as a form of co-singing with the mother, suggesting auditory input of the maternal great call likely triggering the start of a daughter's great call. In fact, Koda et al. (2013) found that gibbon mothers would modify their acoustic patterns when the daughter was co-singing with them, suggesting that mothers might modify their vocal behaviour according to a daughter's responses to her own singing. Although these findings do not show active 'teaching', they suggest that daughters' great call development may acoustically be influenced by those of the mothers.

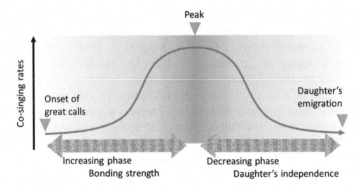

Fig. 15.2 Developmental model based on the mother-daughter co-singing rate. The inverse *U-shaped curve* shows the increase mother-daughter co-singing rate from the onset of great call until daughter full maturation. After the peak, co-singing rate may decrease until cessation corresponding to emigration from the natal group

Soprano Singing as a Unique Vocal Feature

The great loudness of gibbon songs is notably among primate vocalizations. Gibbon songs can reach more than 1 km through the forest, allowing long-distance vocal communication (Mitani 1985). In addition, vocal activity is extremely high (Geissmann 2002). Most gibbons engage in singing activity in the morning, and males may sing for several hours (Brockelman and Srikosamatar 1993), which differs from most other non-human primates who sing loud calls (e.g., Delgado 2006). Considering their small body size, approximately equivalent to a 6-month-old human baby (Némai and Kelemen 1933; Hayama 1970; Hewitt et al. 2002), sound loudness in gibbons is surprising. I propose that this is possible in gibbons through the use of a technique similar to one performed by soprano singers (Koda et al. 2012). Soprano singing is an operatic technique allowing sound propagation to be optimal in a large auditorium. Such a technique has also been named 'formant tuning' in which the soprano increases the fundamental frequency (f_0), determined by vocal fold vibrations, up to the soprano range, while simultaneously tuning the lowest resonance frequency, or first formant frequency, close to the f_0 (Sundberg 1987; Joliveau et al. 2004a, b; Fig. 15.3). The concurrence between the fundamental and the first formant frequencies, which is determined by the shape of the vocal tract, amplifies the loudness of sounds. The formant tuning technique gives greater loudness to the voice for a given effort and may have musical advantages in terms of sound propagation efficiency. In gibbons, the f_0 is normally around 1 kHz, which is close to their predicted lowest formant frequency (Haimoff 1984). According to source-filter theory (Fant 1970), higher audible songs should be performed under such prepared configurations of the vocal apparatus, with a tight coupling between f_0 and formant frequency. In order to investigate

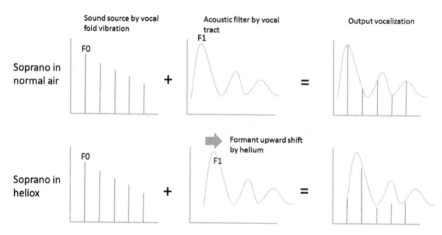

Fig. 15.3 Schematic representations for soprano singing model (*upper*) and heliox effect on soprano singing (*bottom*). All graphs are the illustrations of power spectrum, and the x- and y-axes represent the frequency and sound energy, respectively. The left panels show the sound source generated by the vocal fold vibration, and the sound source possess the harmonic structures, where the lowest and strongest frequency component is the fundamental frequency (f_0). The middle panels show the acoustic resonant filter by the vocal tract, and the lowest and strongest resonant frequency is the first formant frequency (F_1). The right panels are the final output vocalizations, where the sound source and filter are integrated. In soprano singing, the f_0 and F_1 is perfectly matched to amplify the sound intensity, particularly of f_0. In heliox condition, heliox influence only the vocal tract filter without any influence of f_0. Consequently, sound intensity of f_0 is dramatically attenuated due to the mismatch of f_0 and F_1

vocal propagation mechanisms in gibbons, Koda et al. (2012) applied a helium-enriched atmosphere (heliox: Nowicki 1987), an approach which analyzes the acoustic interactions between f_0 and formant frequency with the acoustic simulations of a soprano model. Since the velocity of sound in a heliox should be greater than that in air, the resonance frequencies of the vocal tract are shifted upward by nearly an octave without influencing the f_0 (Fig. 15.3). Remarkably, the heliox experiments with one white-handed gibbon female (*Hylobates lar*) did not show any experimental influence on the f_0, but on formant frequency, indicating an independency between vocal fold vibration and vocal tract movement (Koda et al. 2012; Fig. 15.4). Of particular importance was the evidence that formant frequencies are kept in close position to the f_0. This suggested that gibbons' vocalizations are produced similarly to professional human soprano singers, in which a precise matching of the formant frequency to f_0 amplifies the loudness of songs (Koda et al. 2012).

These findings provide new insights into the evolution of gibbon songs. It is known, for example, that vocal fold and vocal tract shape are genetically determined, and that gibbon songs are high-pitched vocalizations within vertebrates (Fletcher 2004; Barelli et al. 2013). In small body sized primates, such as gibbons, the lengths of both the vocal folds and vocal tract are relatively short which raise the fundamental and formant frequency. The morphology of the vocal folds and vocal

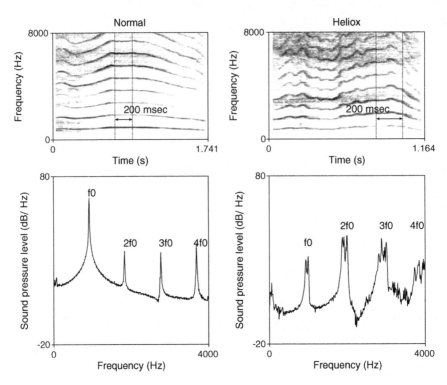

Fig. 15.4 Spectrogram (*top*) and power spectrum (*bottom*) of a sample call of normal air (*left*) and heliox (*right*). *Arrows* show the 200-ms frame used to generate power spectrum. In normal air, the f_0 was amplified distinctively from the other upper harmonics ($2 f_0$, $3 f_0$, $4 f_0$). In heliox, f_0 was dramatically suppressed, as predicted by the soprano model. Adapted from Koda et al. (2012, Fig. 5). Copyright © 2012 by Wiley. Reprinted by permission of Wiley

tract length in gibbons would predominantly determine the position of the f_0 as being very close to the formant frequency in their songs. In addition to morphological conditions, gibbons modify the shape of the vocal tract for a precise matching of the formant frequency to the f_0. Therefore, elaborate rhythmical control of the shapes of the vocal tract, like jaw or lip movements in humans, are required for soprano singing in gibbons. Interestingly, the cerebellum, a brain region responsible for the regulation of rhythmic motor control, is more expanded in gibbons than it is in other non-human primates (MacLeod et al. 2003). Although cerebellum expansion is generally linked to brachiation, a specific locomotion pattern adapted to the almost exclusively arboreal life of gibbons, the neurological modifications perhaps also allowed gibbons to control lip and jaw movements in an elaborated rhythmic manner. Thus, the gibbon soprano may be a result of morphological (small vocal apparatus) and neurological (cerebellum expansion) modifications.

What Makes Gibbon Song so Unique?

The recent studies of Koda et al. (2012, 2013) emphasize how possible adaptations to ecological environments may determine the emergence of peculiar and unique vocal performance in gibbons that are not present in other non-human primates. These findings suggest a novel scenario in which a common gibbon ancestor first pre-adapted to its ecological environment (Reichard et al. 2016), which in turn shaped their social structure, and later shifted their communication towards the development of unique songs.

Presumably, evolutionary processes leading to the achievement of co-singing as well as soprano singing in the gibbon lineage may not have been linear. Different biological foundations, which are originally adapted to several environmental factors, shaped gibbon songs. For example, hylobatids small vocal apparatus represents a primary biological foundation for soprano singing as discussed above. However, a small vocal apparatus probably originally evolved as a consequence of small body size, perhaps even reduction in body size compared to stem hominoids (Reichard et al. 2016), which may have been the result of a set of adaptations in response to a changing environment. Hylobatids small body size might be advantageous during rapid movement through the canopy by brachiation (Preuschoft et al. 2016). Brachiation as well as soprano singing, are shared characteristics in terms of rapid and rhythmic movements, which both could have also resulted from adaptations to an arboreal lifestyle.

I suggest that arboreality may have potentially influenced the style of gibbon vocal communication, and rapid movements through a dense canopy, which would have given rise to an emphasis on auditory rather than visual signals (Bradbury and Vehrencamp 1998). Aside from ecological factors, social factors may also have been important in the evolution of loud, complex vocalizations, because mother-offspring bonds are long lasting with dispersal of offspring occurring after many years at maturity (Brockelman et al. 1998; Reichard and Barelli 2008), creating an environment rich in mother-offspring communication. The development of loud calls and mother-daughter co-singing could have been facilitated from long-lasting bonds, and this social characteristic of hylobatid social relationships may yet have been an integral factor in the emergence of the unique forms of vocal communication that are so characteristic of gibbons.

In conclusion, several biological traits, i.e., small body size, arboreality, and a high degree of sociality, perhaps evolved before the emergence of gibbons elaborate songs. I suggest that the aforementioned biological features, which seem superficially unrelated to gibbon songs, need to be investigated in more detail in future studies in order to better understand the origin and evolution of gibbon songs. Recently, a genome-wide analysis of the four gibbon genera has begun to elucidate the family's evolutionary history and diversification. In particular, positive selection of genes important in forelimb development and connective tissues may have been involved in adaptations to their specialized brachiation locomotion (Carbone et al. 2014). These genes are not directly involved with songs, however, since brachiation is partly shared with singing in terms of rhythmic control of the body, the enlargement of the cerebellum might indirectly have influenced of gibbons to perform rhythmic control of songs as well. The integrated understanding of unique ecological, morphological, and social traits helps to answer why gibbon songs are so distinctive. Future research using interdisciplinary approaches integrating neurological, behavioural, ecological, and genomic methodologies are necessary for an even deeper understanding of the evolution of gibbon songs.

Acknowledgements I am grateful to Claudia Barelli, Ulrich Reichard and Sofia Bernstein for their comments, and Hirohisa Hirai for his encouragements. This review work is partly supported by JSPS KAKENHI (15K00203, 25285199 to HK as PI or co-PI) and by the SPIRITS program from Kyoto University (to HK as co-PI).

References

Barelli C, Mundry R, Heistermann M, Hammerschmidt K (2013) Cues to androgens and quality in male gibbon songs. PLoS ONE 8:e82748

Bradbury JW, Vehrencamp SL (1998) Principles of animal communication. Sinauer Associates, Massachusetts

Brockelman WY, Schilling D (1984) Inheritance of stereotyped gibbon calls. Nature 312:634–636

Brockelman WY, Srikosamatar S (1993) Estimation of density of gibbon groups by use of loud songs. Am J Primatol 29:93–108

Brockelman WY, Reichard U, Treesucon U, Raemaekers JJ (1998) Dispersal, pair formation and social structure in gibbons (*Hylobates lar*). Behav Ecol Sociobiol 42:329–339

Carbone L, Alan Harris R, Gnerre S et al (2014) Gibbon genome and the fast karyotype evolution of small apes. Nature 513:195–201

Clarke E, Reichard UH, Zuberbühler K (2006) The syntax and meaning of wild gibbon songs. PLoS ONE 1:e73

Cowlishaw G (1992) Song function in gibbons. Behaviour 121:131–153

Delgado RA (2006) Sexual selection in the loud calls of male primates: signal content and function. Int J Primatol 27:5–25

Fant G (1970) Acoustic theory of speech production. Walter de Gruyter

Farabaugh SM (1982) The ecological and social significance of duetting. Acoust Commun Birds 85–124

Fletcher NH (2004) A simple frequency-scaling rule for animal communication. J Acoust Soc Am 115:2334

Geissmann T (1984) Inheritance of song parameters in the Gibbon song, analysed in 2 hybrid Gibbons (*Hylobates pileatus* × *H. lar*). Folia Primatol 42:216–235

Geissmann T (1999) Duet songs of the siamang, *Hylobates syndactylus*: II. Testing the pair-bonding hypothesis during a partner exchange. Behaviour 136:1005

Geissmann T (2000) Gibbon songs and human music from an evolutionary perspective. In: Wallin NL, Merker B, Brown S (eds) The origins of music. The MIT Press, Cambridge, pp 103–123

Geissmann T (2002) Duet-splitting and the evolution of gibbon songs. Biol Rev Camb Philos Soc 77:57–76

Geissmann T, Orgeldinger M (2000) The relationship between duet songs and pair bonds in siamangs, *Hylobates syndactylus*. Anim Behav 60:805–809

Gittins SP (1984) The vocal repertoire and song of the agile gibbon. In: Preuschoft H, Chivers D, Brockelman W, Creel N (eds) The lesser apes: evolutionary and behavioural biology. Edinburgh University Press, Edinburgh, pp 354–375

Haimoff EH (1984) Acoustic and organizational features of gibbon songs. In: Preuschoft H, Chivers D, Brockelman W, Creel N (eds) The lesser apes: evolutionary and behavioural biology. Edinburgh University Press, Edinburgh, pp 333–353

Hayama S (1970) The Saccus laryngis in primates. J Anthropol Soc Nippon 78:274–298 (in Japanese with English abstract)

Hewitt G, MacLarnon A, Jones KE (2002) The functions of laryngeal air sacs in primates: a new hypothesis. Folia Primatol 73:70–94

Joliveau E, Smith J, Wolfe J (2004a) Vocal tract resonances in singing: the soprano voice. J Acoust Soc Am 116:2434–2439

Joliveau E, Smith J, Wolfe J (2004b) Acoustics: tuning of vocal tract resonance by sopranos. Nature 427:116

Koda H, Nishimura T, Tokuda IT et al (2012) Soprano singing in gibbons. Am J Phys Anthropol 149:347–355

Koda H, Lemasson A, Oyakawa C et al (2013) Possible role of mother-daughter vocal interactions on the development of species-specific song in gibbons. PLoS ONE 8:e71432

MacLeod CE, Zilles K, Schleicher A et al (2003) Expansion of the neocerebellum in Hominoidea. J Hum Evol 44:401–429

Marshall JT, Marshall E (1976) Gibbons and their territorial songs. Science 193:235–237

Merker B, Cox C (1999) Development of the female great call in *Hylobates gabriellae*: a case study. Folia Primatol 70:97–106

Mitani JC (1985) Gibbon song duets and intergroup spacing. Behaviour 92:59–96

Némai J, Kelemen G (1933) Beiträge zur Kenntnis des Gibbonkehlkopfes. Anat Embryol 100:512–520

Nowicki S (1987) Vocal tract resonances in oscine bird sound production: evidence from birdsongs in a helium atmosphere. Nature 325:53–55

Oyakawa C, Koda H, Sugiura H (2007) Acoustic features contributing to the individuality of wild agile gibbon (*Hylobates agilis agilis*) songs. Am J Primatol 69:777–790

Preuschoft H, Chivers D, Brockelman WY, Creel N (1984) the lesser apes: evolutionary and behavioural biology. Edinburgh University Press, Edinburgh

Raemeakers JJ, Raemeakers PM (1985) Field playback of loud calls to gibbons (*Hyiobates lar*): territorial, sex-specific and species-specific responses. Anim Behav 33:481–493

Reichard UH (2003) Social monogamy in gibbons: the male perspective. In: Reichard UH, Boesch C (eds) Monogamy: mating strategies and partnerships in birds. Cambridge University Press, Humans and other Mammals, pp 190–213

Reichard UH, Barelli C (2008) Life history and reproductive strategies of Khao Yai *Hylobates lar*: implications for social evolution in apes. Int J Primatol 29:823–844

Sundberg J (1987) The science of the singing voice. Northern Illinois University Press

Tenaza J (1985) Songs of hybrid gibbons (*Hylobates lar × H. muelleri*). Am J Primatol 8: 249–253

Todd NPMA, Merker B (2004) Siamang gibbons exceed the saccular threshold: intensity of the song of *Hylobates syndactylus*. J Acoust Soc Am 115:3077

Index

9 781493 956128